Software Tools for the Simulation of Electrical Systems

Theory and Practice

Software Tools for the Simulation of Electrical Systems

Theory and Practice

Ashok Kumar L.
Dept. of Electrical & Electronics Engineering,
PSG College of Technology, Coimbatore,
Tamil Nadu, India

Indragandhi V.
School of Electrical Engineering,
Dept. of Energy & Power Electronics, VIT, Vellore,
Tamil Nadu, India

Uma Maheswari Y.
Pramura Software (P) Ltd., Coimbatore,
Tamil Nadu, India

Academic Press is an imprint of Elsevier
125 London Wall, London EC2Y 5AS, United Kingdom
525 B Street, Suite 1650, San Diego, CA 92101, United States
50 Hampshire Street, 5th Floor, Cambridge, MA 02139, United States
The Boulevard, Langford Lane, Kidlington, Oxford OX5 1GB, United Kingdom

British Library Cataloguing-in-Publication Data
A catalogue record for this book is available from the British Library

Library of Congress Cataloging-in-Publication Data
A catalog record for this book is available from the Library of Congress

ISBN: 978-0-12-819416-4

For Information on all Academic Press publications
visit our website at https://www.elsevier.com/books-and-journals

Publisher: Brian Romer
Acquisitions Editor: Lisa Reading
Editorial Project Manager: Naomi Robertson
Production Project Manager: Nirmala Arumugam
Cover Designer: Christian J. Bilbow

Typeset by MPS Limited, Chennai, India

ELSEVIER
Book Aid
International
Working together
to grow libraries in
developing countries
www.elsevier.com • www.bookaid.org

Contents

About the authors

Dr. Ashok Kumar L. was a Postdoctoral Research Fellow at San Diego State University, California. He is a recipient of the BHAVAN fellowship from the Indo-US Science and Technology Forum and SYST Fellowship from DST, Govt. of India. His current research focuses on integration of Renewable Energy Systems in the Smart Grid and Wearable Electronics. He has 3 years of industrial experience and 19 years of academic and research experience. He has published 187 technical papers in International and National journals and presented 197 papers in National and International Conferences. He has completed 27 Government of India funded projects, and currently seven projects are in progress. His PhD work on wearable electronics earned him a National Award from ISTE, and he has received 24 awards at the national level. Ashok Kumar has seven patents to his credit. He has guided 98 graduate and postgraduate projects. He is a member of, and in prestigious positions in, various national forums. He has visited many countries for institute–industry collaboration and as a keynote speaker. He has been an invited speaker in 178 programs. Also he has organized 79 events, including conferences, workshops, and seminars. He completed his graduate program in Electrical and Electronics Engineering from the University of Madras and his postgraduate from PSG College of Technology, India, and Masters in Business Administration from IGNOU, New Delhi. After completion of his graduate degree, he joined Serval Paper Boards Ltd., Coimbatore (now ITC Unit, Kovai) as a project engineer. Presently, he is working as a Professor and Associate HoD in the Department of EEE, PSG College of Technology and also doing research work in wearable electronics, smart grid, solar PV, and wind energy systems. He is also a Certified Charted Engineer and BSI Certified ISO 500001 2008 Lead Auditor. He has authored the following books in his areas of interest: (1) *Computational Intelligence Paradigms for Optimization Problems Using MATLAB/SIMULINK*, CRC Press; (2) *Solar PV and Wind Energy Conversion Systems—An Introduction to Theory, Modeling with MATLAB/SIMULINK, and the Role of Soft Computing Techniques*, Green Energy and Technology, Springer, United States; (3) *Electronics in Textiles and Clothing: Design, Products and Applications*, CRC Press; (4) *Power Electronics with MATLAB*, Cambridge University Press, London; (5) *Automation in Textile Machinery: Instrumentation and Control System Design Principles*, CRC Press, Taylor &

Francis Group, United States; (6) monograph on *Smart Textiles*; (7) monograph on *Information Technology for Textiles*; *(8) Deep Learning Using Python*, Wiley India Publications; (9) monograph on *Instrumentation & Textile Control Engineering*; and (10) *Computational Paradigm Techniques for Enhancing Electric Power Quality*, CRC Press, Taylor & Francis Group, United States.

Dr. Indragandhi V. has a Power Electronics background from her studies as follows: She received an M.E in Power Electronics and Drives from Anna University and was awarded a Gold Medal for the achievement of University first rank. Subsequently, she was awarded a doctorate of philosophy in the field of Power Converters for Renewable Applications at Anna University, Chennai in the year 2015. At present, she is serving as an Associate Professor in the School of Electrical Engineering, Energy, and Power Electronics Department, Vellore Institute of Technology, Vellore, Tamil Nadu. She has been engaged in teaching cum research work for the past 12 years. Moreover, she has organized 25 events, including conferences, expert talks, workshops, and value-added programs. She is an active senior member in IEEE and member of IET.

She has authored more than 60 research articles and 4 book chapters in leading peer-reviewed international journals and published articles in refereed impact factor journals such as IET, Elsevier, and Springer publications. She has visited many countries and presented her technical research work in international forums. Also she has two patents to her credit. Currently, six PhD scholars are pursuing their research under her guidance. She has guided more than 50 UG and PG projects in the area of power electronics, drives, and renewable energy sources. Currently, she serves as a reviewer for reputed journals such as IEEE Transactions on Power Delivery and IEEE Transactions on Power Electronics.

Uma Maheswari Y. is a Technology Manager at Pramura Software Private Limited, Coimbatore. She has around 16 years of industrial experience. She completed her graduate program in Electrical and Electronics Engineering at Amrita College of Engineering, Coimbatore and her postgraduate program in Embedded System and Technologies at Anna University, Coimbatore. She has authored the book titled *Power Electronics with MATLAB*. Her expertise is in the design of PCB and simulation software.

Preface

Simulation is the stepping stone for implementation

An interactive environment for numerical computation, visualization, and programming is a key positive point about learning the software tools related to electrical sciences. Also students are required to work with different circuit topologies and design based on the applications. They face difficulties in choosing software tools for different applications and they need a start-up guide to help them find solutions to the problems.

This book definitely helps students to come up with new ideas in the simulation of electrical, electronics, and instrumentations systems. This book, on a very topical subject, is aimed at engineering students who either practice or implement simulation software tools of switches, circuits, controllers, instruments, and automation system design. Further this book covers power electronic switches, FACTS controller devices, and simulation model building using all of the possible software, that is, MATLAB, PSIM, Pspice, and PSCAD. Also the software Labview and PLC that are used in industrial automation, process control, monitoring, and measurement systems are discussed with a number of examples and step-by-step procedures. Moreover, the photovoltaic software PVSyst is presented in this book which is helpful to readers doing research in renewable energy systems. In addition every chapter contains complete simulated circuits of practical problems with the procedure and unsolved problems for practice as well.

The book has been divided into 10 chapters. Chapter 1, MATLAB/Simulink, covers the simulation of power transistor thyristor simulation using MATLAB. Chapter 2, PSIM, deals with the simulation of power electronic circuits using PSIM software, and Chapter 3, PSpice, deals with the simulation of IGBT and TRIAC performance characteristics using Pspice software. The Multisim software toolbox is presented in Chapter 4, it includes Multisim, along with the simulation of converters. The PCB Design Tool—Design Spark—and design examples are discussed in Chapter 5. Chapter 6, Simulation of Hydraulic and Pneumatic Valves—PLC, presents PLC software and its usefulness in designing pneumatic circuits. Chapter 7, LabVIEW, introduces the LabVIEW and Virtual Instrumentation concepts. Chapter 8,

PSCAD, provides an overview of the power system simulation software PSCAD and Chapter 9, PVSYST, presents the simulation of photovoltaic systems using PVSyst. The last chapter (Chapter 10: Applications of Software Tools) describes how to utilize the knowledge that has been gained from the previous chapters for real-time applications.

Acknowledgments

Dr. Ashok Kumar L. would like to take this opportunity to acknowledge those people who helped in completing this book. He is thankful to his wife, Uma Maheswari Y., for her constant support during writing. Without her, all these things would not be possible. He would like to express his special gratitude to her daughter A.K. Sangamithra who initiated the process by asking a simple question of, "What is Simulation?" and for her smiling face and support; it helped a lot in completing this work.

Dr. Indragandhi V. would like to take this opportunity to thank her husband Arunachalam and daughter Subiksha for their constant support and time during writing. She would like to express her special gratitude to her family members Jai Girish, Rajabrindha, and Subramaniyaswamy for showing their heartfelt love and caring words. She dedicates the book to her father Mr. Vairavasundaram and mother Ms. Chellammal who is the backbone of all her successes.

Uma Maheswari Y. wishes to thank her daughter A.K. Sangamithra in being patient and giving her all the love, time, and space to finish her work. She wishes to acknowledge her husband Aski's support in the successful completion of this book. She dedicates the book to her father Mr. Yuvaraj and mother Ms. Kalavathi who laid the foundation for all her successes and special thanks to her brother Mr. Dhayaneswaran Y.

Chapter 1

MATLAB®/Simulink

Chapter Outline

1.1 Introduction

1.1.1 Basics of MATLAB®

A simulation is the imitation of functioning over time of a real-world mechanism or a system. The simulation involves the creation of a model, which describes the key features, actions, and functions of a physical or abstract structure or mechanism selected. The mechanism itself is defined by the model while the simulation is hierarchical over time.

Modeling is used in a variety of settings, such as performance optimization system modeling, software development, research, teaching, schooling, and video games. Software simulations are often used for the analysis of models for simulation. Simulation is being used in the theoretical simulation of natural systems or human processes to offer insight through operations and economy. Simulation can be used to illustrate the potential real consequences of alternate environments and action courses. Simulation is also used when it is impossible to implement the system in real time.

The primary simulation concerns include collection of accurate source information about the appropriate set of key features and behaviors, use of simplistic simulation methods and conclusions, and consistency and validation of simulation tests. Model verification and validation procedures and protocols are a continuing field in academic study, refined study, research, and development, especially in computer simulation technology or practice.

Simulation software: It is based on the simulation method with a variety of mathematical formulas for a real phenomenon. It is basically a program that allows the user to control a simulation process without executing this. Simulation software is widely used for designing equipment to ensure that the final product is as close to design specifications as possible without costly changes in processes. Real-time modeling software is commonly used in sports, but also has major industries. If the penalty for improper operation is expensive, such as aircraft pilots, power plant operators or chemical

Software Tools for the Simulation of Electrical Systems. DOI: https://doi.org/10.1016/B978-0-12-819416-4.00001-6

facilities operators, the actual control panel is mocked, and the physical response is simulated in real time and gives valuable trainings experiences without fear of disaster.

Advanced computer programs can simulate power system behavior, weather conditions, electronic circuits, chemical reactions, mechatronics, heat pumps, feedback control systems, atomic reactions, and even complex biological processes. By theory, all phenomena can be replicated on a machine that can be simplified to statistical data and equations. Simulation can be hard because the majority of natural phenomena are influenced by nearly endless numbers. One of the techniques to create effective simulations is to evaluate the key factors influencing the simulation goals.

Simulations are also used for testing new hypotheses in addition to imitating the mechanism for analyzing how they work under different conditions. The theoretician can then codify the associations in the context of a computer system, using a philosophy of causal relations. If the system then implements the real process, the proposed relationships are likely to be correct.

Electronics simulation: To simulate the behavior of a particular device or circuit, the app uses mathematical equations. Essentially, it is a software application that transforms a computer into a fully operating electronic laboratory. To make it easy and smooth to add a Schedule Editor, a SPICE, and OnScreen Waveforms, an interactive emulator is added. Through simulating the behavior, it increases significantly efficiency and provides insight into the comportment and reliability of electronic circuit structures before actually building them. Some devices use a SPICE motor to mimic the exceptional performance and precision of analog, optical, or hybrid A/D circuits. These usually contain large collections of models and tools. Although such simulators typically have the capabilities to export the printed circuit-board, they are not important to the design and testing of electronic circuit-related circuits.

Simulators with analog and event-driven digital simulation capabilities, which are known as mixed-mode simulators, operate as purely analog electronic circuit simulators. This requires emulation, which involves the synthesis of the two analog, event-driven elements (digital or sampled). One optimized schematic can power a full mixed-signal analysis. Both digital models in mixed-mode simulators have accurate time and time delays for propagation.

The mixed-mode simulator event-guided algorithm is a general purpose that supports nondigital data types. For example, elements may simulate digital signal processing features or sampled data filters by using real or integer values. Due to the fact that the event-led algorithm is quicker than the traditional SPICE matrix solution simulation time for circuits using event-driven models rather than analog models is decreased considerably.

The simulation mixed mode is performed on three levels: (1) primitive digital components, which use time models and an embedded 12 or 16 state digital logic simulator, (2) subcircuit models that use the real integrated circuit transistor topology, and (3) in-line Boolean logic terminology.

The exact description of IC's I/O features is primarily used when examining transmission line and signal completeness concerns. Boolean logic expressions are delay-free functions used to effectively process logic signals in an analog environment. Both two computational methods use SPICE to solve a problem when mixed-mode capacities are used in the third process, computer primitives. Nonetheless, a lot of simulations (especially those using A/D technology) allow all three methods to be combined. There is no reasonable solution alone.

MATLAB: It is a high-level mathematical programming terminology. It combines calculation, simulation, and programming in a simple-to-use setting where problems and solutions are presented in common mathematical notations. Typical applications include development of math and calculation software, modeling, concept analysis, testing, and visualization; science and engineering visualization; and software development including the construction of MATLAB Visual User Interface, an interactive interface where basic data is an array of unbalanced data. In a fraction of the time it would take you to type a program in a scale noninteractive language, such as C or Fortran, you solve many technical computer problems, particularly those with matriximum and vector formulations.

MATLAB is an acronym for "matrix laboratory." MATLAB was originally developed to provide easy access to LINPACK and EISPACK matrix software projects that together represent the best in matrix calculation tools.

Over the years, MATLAB has improved and many people have contributed to it. It is the standard tool for beginner and comprehensive courses in math, engineering, and science in the university environment. MATLAB is the industry's preferred method for study, growth, and analysis in high productivity. MATLAB delivers a collection of applications for toolboxes. To most MATLAB users, toolboxes are very necessary for knowledge and use of the technology. The toolboxes are a comprehensive set of MATLAB (M-files) functions that extend the MATLAB framework to solve different problem classes.

1.1.1.1 Design and simulation of power converter

Single-phase half-controlled converter

When converting AC to DC power conversion, a single stage half-wave thyristor converter circuit is used. The input AC is given from a transformer to the thyristor converter with the necessary AC voltage, which is centered on the required DC voltage. By adding the correct pulse gate signal to the thyristor port terminal, the thyristor is triggered by a time angle of $\omega t = \alpha$ (Figs. 1.1–1.3).

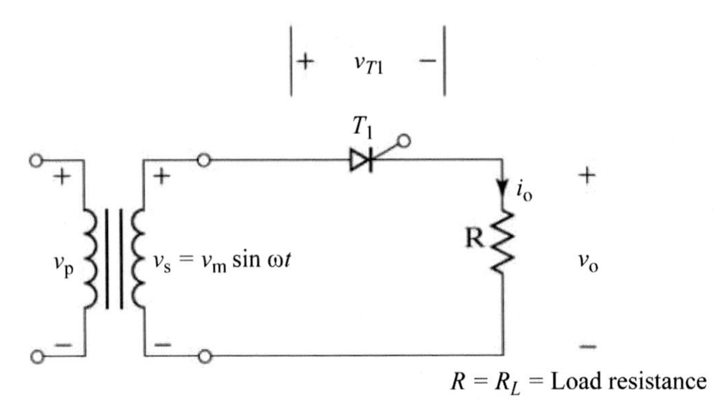

FIGURE 1.1 Single-phase half-controlled converter.

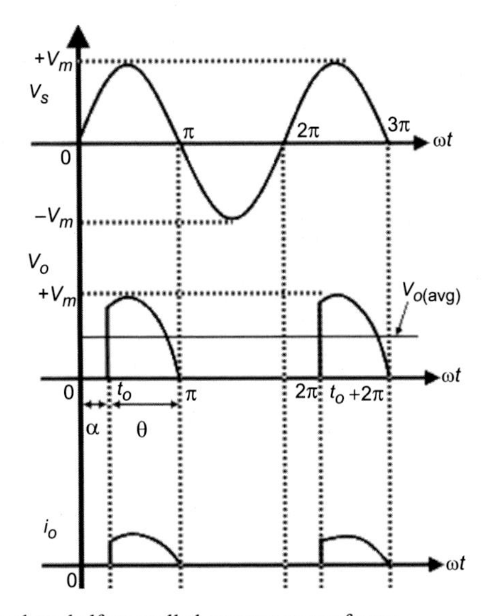

FIGURE 1.2 Single-phase half-controlled converter waveforms.

In the silicon controlled rectifier (SCR)-positive half cycle, SCR begins the lead at shooting angles—alpha—and the boxes include the signal processing, control systems, neural networks, FUZZY logic, wavelets, and simulations. Small drop over SCR is ignored so that output voltage equal to voltage supply. The load of _RL “triggers a slow rise of the current through SCR. At _π”, it is at zero where the maximum value is load current. The inductor retains energy and generates the voltage in a positive half loop.

The voltage formed across the inductor in a negative half cycle, SCR will forward biases and retain its steering. It basically denies change in current with the inductance property. Current output and input flows in the same loop, and $i0 = 0$ all the time. After π the inductor energy is provided to the

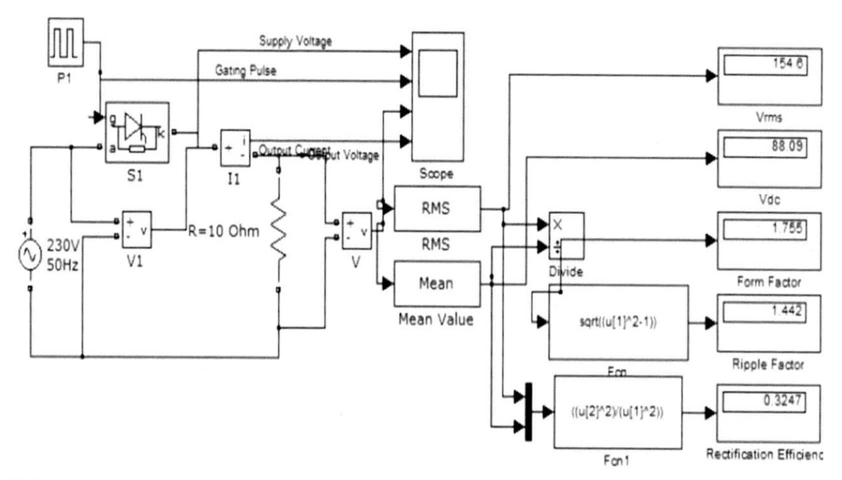

FIGURE 1.3 Simulation model: single-phase half-controlled converter.

hands, and the flow of _*io*‘ is accomplished. The electricity is decreased as if the circuit is used, which decreases the current. At _β‘ energy,’ is zero, and’ alternative energy,’ shuts off. _β‘ is 0 from _β‘ to _$2\pi + \alpha$ oscillations, therefore the conduction is discontinuous (Figs. 1.4–1.7).

Single-phase fully controlled rectifier

This type of power electronics–based rectifier circuit is widely used in controlling the speed of DC motors. This circuit is obtained by replacing all the diodes used in uncontrolled or half-controlled rectifiers with thyristors, as shown in Fig. 1.8. From the circuit, we can observe that one thyristor from a top group (T1, T3) and one thyristor from the bottom group (T2, T4) must conduct for load current flow. Yet T1T3 or T2T4 are not able to operate at the same time (Figs. 1.9–1.17).

Three phase converters

Three phase converters provide higher average output voltage. Frequency of ripples on output voltage is higher compared with that of the single-phase converter. Thus the filtering requirements for smoothing out load current and load voltage are simpler. For these reasons, three phase converters are extensively used in high-power variable-speed drivers.

Three single-phase half-wave rectifiers are integrated in one circuit to feed a typical load with a three-phase semisoft. Thyristor S1 is a half-wave rectifier array, with one input step winding _*a-n*‘. The second S2 thyristor in series is the second half-wave operated rectifier with the _*b-n*‘ supply process. In the third half-wave rectifier, the thyristor S3 in series with the winding of the supply step. Fig. 1.18 shows three phase fully controlled rectifier.

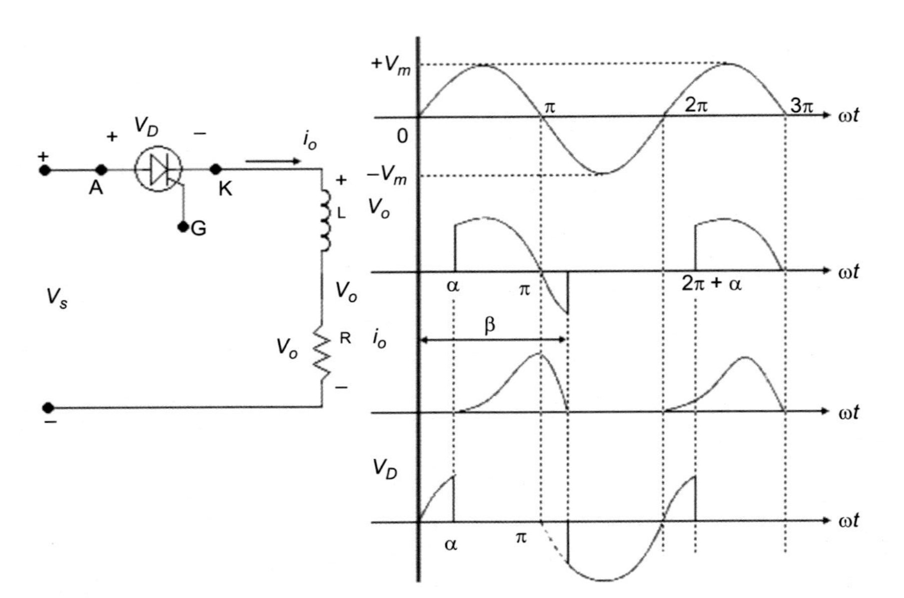

FIGURE 1.4 Single-phase half-controlled converter-RL load.

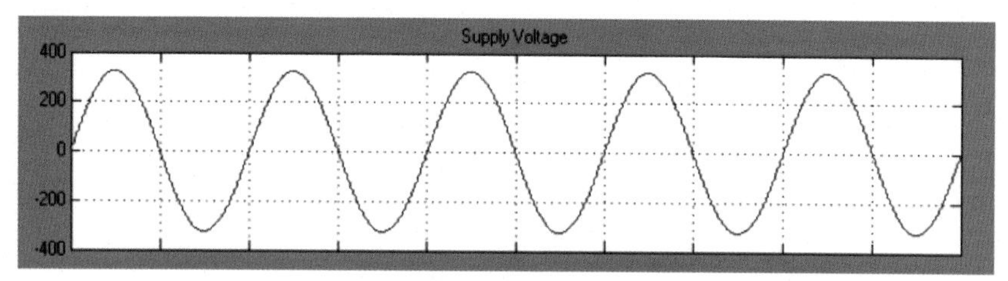

FIGURE 1.5 Waveform of supply voltage.

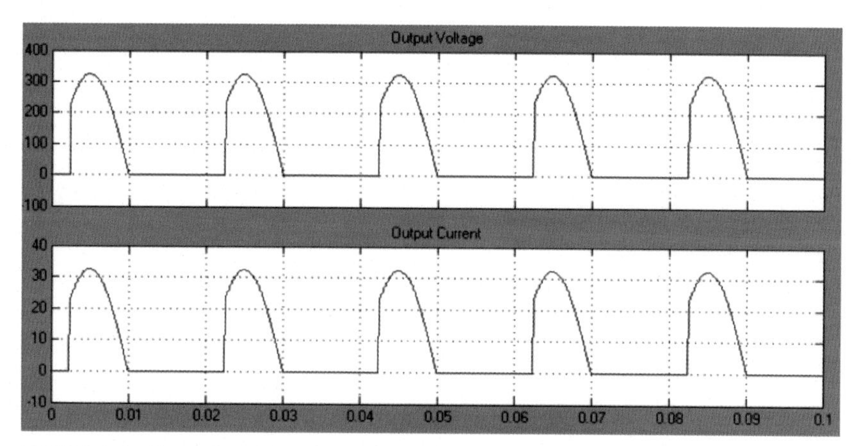

FIGURE 1.6 Simulation waveforms: single-phase half-controlled converter.

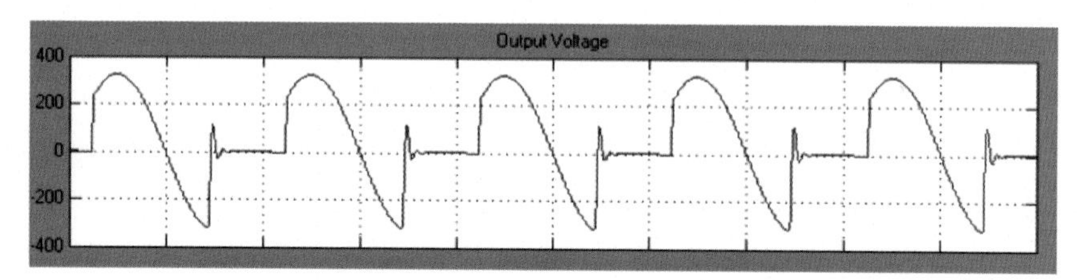

FIGURE 1.7 Simulation waveforms: single-phase half-controlled converter-RL load.

In industrial applications, up to 15 kW output power is used extensively for three phases: half-controlled bridge converters and complete-controlled bridge converters (Figs. 1.19 and 1.20). A fully controlled bridge-controlled rectifier is a three-phase direct transformer with six thyristors in the shape of full-wave bridge configuration. All six thyristors are regulated switches which are triggered by the application of specific trigger signals at appropriate times.

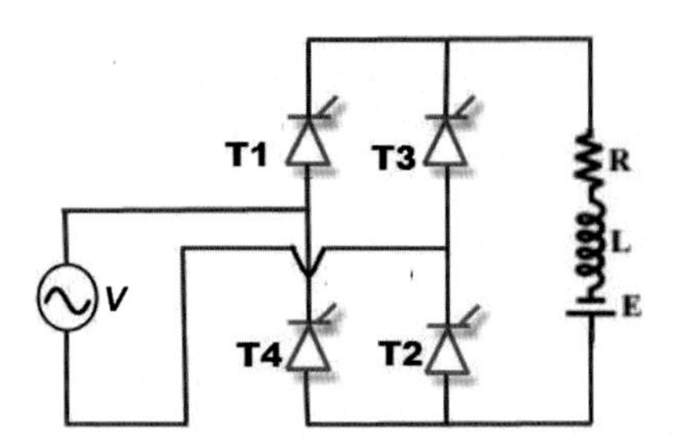

FIGURE 1.8 Single-phase fully controlled rectifier.

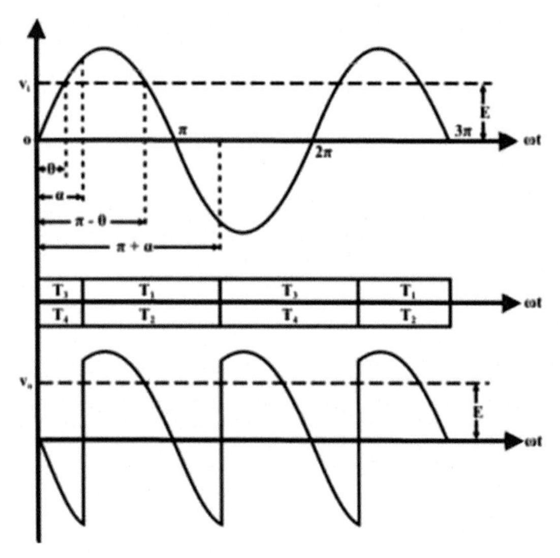

FIGURE 1.9 Single-phase fully controlled rectifier waveforms.

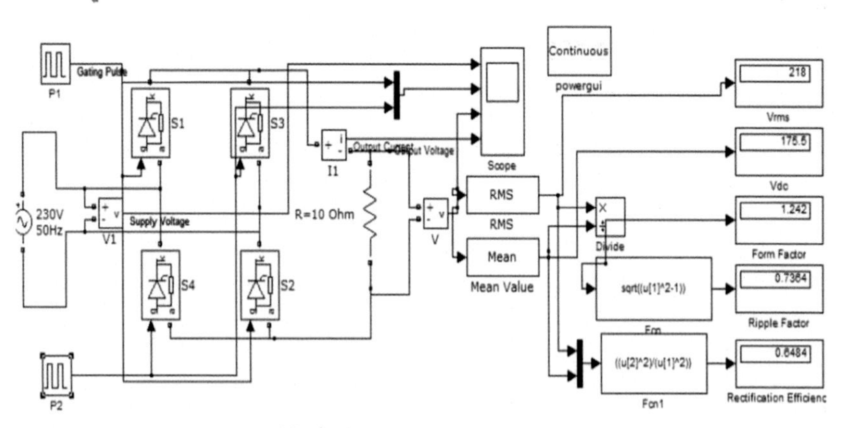

FIGURE 1.10 Simulation model: single-phase fully controlled converter.

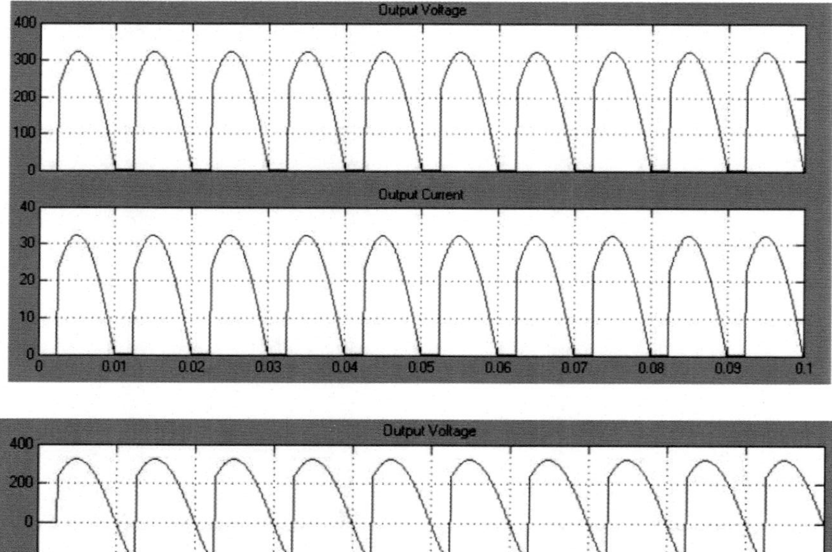

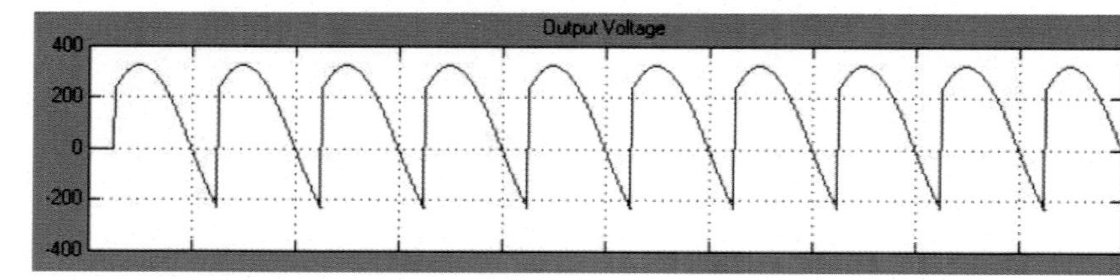

FIGURE 1.11 Simulation waveforms: single-phase fully controlled converter.

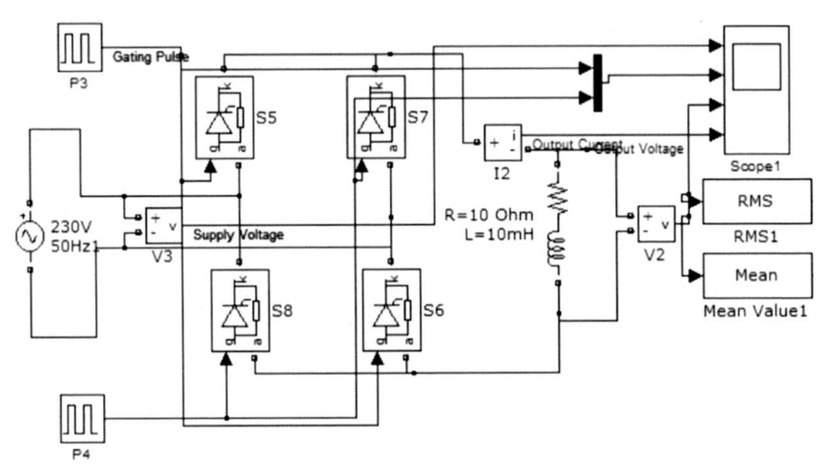

FIGURE 1.12 Simulation model: single-phase fully controlled converter with RL load.

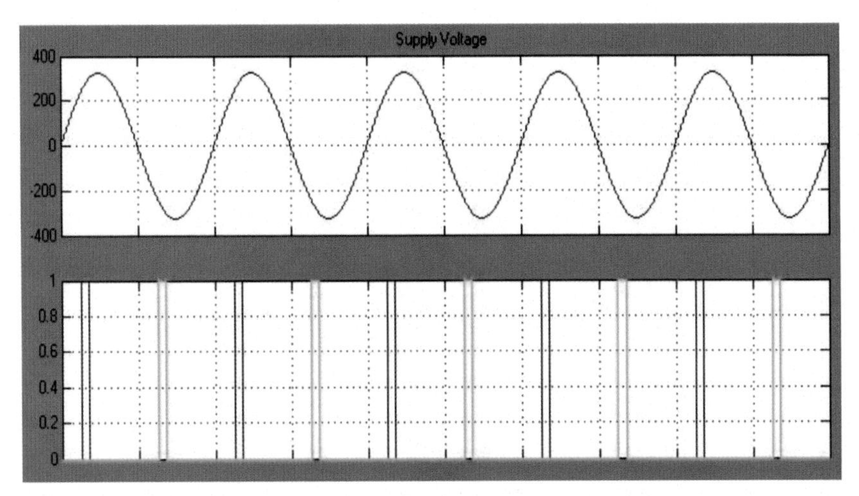

FIGURE 1.13 Simulation waveforms: supply voltage.

Buck converter

The conceptual-type buck converter is best understood in reference to the current and the voltage of the inductor. The present in the circuit is null, starting with the turn open (off-state). The current starts to increase when the switch is first closed (on-state) and the induction generates a voltage opposite throughout the terminals, as a result of the change in current. This reduction of the voltage counteracts the source voltage and reduces the net voltage throughout the charge. The rate of change decreases over time and then the tension on the whole of the inductor decreases and the voltage increases at

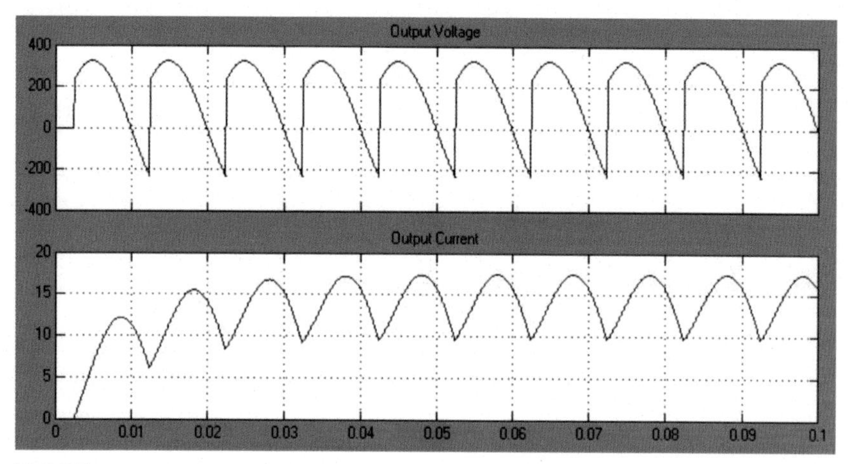

FIGURE 1.14 Simulation waveforms: single-phase fully controlled converter with RL load.

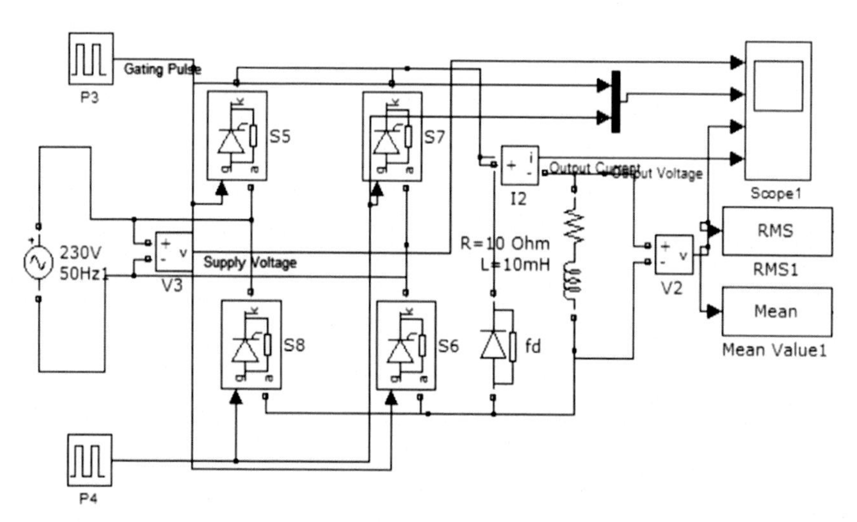

FIGURE 1.15 Simulation model: single-phase fully controlled converter with freewheeling diode.

load. The inductor retains energy as a magnetic field during this period. If the valve is opened while the current is in motion, the voltage will collapse across the inductor so that the total tension at the load is always less than the input tension source. The voltage source is removed from the circuit when the switch is open again (off-state) and the current flow decreases. The rising current induces a voltage decrease over the inductor (contrary to a drop on the state) and the inductor is now a current source (Fig. 1.21).

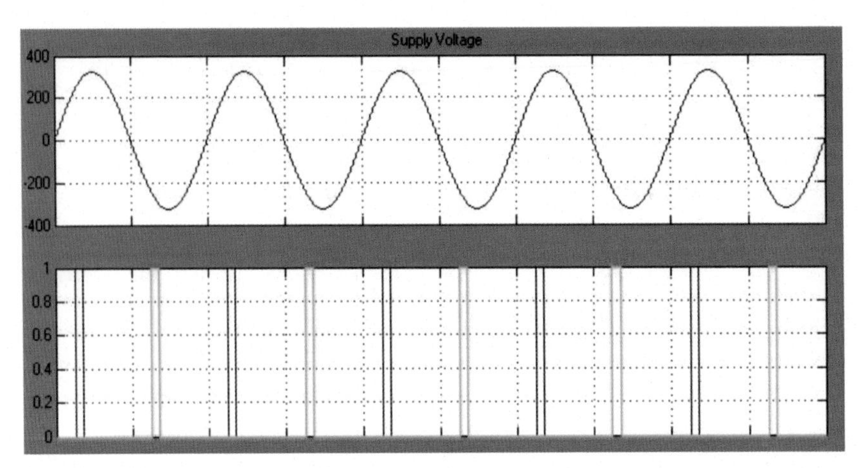

FIGURE 1.16 Simulation waveforms: supply voltage.

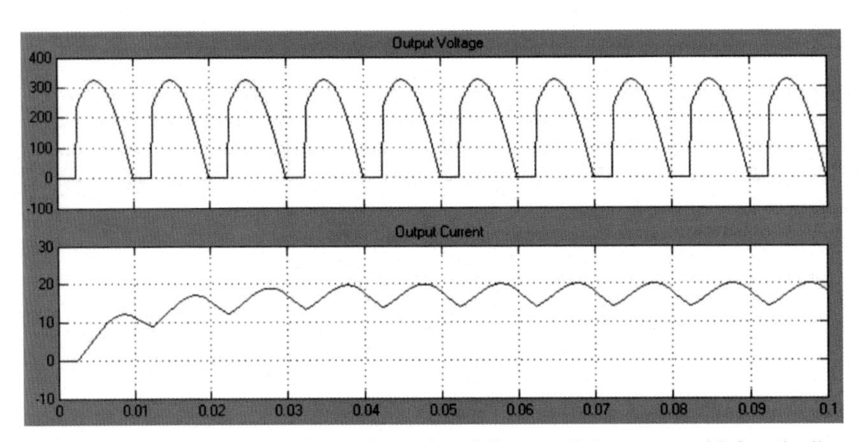

FIGURE 1.17 Simulation waveforms: single-phase fully controlled converter with freewheeling action.

The energy contained in the magnetic field of the inductor supports the current flow through the charge. When the current is attached to the current, which flows on-state, the current flows while the input voltage is removed, and the current reaches greater than the average input current. The average "increase" means the voltage is lower and preserves ideally the power supplied to the charge. The driver discharges the energy into the remainder of the circuit during the off-state process. The voltage at the charge is always higher than zero if the switch is closed again before the inductor is completely discharged (state) (Fig. 1.22).

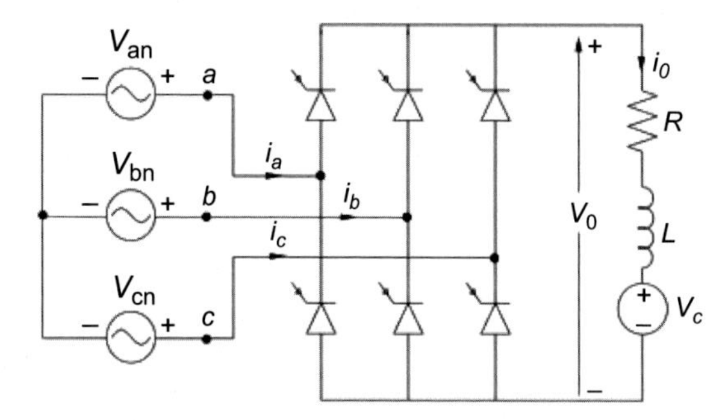

FIGURE 1.18 Three-phase fully controlled rectifier.

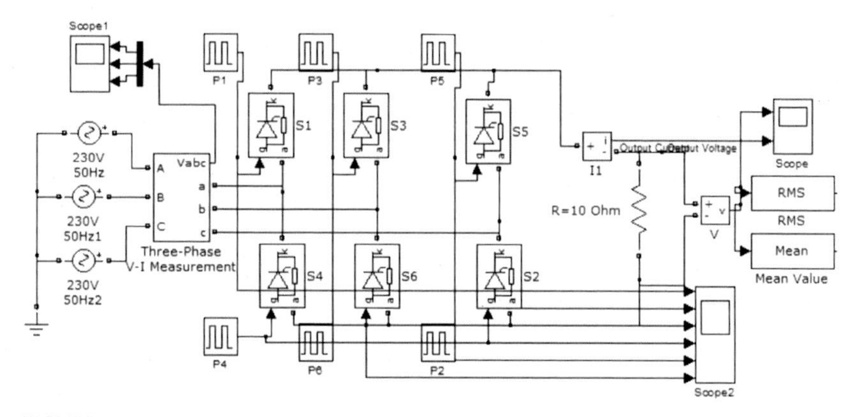

FIGURE 1.19 Simulation model: three-phase fully controlled rectifier.

Boost converter

An inductor's potential to combat current changes by generating and withdrawing a magnetic field is the main driver of the boost converter. The output voltage of the boost converter is always higher than the input voltage.

1. When the clockwise door closes and the inductor consumes voltage, producing a magnetic field, the current flows through the inductor. On the left side of the inductor, polarity is positive.
2. The current reduces as an impedance increases when the door is opened. To keep the current toward the load, the previously created magnetic field is removed. The polarity is therefore inverted (that is to say that the left side of the inductor is negative). Therefore two outlets in series charge the condenser via diode D by higher voltage.

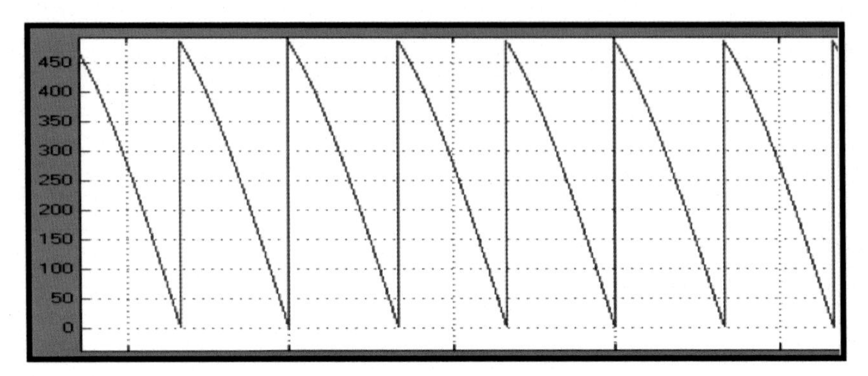

FIGURE 1.20 Simulation waveforms: three-phase fully controlled rectifier.

If the switch is cycled quickly enough, the inductor will not fully load between charging points and the load will always see a load higher than the load of the source when the switch has been opened. The condenser is also powered to this maximum voltage while the valve is opened in parallel with the load. The capacitor is able to provide the load with voltage and electricity when the door is then closed and the right side is separated from the left side. The blocking diode stops the condenser from discharging the circuit during this process. Of course the door must be opened again fairly quickly to prevent the condenser from too much discharge (Figs. 1.23–1.26).

The buck boost converter is a DC-to-DC converter. The DC-to-DC converter output voltage is lower or larger than the input voltage. The magnitude output voltage depends on the working cycle. These converters are also known as transformers step up and down, and these names are derived from the analog transformer step up and down. The input voltages are more or less than the input voltage step-up/-down to some degree. The input power is equal to the output capacity with the low storage energy.

It is a kind of DC-to-DC converter with an output voltage value. The input voltage may be more or less the same. The buck boost converter is identical to the fly back circuit and the transformer is replaced by a single inductor. Several kinds of converters are included in the buck boost converter, the other one being the buck converter. The output voltage range of these transformers equals the input voltage (Figs. 1.27–1.31).

1.1.1.2 Simulation of different transformerless inverters

H-bridge with unipolar modulation

Fig. 1.32 shows a typical full bridge topology. Generally, two pulse width modulation (PWM) techniques are used for full bridge inverter unipolar and bipolar modulation. Unipolar modulation is also known as three-level modulations (Fig. 1.33). It generates three-level output voltage: + Vdc, 0, − Vdc,

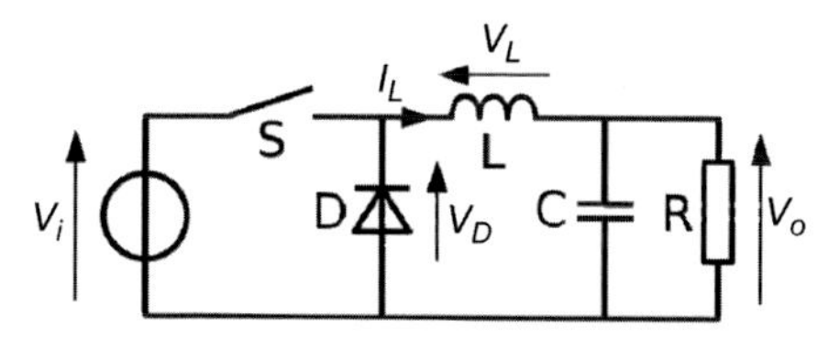

FIGURE 1.21 Circuit diagram of a buck regulator.

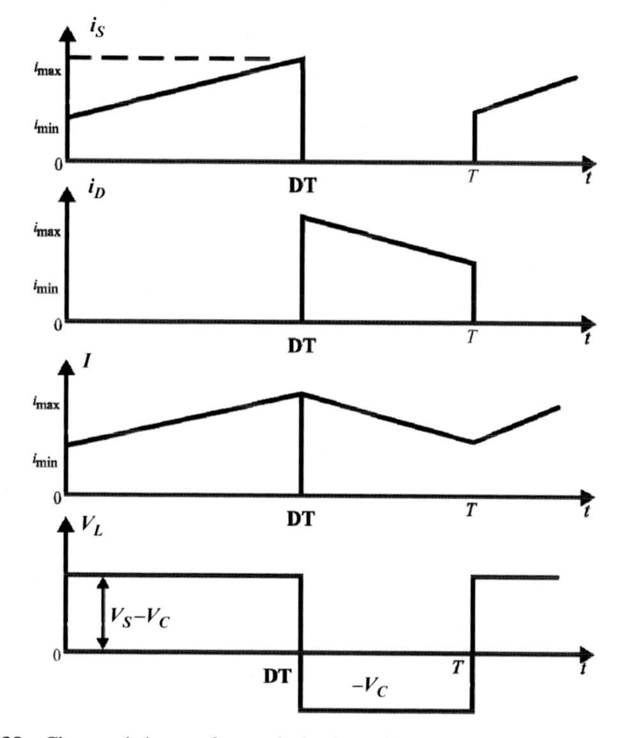

FIGURE 1.22 Characteristic waveforms of a buck regulator.

with double of the switching frequency as shown in Fig. 1.34. In every switching transition, the voltage changes across the inductor by Vdc. Thus unipolar PWM reduces change of voltage (dv/dt), ripple current, filter size and losses in both switches and inductors. Voltage oscillating with switching frequency energizes stray capacitances which generates high frequency common mode voltage as shown in Fig. 1.35. This high-frequency, common-mode voltage generates dangerous leakage current up to a few amperes. Hence, unipolar PWM is not suitable for transformerless full-bridge topology.

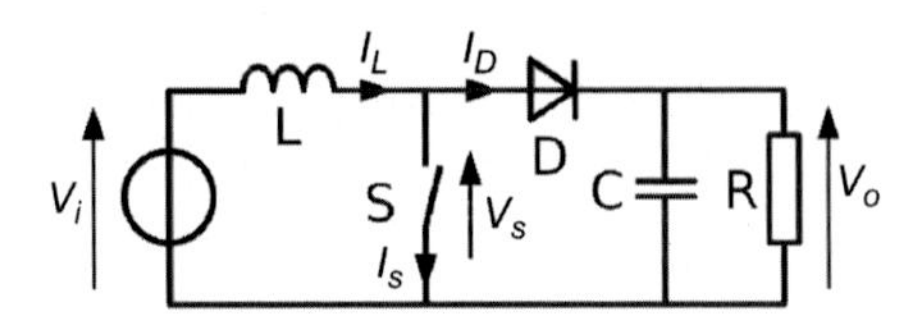

FIGURE 1.23 Circuit diagram of the boost converter.

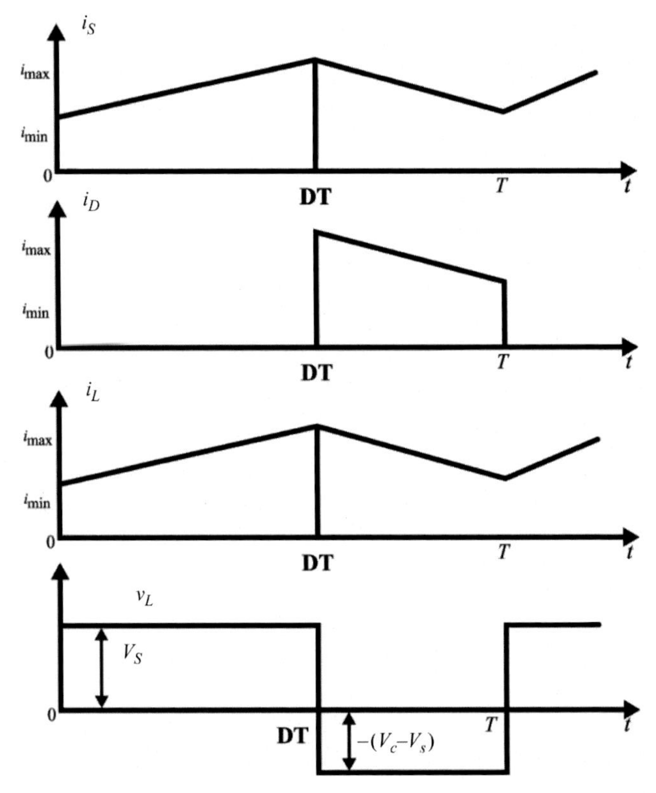

FIGURE 1.24 Characteristic waveform of the boost converter.

H-bridge with bipolar modulation

Fig. 1.36 shows a typical full bridge topology. Fig. 1.37 shows bipolar pulse pattern for full bridge inverter circuit. Bipolar modulation, also known as two level modulations, generating two level output voltage: + Vdc and − Vdc, as shown in Fig. 1.38. In every switching transition, the voltage changes across the inductor by twice of input voltage, 2Vdc. Such modulation technique reduces the overall efficiency of the inverter due to large current ripple across the inductors and high switching losses. Compared to

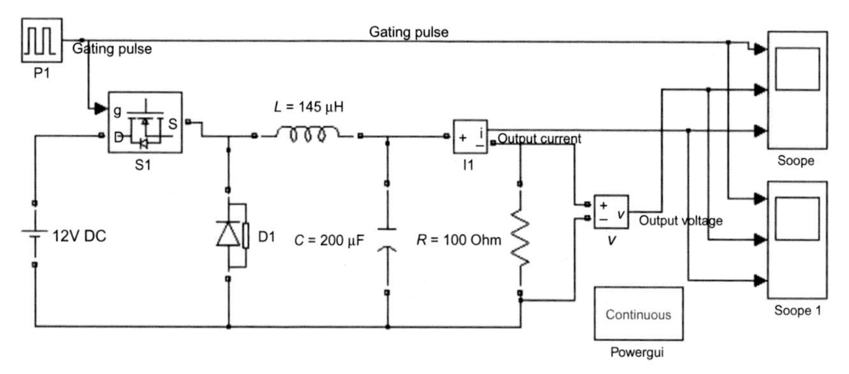

FIGURE 1.25 Simulation model: buck converter.

unipolar PWM, leakage current is significantly reduced due to constant common mode voltage as shown in Fig. 1.39.

H-bridge with DC bypass (H5)

H5 topology is modified by adding an extra switch S1 in the input side (DC) of the full bridge inverter as shown in Fig. 1.40. The upper pair of switches S3 and S5 is operated at grid frequency while the lower pair of switches S4 and S6 is operated at switching frequency. S1 and S6 are simultaneously turned on at high frequencies in the positive half period, while S3 is constantly disabled. Current flows through S1, S3, and S6. During the positive zero voltage vector, S1 is turned off, disconnecting the photovoltaic (PV) from the grid. Present S3 freewheels and S5 antiparallel diode (Figs. 1.41 and 1.42).

S1 and S4 are turned at high frequency simultaneously, while S5 is turned on continuously in the negative half cycle. The route via S1, S5, and S4 is now open. During the negative zero voltage vector, S1 is turned off and current freewheels through S5 and the antiparallel diode of S3. Hence, H5 topology is also known as DC decoupling topology where PV is disconnected from the grid by disconnecting S1 in the DC side of the inverter. SMA is proprietary for H5 topology.

H-bridge with AC bypass highly efficient and reliable inverter concept

HERIC topology is designed by modification of full bridge topology with two extra switches S5 and S6 connected across the output side (AC side) of the inverter as shown in Fig. 1.43. At switching frequency, a half-wave grid voltage is operated on each pair of diagonal switches. During the zero voltage vectors, S5 is turned on during the positive half cycle. Current freewheels through the antiparallel diode of S6, S5 and the grid.

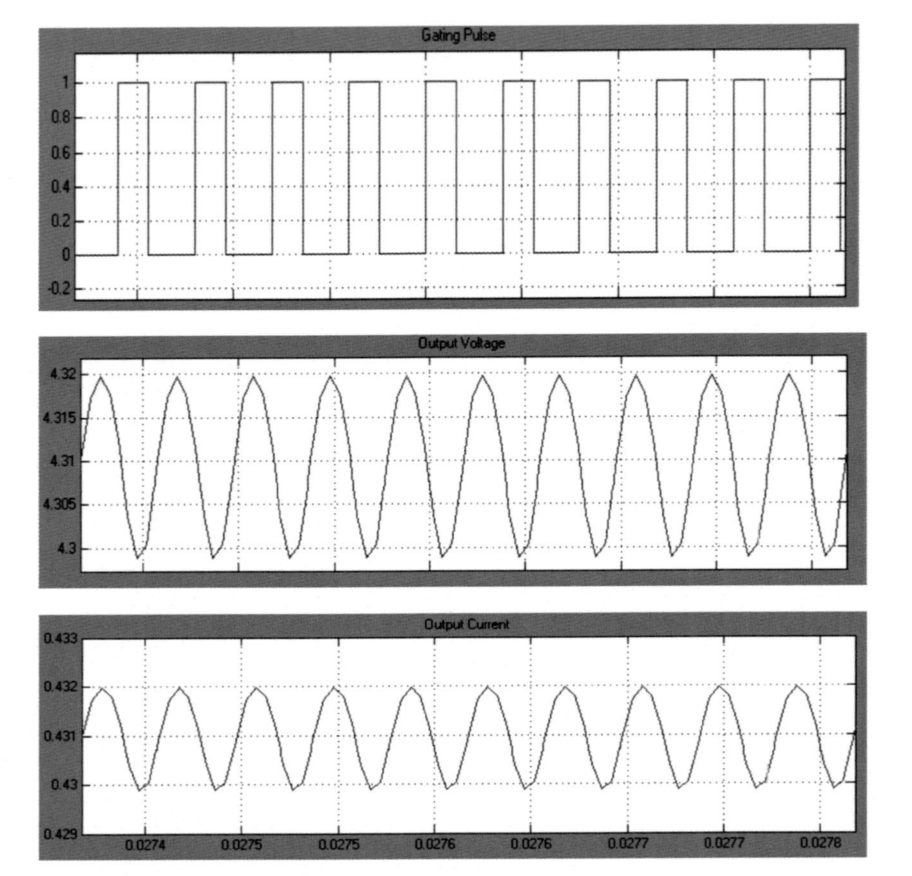

FIGURE 1.26 Simulation waveforms: gating pulse, output voltage, and current of buck converter.

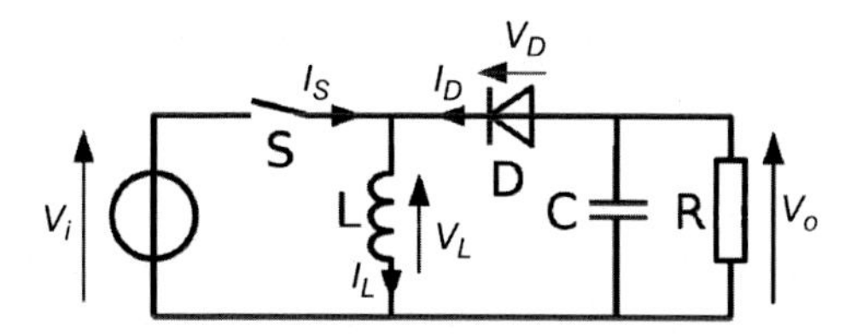

FIGURE 1.27 Circuit diagram of buck boost converter.

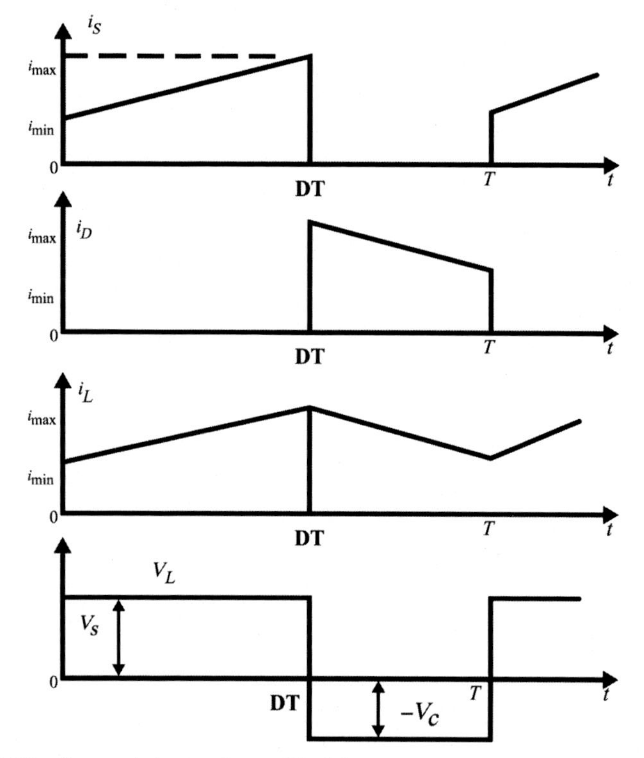

FIGURE 1.28 Characteristic waveforms of buck boost converter.

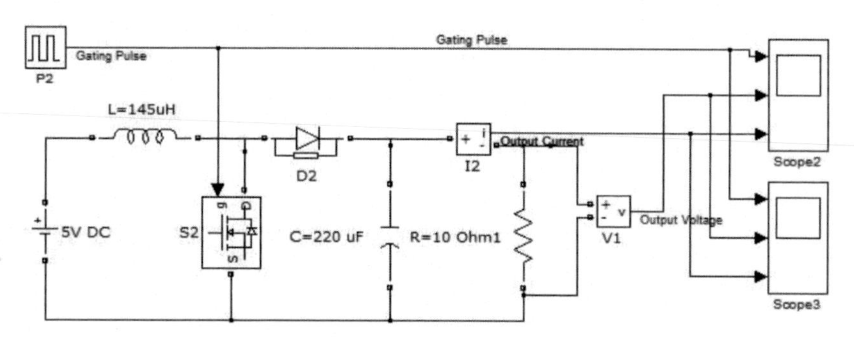

FIGURE 1.29 Simulink model: boost converter.

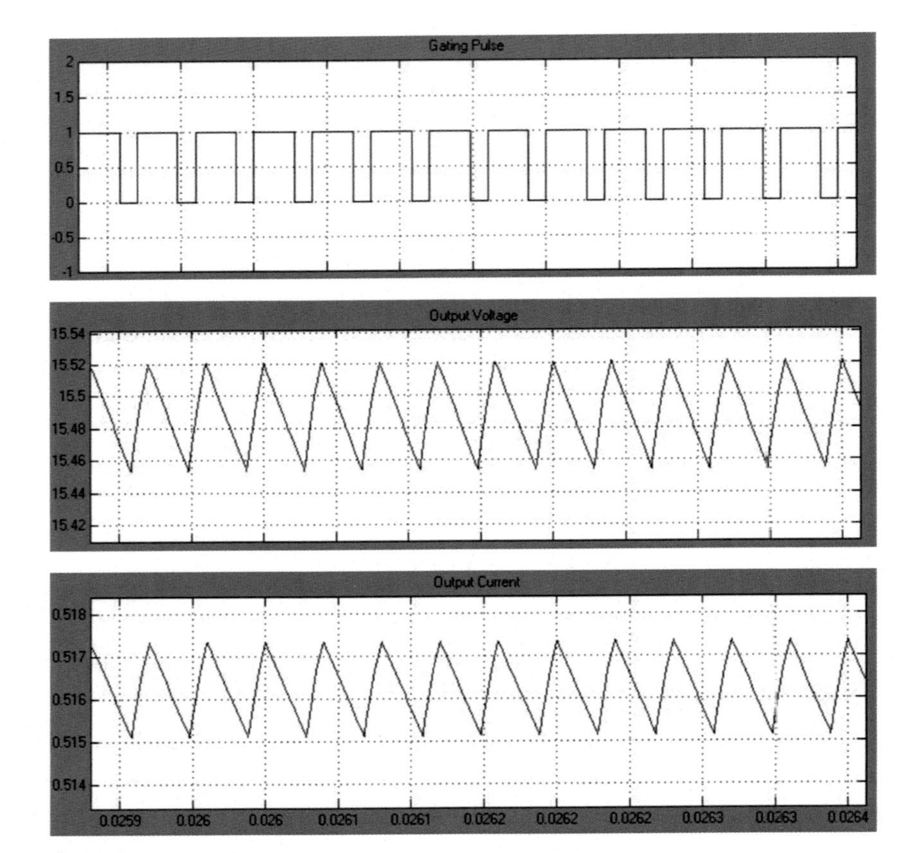

FIGURE 1.30 Simulation waveforms: boost converter.

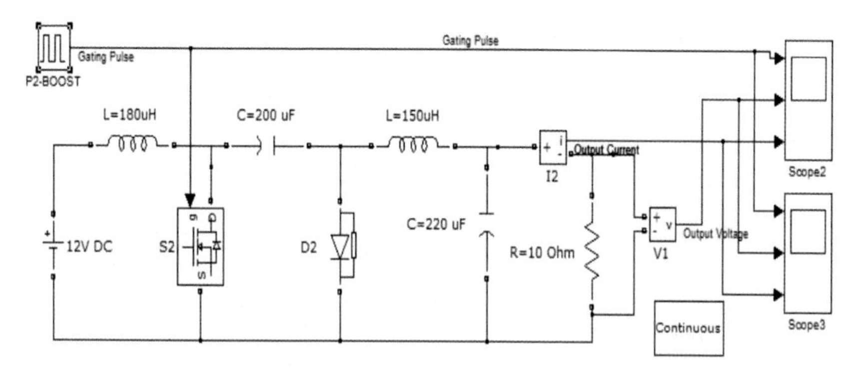

FIGURE 1.31 Simulation model: buck boot converter.

On the other hand, S6 is turned on during the negative half cycle and current freewheels through antiparallel diode of S5, S6 and the grid. PV is disconnected from the grid during the zero voltage vectors because S1, S4 or S2, S3 are all in off state. Hence, highly efficient and reliable inverter concept (HERIC) topology is also known as AC decoupling topology because PV is disconnected from the grid by short circuiting the AC side of the inverter. As shown in Fig. 1.44, the output voltage of the HERIC topology consists of three levels and the voltage to ground of stray capacitance is sinusoidal in shape. Leakage current is very small due to constant common mode voltage as shown in Fig. 1.45.

1.1.1.3 H6 topology

According to HERIC inverter, high frequency common mode voltage can be totally avoided if the voltage of the freewheeling path is clamped to the half of the input voltage. H6 topology, for this reason, is therefore planned. Fig. 1.46 shows the H6 inverter topology. Two additional switches and diodes, S5, S6, and D5, D6, are added to the conventional full bridge inverter. In the positive half cycle, S1 and S4 are turned on continuously. S5 and S6 commutate simultaneously at switching frequency while S2 and S3 commutate together and complementarily to S5 and S6. During the positive zero voltage vector, S5 and S6 are turned off and S2 and S3 are turned on. Hence, freewheeling current finds its path in two ways: S1 and the antiparallel diode of S3, and S4 and the antiparallel diode of S2.

On the other hand, in the negative half cycle, S2 and S3 are turned on continuously. S5 and S6 commutate simultaneously at switching frequency while SI and S4 commutate together and complementarily to S5 and S6. During the negative zero voltage vector, S5 and S6 are turned off and S1 and S4 are turned on. Hence, freewheeling current finds its path in two ways: S3 and the antiparallel diode of S1, and S2 and the antiparallel diode of S4. PV is disconnected from grid by the use of S5 and S6.

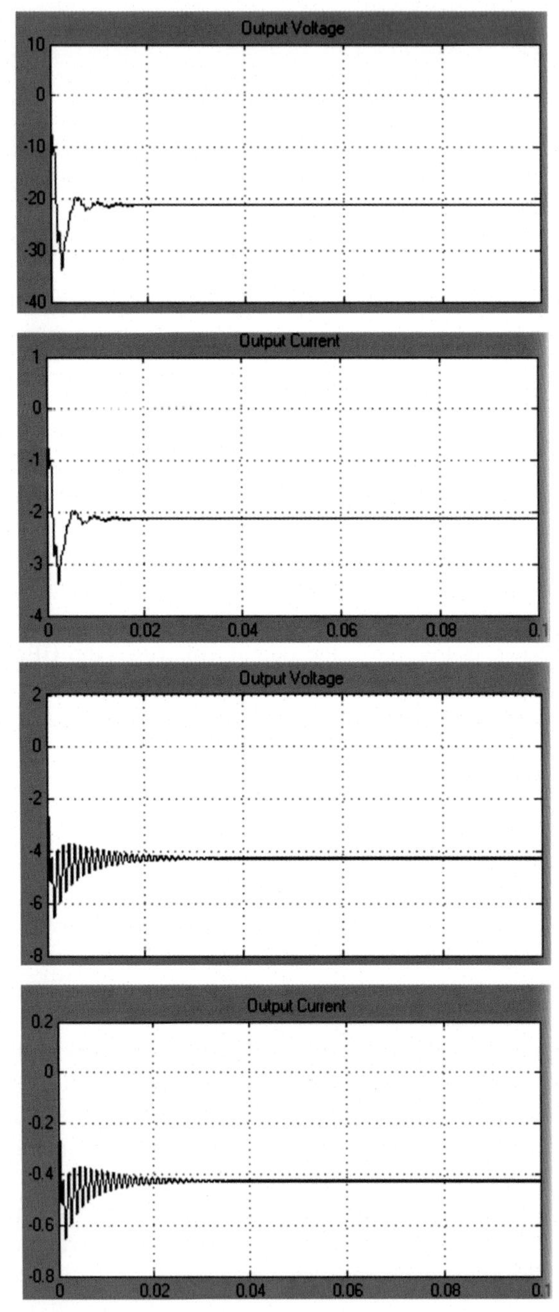

FIGURE 1.32 Simulation waveforms: buck boost converter.

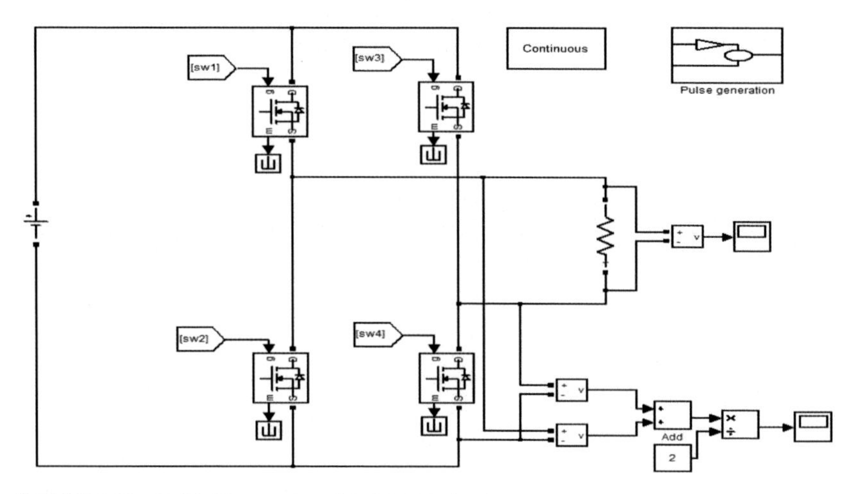

FIGURE 1.33 Full bridge topology SIMULINK circuit.

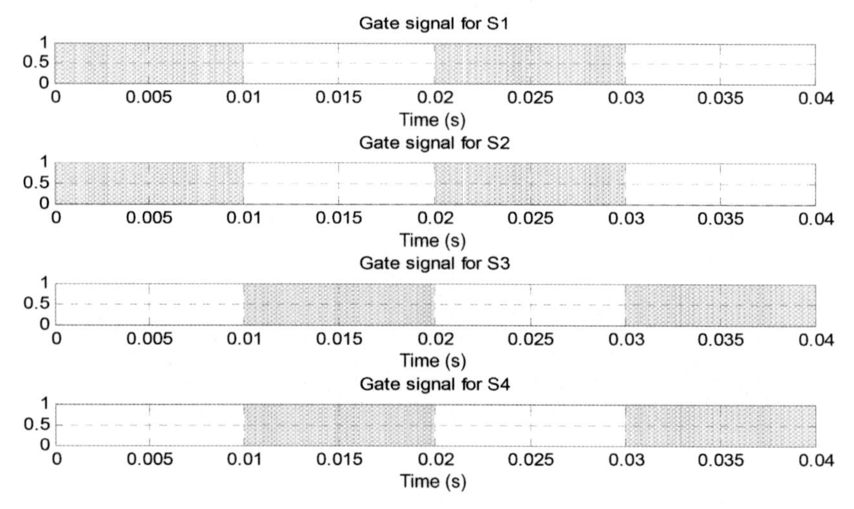

FIGURE 1.34 Unipolar pulse pattern for full bridge inverter.

As shown in Fig. 1.47, the output voltage of the H6 topology consists of three levels. The topology and control strategy guarantees constant common mode voltage which generates very small leakage current as shown in Fig. 1.48. High efficiency is achieved without compromising the common mode behavior. In conclusion, H6 topology is suitable for transformerless grid connected PV system (Fig. 1.49).

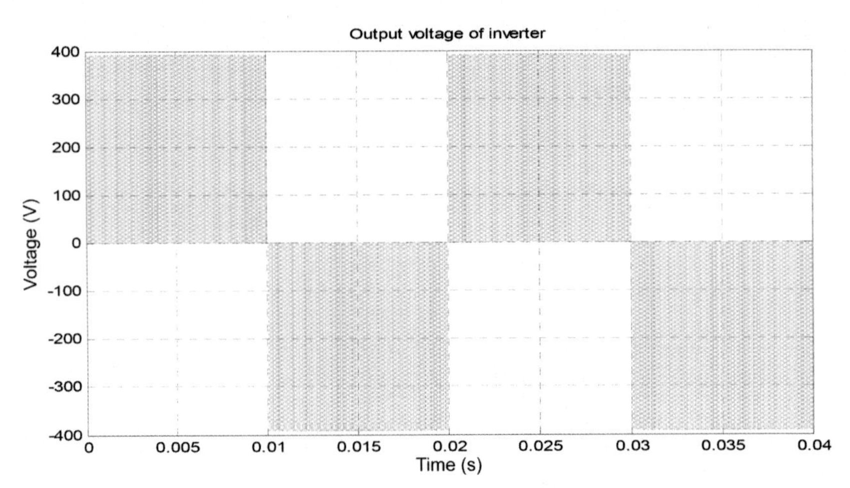

FIGURE 1.35 Output voltage of full bridge inverter with unipolar PWM. *PWM*, Pulse width modulation.

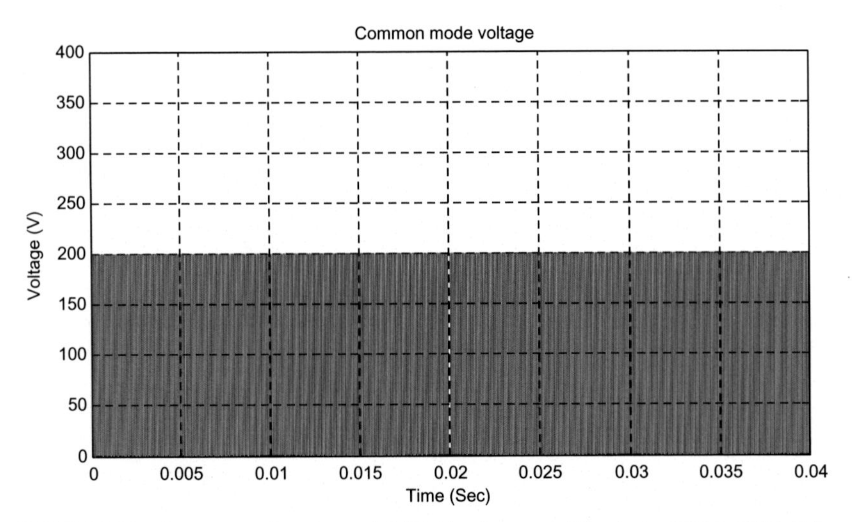

FIGURE 1.36 Common mode voltage of full bridge inverter with unipolar PWM. *PWM*, Pulse width modulation.

1.1.1.4 oH5 topology

Compared to H6, oH5 is used to set freewheeling path voltage to half the input voltage to keep the normal mode voltage from happening entirely. S5 is continuously allowed in the positive half cycle. S1 and S6 switch at frequency and in addition to S2 and S5 simultaneously. S1, S3, and S6 flow then. Current freewheels via S3 and S5 antiparallel diode are possible during

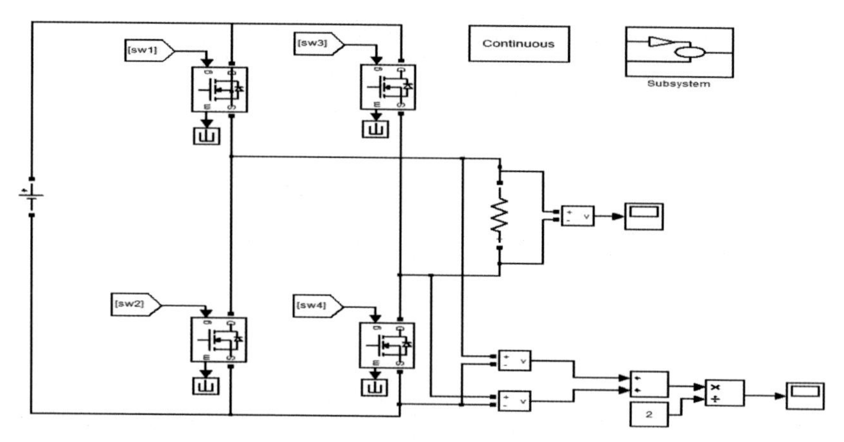

FIGURE 1.37 Full-bridge topology SIMULINK circuit.

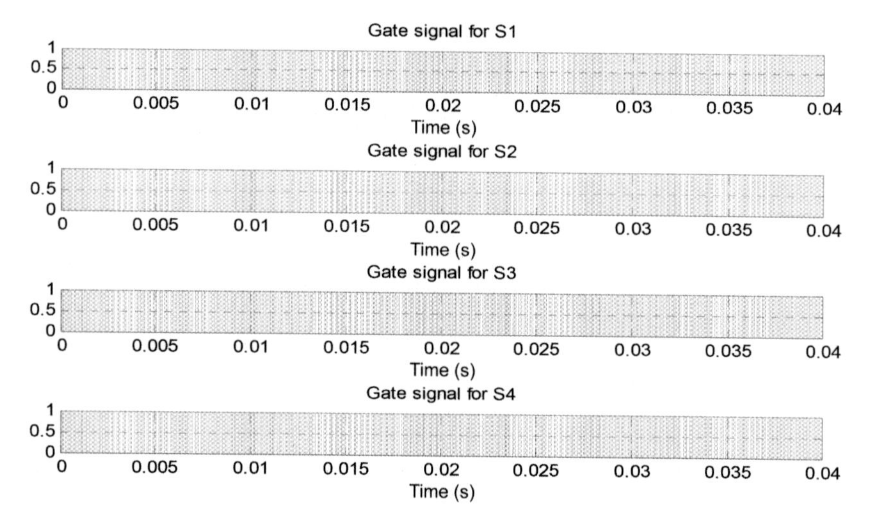

FIGURE 1.38 Bipolar pulse pattern for full bridge inverter.

the positive zero voltage vector. To insure the freewheel direction is maintained at half the input voltage, S2 is turned on.

S3 is activated constantly, on the other hand, in the negative half cycle. At the frequency switching and in addition to S2 and S3, S1 and S4 switch concurrently. S1, S5 and S4 are currently in circulation. Present freewheels by S5 and antiparallel S3 diodes are used during the adverse zero voltage path. S2 is allowed to ensure that half of the input voltage is fastened to the freewheeling path.

The current tension of oH5 topology is in three stages, as shown in Fig. 1.50, and is ground voltage. Due to the constant typical voltage mode

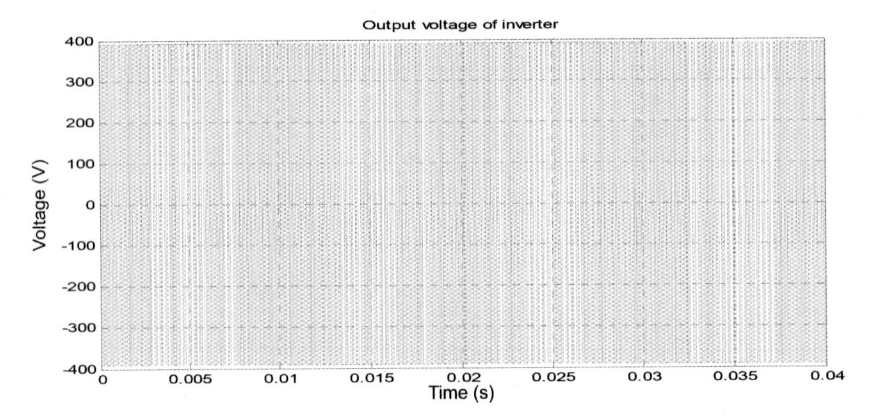

FIGURE 1.39 Output voltage of full bridge inverter with bipolar PWM. *PWM*, Pulse width modulation.

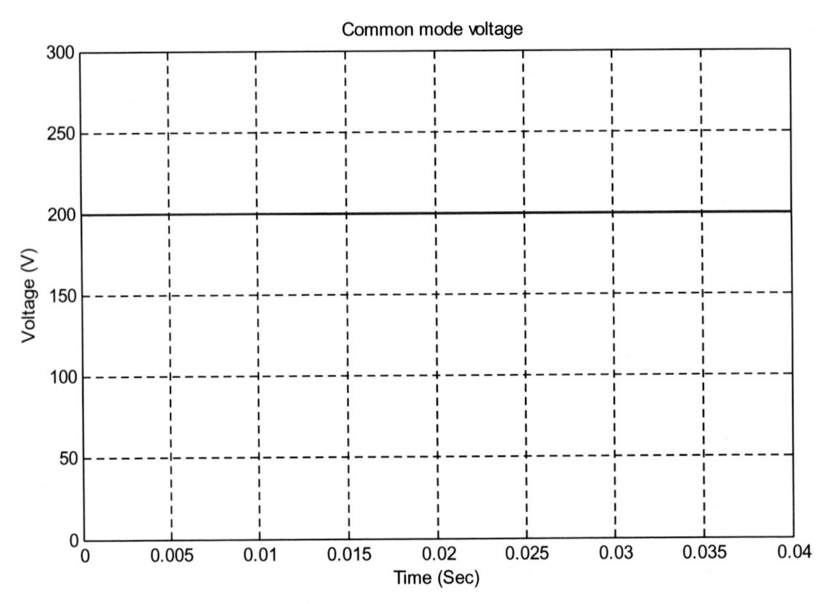

FIGURE 1.40 Common mode voltage of full bridge inverter with bipolar PWM. *PWM*, Pulse width modulation.

shown in Fig. 1.51, leakage current was very high. This topology brings together both the benefits of unipolar and bipolar modulation. Without violating typical mode behavior, high efficiency is achieved. Eventually, oH5 is suitable for the PV network attached to the transformerless grid.

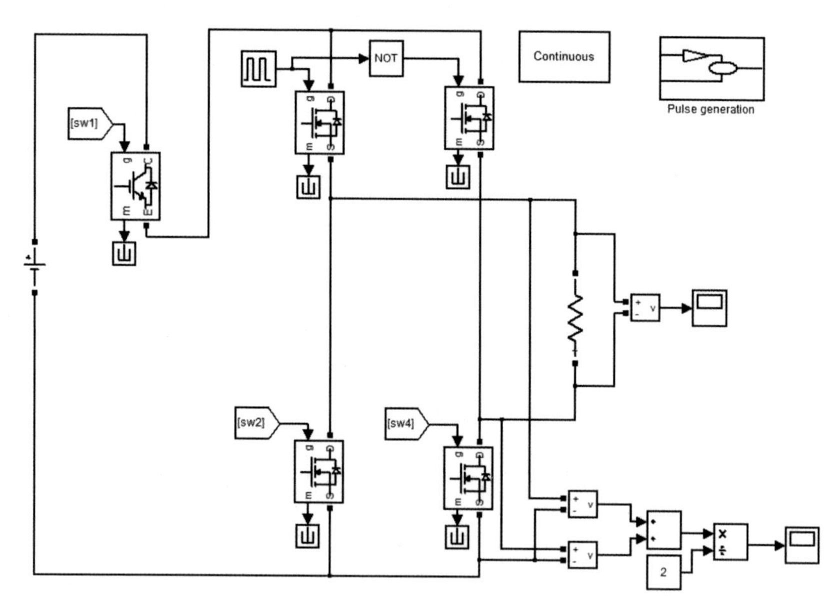

FIGURE 1.41 H5 topology SIMULINK circuit.

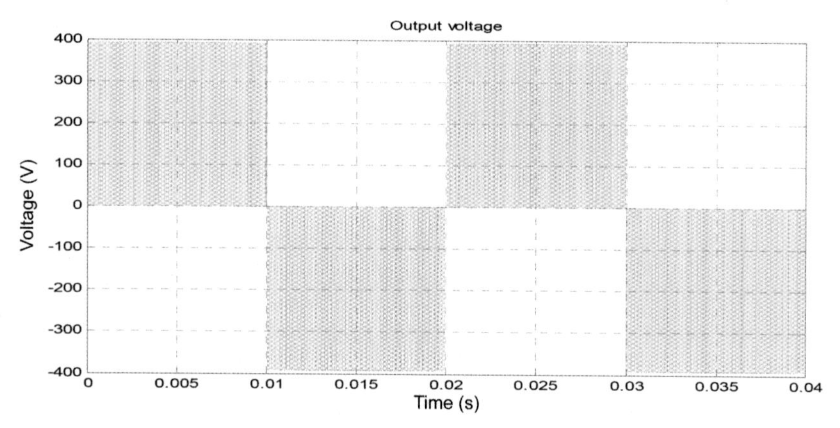

FIGURE 1.42 Output voltage of H5 inverter.

Multilevel inverter (MLI) is a power electronic system that offers the desired voltage alternation at the output with multiple lower voltages DC as input. In order to generate the AC voltage from DC voltage, a two-style inverter is usually used. The problem now emerges what is needed when we have a two-story converter with multistandard inverters. We must first look

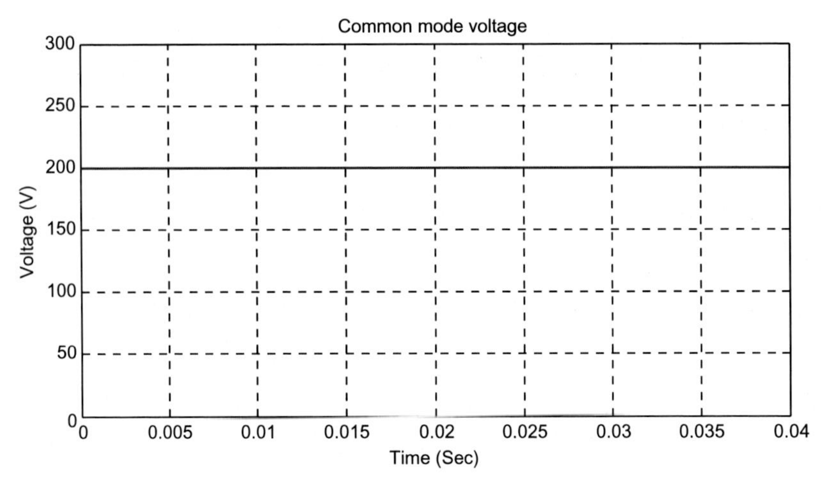

FIGURE 1.43 Common mode voltage of H5 inverter.

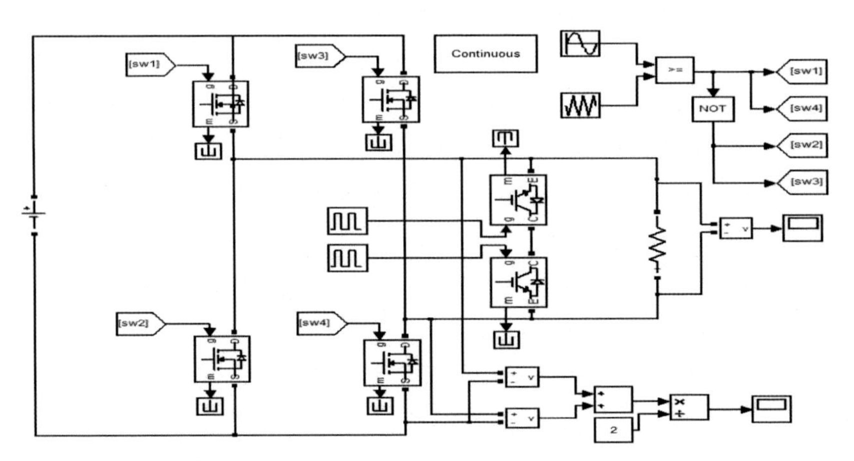

FIGURE 1.44 Highly efficient and reliable inverter concept topology SIMULINK circuit.

at the idea of multidecker inverters in order to answer this issue (Figs. 1.52 and 1.53).

Introduction

Consider a two-tier inverter case first. A double-level inverter produces two separate voltages, that is, supposing that we provide Vdc as a two-level

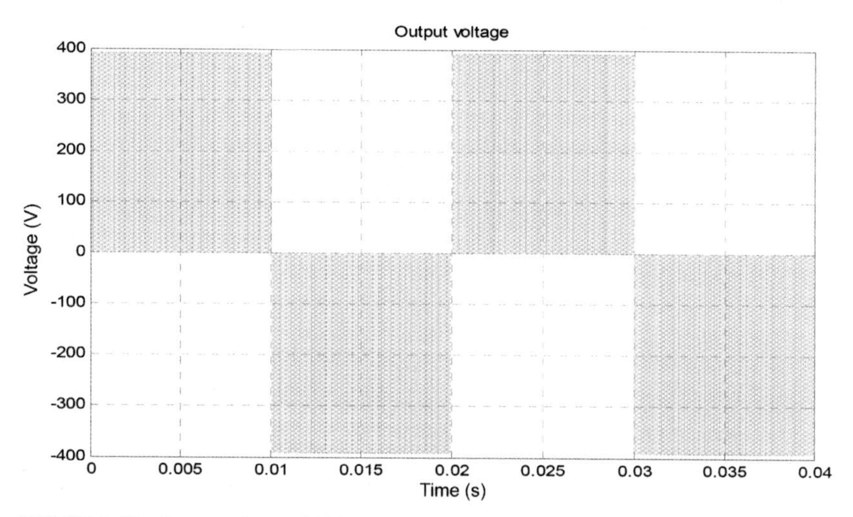

FIGURE 1.45 Output voltage of highly efficient and reliable inverter concept inverter.

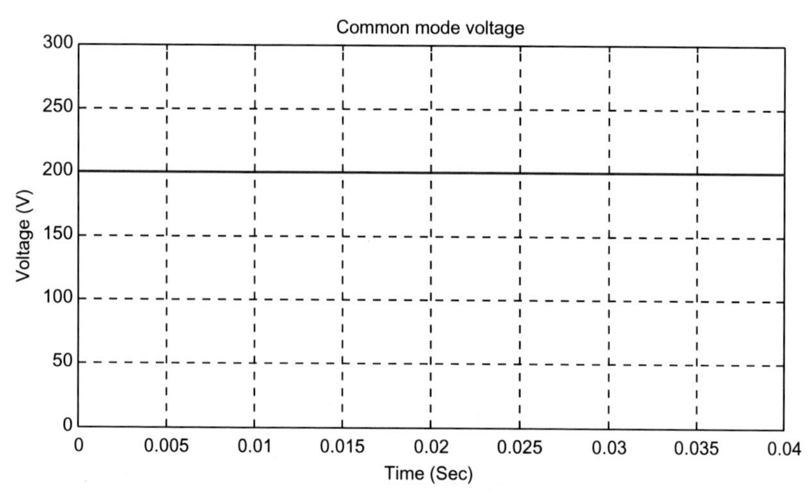

FIGURE 1.46 Common mode voltage of Highly efficient and reliable inverter concept inverter.

inverter input, the output is + Vdc/2 and − Vdc/2. Such two newly generated voltages are normally exchanged for the creation of an AC voltage. The reference wave in the dashed blue line is used for switching mostly PWM as shown in the Fig. 1.54. While this AC generation approach is successful, it does have little downside as it causes harmonic voltage distortions and has high d*v*/d*t* compared to a multiple-level inversor. This method works

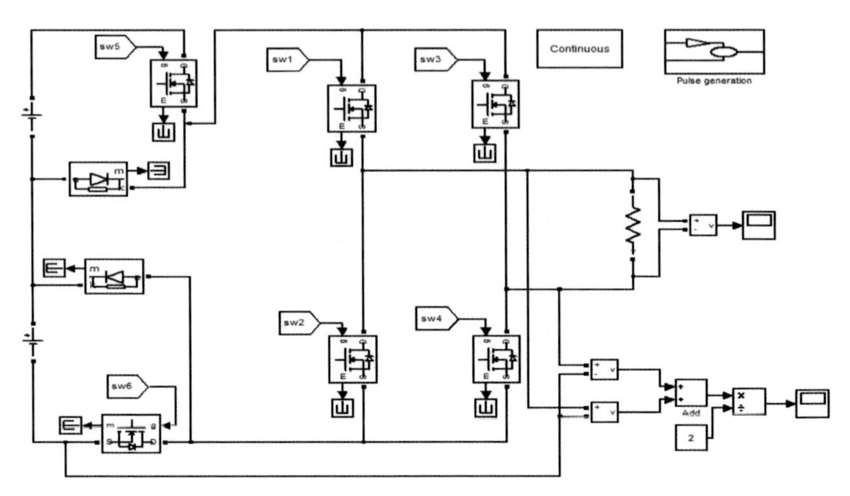

FIGURE 1.47 H6 topology SIMULINK circuit.

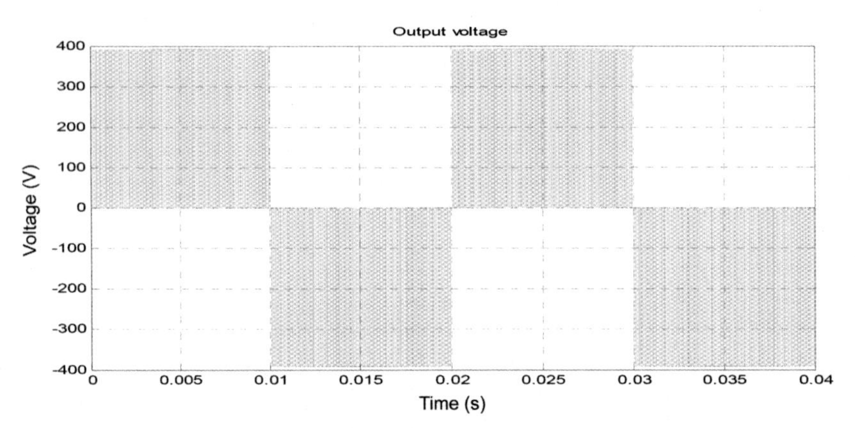

FIGURE 1.48 Output voltage of H6 inverter.

normally, but it creates problems in a few applications, particularly in those where low output voltage distortion is required.

Concept

The concept of a MLI is a modification of a two-level inverter. In MLIs we do not deal with the two level voltage instead to create a smoother stepped output waveform; more than two voltage levels are combined together and the output waveform obtained in this case has lower d*v*/d*t* and also lower harmonic distortions. Smoothness of the waveform is proportional to the voltage levels; as we increase the voltage level the waveform becomes

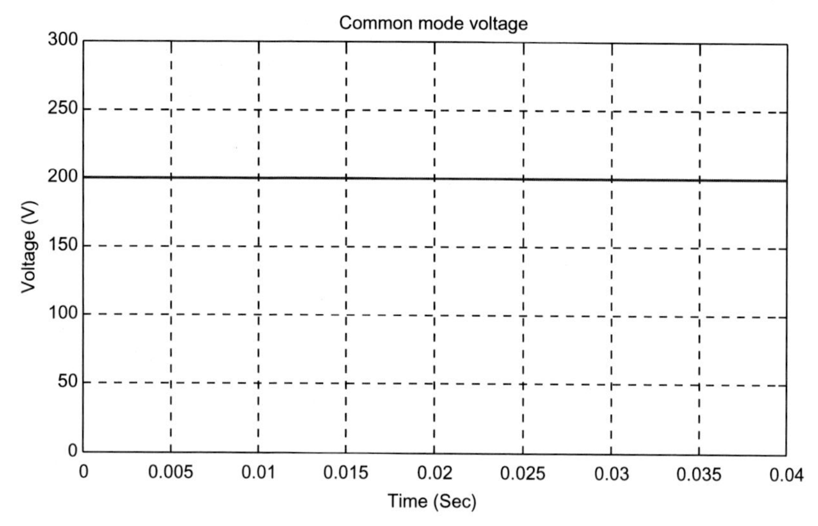

FIGURE 1.49 Common mode voltage of H6 inverter.

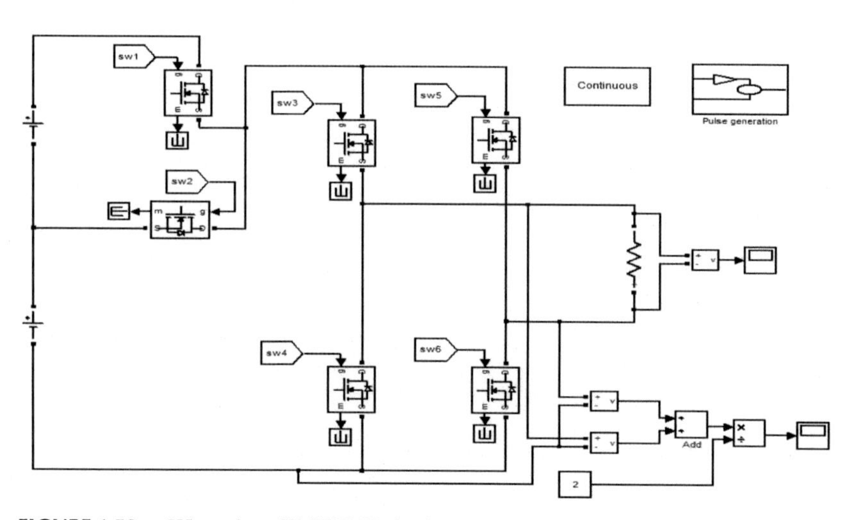

FIGURE 1.50 oH5 topology SIMULINK circuit.

smoother, but the complexity of controller circuit and components also increases along with the increased levels. The waveform for the three-, five-, and seven-level inverters is shown in Fig. 1.54, where we clearly see that as the levels are increasing, waveform becoming smoother.

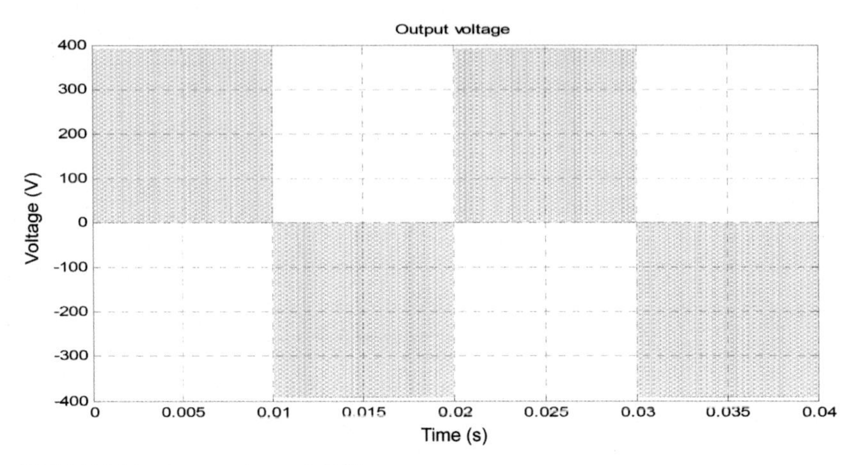

FIGURE 1.51 Output voltage of oH5 inverter.

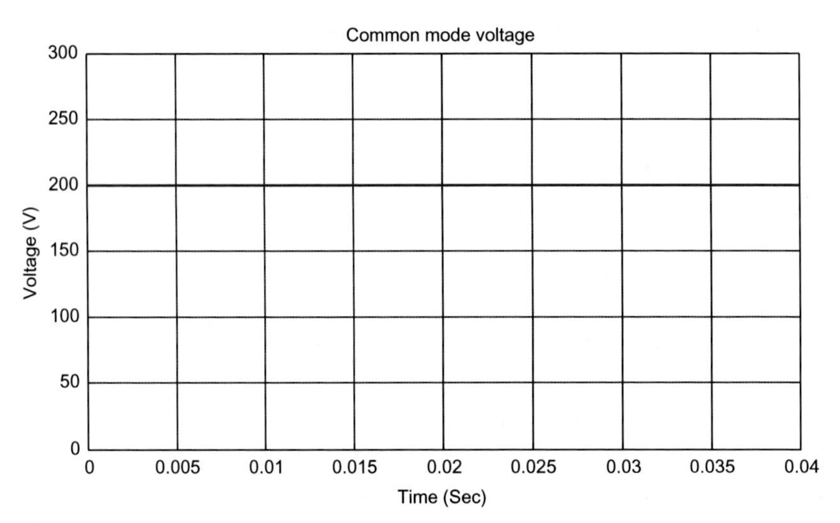

FIGURE 1.52 Common mode voltage of oH5 inverter.

Types of multilevel inverters

MLI is a type of two-level inverter modification. In the case of MLIs, then, we do not work with two levels of voltage to produce a more linear step-by-start waveform and to add more than two voltage levels, and there are less d*v*/d*t* and lower harmonic fluctuations on the output waveform produced. The waveform smoothness is proportional to the voltage level since the waveform increases its voltage level, but the control circuit complexity and the components also increases in conjunction with the increase in the

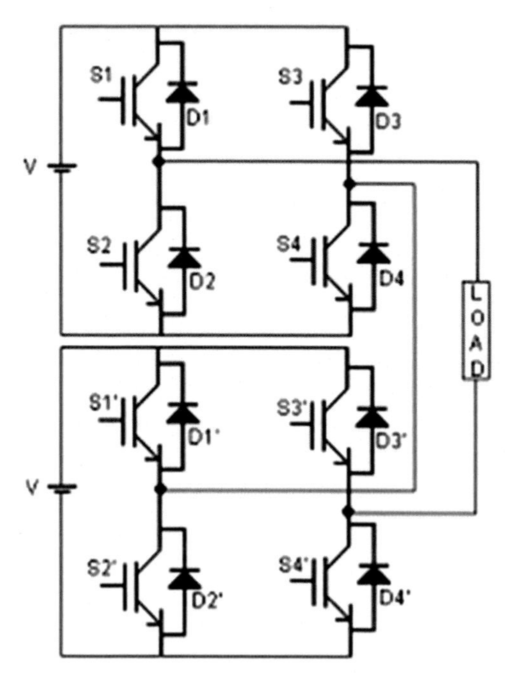

FIGURE 1.53 Cascaded H-bridge multilevel inverter.

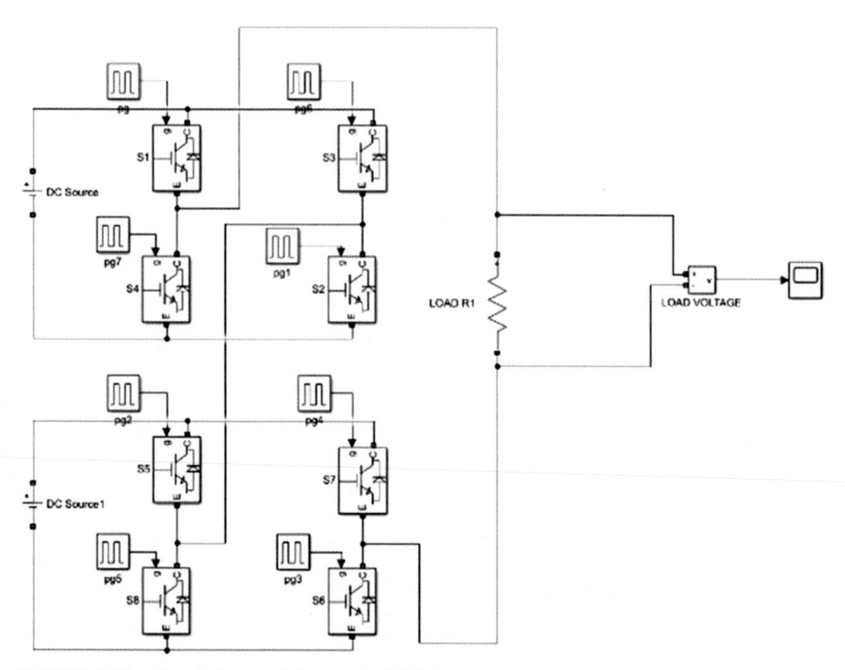

FIGURE 1.54 Simulation model: cascaded H-bridge inverter.

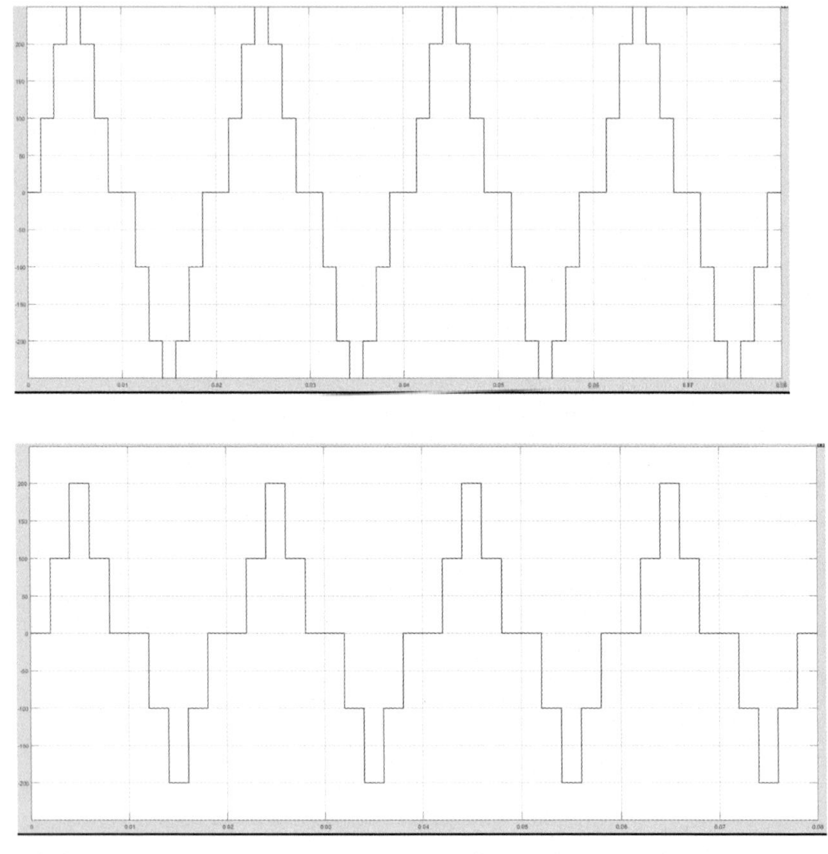

FIGURE 1.55 Simulation waveforms: cascaded H-bridge inverter.

voltage. In Fig. 1.54, we clearly see that the waveform for three, five, and seven levels of the inverter becomes smoother when the levels increase.

MLI topologies are given. The distinction is the switching function and the input voltage source for the MLIs. Three most widely used multilevel reverse geometry topologies include cascaded multilevel reverse-cutter H-bridge, diode MLI clamped, and flying capacitor multilevel cascaded H-bridge MLIs. The output voltages of the single H-bridge cells are the total voltages generated by each cell, for example, if there are k cells of the multilevel H-bridge inverter, the output voltage numbers will be $2k + 1$. Each cell contains one H-bridge. The inverter form is desirable in contrast with the other two types of inverters because it needs less components than the other two and therefore has a smaller overall weight and size. The K level cascaded H-bridge inverter is illustrated in Fig. 1.52.

Each phase is connected with a single DC source in a single phase inverter. Each degree produces three positive, negative, and zero voltages.

This can be accomplished by connecting the AC source to the DC output and by using different four switch combinations. If two switches in opposite positions are on, the inverter must stay on. When all the inverters switch on or off, it will turn off. Switching angles are specified and applied to minimize the total harmonic distortion. The equations for the moving angle estimation will be similar. Two H-bridge inverters are cascaded in five stages with the cascading H-bridge MLIs. It has five output levels and uses eight switching devices to power, while four H-bridge inverters are cascaded on nine levels. There are 9 performance stages, and 16 instruments are used and tested.

The picture is shown in the diagram. Cascade H-bridge MLI are mostly used for static var applications in renewable energy sources as well as battery-driven applications. Cascading MLI cascading H-bridge MLIs multilevel cascaded H bridge inverters can be used as a delta or wye. The inverter is regulated by controlling the power factor. It is safest when used as a solar cell or fuel cell.

Cascade H-bridge MLI benefits include the following: (1) outcome voltages are from twice as many sources; (2) manufacturing is easy and quick; (3) packaging and layout is modularized; and (4) easily adjustable with cascade H-bridge MLIs. However, disadvantages also remain: Every H-bridge needs a separate DC source for application limited by large numbers of sources.

1.1.1.5 Simulation circuit and results

Simulation circuit and results of cascaded H-bridge inverter (Fig. 1.55).

Chapter 2

PSIM Simulation Practices

Chapter Outline

2.1 Introduction to PSIM

2.1.1 Introduction

Power simulations (PSIM) is a simulation package specifically designed for power electronics and motor control. With fast simulation and a friendly user interface, PSIM provides a powerful simulation environment for power electronics, analog and digital control, and motor drive system studies. The PSIM simulation package consists of three programs: circuit schematic program PSIM, PSIM simulator, and a waveform processing program, SIMVIEW. It has the following add-on modules; the overall environment is shown in Fig. 2.1.

Software Tools for the Simulation of Electrical Systems. DOI: https://doi.org/10.1016/B978-0-12-819416-4.00002-8

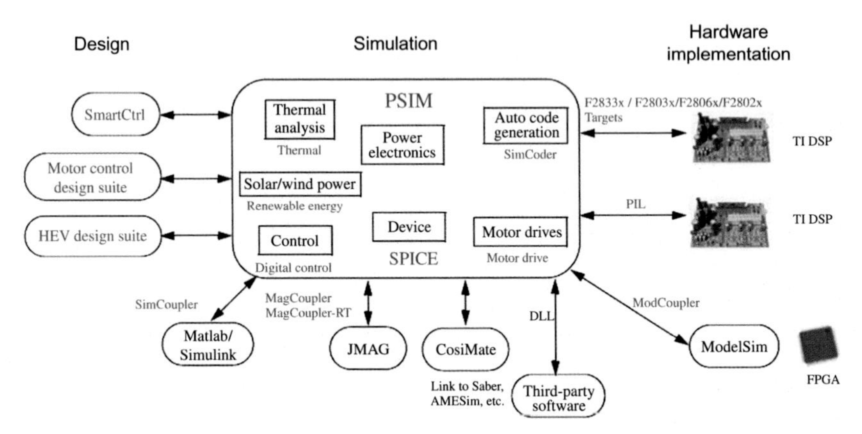

FIGURE 2.1 PSIM overall environment.

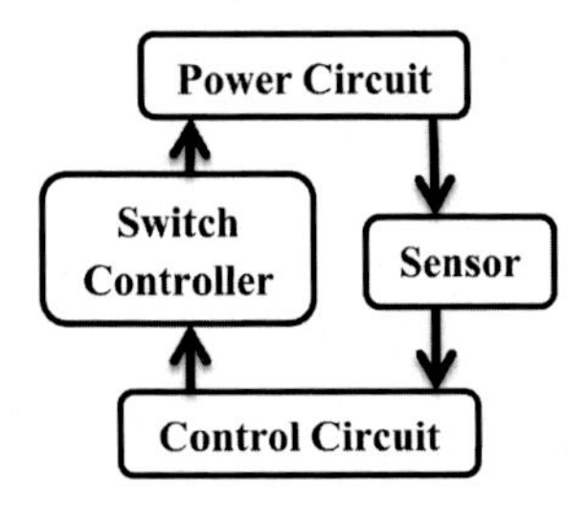

FIGURE 2.2 PSIM circuit structure.

2.1.2 Circuit structure

A circuit is represented in PSIM in four blocks: power circuit, control circuit, sensors, and switch controllers. The figure below shows the relationship between these blocks (Fig. 2.2). The power circuit consists of switching devices, RLC branches, transformers, and coupled inductors. The control circuit is represented in a block diagram. Components in the s domain and z domain, logic components (such as logic gates and flip flops), and nonlinear components (such as multipliers and dividers) are used in the control circuit. Sensors measure power circuit voltages and currents and pass the values to the control circuit. Gating signals are then generated from the control circuit and sent back to the power circuit through switch controllers to control switches.

2.1.3 Software/Hardware requirement

PSIM runs in Microsoft Windows 7/8/10 on personal computers. The minimum RAM memory requirement is 1 GB.

2.1.4 Installing the program

A quick installation guide is provided in the flier "PSIM – Quick Guide" and on the CD-ROM.

Some of the files in the PSIM directory are shown in Tables 2.1 and 2.2.

2.1.5 Simulating a circuit

To simulate the sample one-quadrant chopper circuit "chop.sch":

1. Start PSIM. Choose *Open* from the *File* menu to load the file "chop.sch."
2. From the *Simulate* menu, choose *Run PSIM* to start the simulation. The simulation results will be saved to File "chop.txt." Any warning messages occurring in the simulation will be saved to File "message.doc."

TABLE 2.1 PSIM directory.

Files	Description
psim.dll	PSIM simulator
psim.exe	PSIM circuit schematic editor
Simview.exe	Waveform processor SIMVIEW
Psim.lib, psimimage.lib	PSIM libraries
*.hlp	Help files
*.sch	Schematic files
SetSimPath.exe	Program to set up the SimCoupler Module
PcdEditor.exe	Device database editor

TABLE 2.2 PSIM file extensions.

*.psimsch	PSIM schematic file (binary)
*.psimpjt	PSIM project file
*.txt	PSIM simulation output file (text)
*.fra	PSIM ac analysis output file (text)
*.smv	PSIM waveform file (binary)
*.lib	PSIM library file
*.dev	Device database file

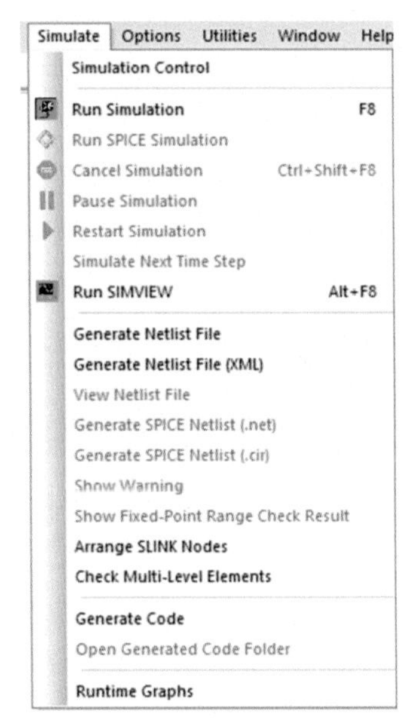

FIGURE 2.3 Simulate menu.

3. If the option *Auto-run SIMVIEW* is not selected in the *Options* menu, from the *Simulate* menu, choose *Run SIMVIEW* to start SIMVIEW. If the option *Auto-run SIMVIEW* is selected, SIMVIEW will be launched automatically. In SIMVIEW, select curves for display as shown in Fig. 2.3.

2.1.5.1 Running the simulation

There are two options to run a PSIM simulation: using the Simulate menu or with a Command Line (Table 2.3).

To view the simulation results in the middle of the simulation, one can either go to Simulate >> Runtime Graphs to select the waveforms, or use the voltage/current scopes (under Elements >> Other >> Probes) to view the waveforms (Fig. 2.4).

The difference between the runtime graphs and the voltage/current scopes is that only waveforms that are saved for display in SIMVIEW (such as voltage probes, current probes, current flags, etc.) are available for the runtime graphs. In addition, a runtime graph displays the waveform in its entirety, from the beginning to the final study time. Because of this, the runtime graphs are disabled in the free-run mode as the final study time is undetermined.

On the other hand, voltage/current scopes can be used in either the one-time simulation mode or in the free-run mode. Voltage scopes can be connected to any nodes, and will display the node-to-ground voltage waveforms.

TABLE 2.3 Simulate menu parameters.

Simulation Control	To set the simulation parameters, such as time step, total time, etc. When this is selected the cursor will change to the image of a clock. Place this clock on the schematic, and double-click to display the property window
Run Simulation	To run the simulation
Cancel Simulation	To cancel the simulation that is currently in progress
Pause Simulation	To pause the simulation that is currently in progress
Restart Simulation	To resume a paused simulation
Simulate Next Time Step	To run the simulation to the next time step, and pause
Run SIMVIEW	To launch the waveform display program SIMVIEW
Run Parameter Sweep	To run parameter sweep simulation
Generate Netlist File	To generate the netlist file from the schematic
Generate Netlist File (xml)	To generate the netlist file in xml format from the schematic
View Netlist File	To view the generated netlist file
Show Warning	To show the warning messages, if any, from the simulation
Show Fixed-Point Range Check Result Arrange SLINK Nodes	To display the fixed-point range check result
	To rearrange the SLINK nodes. This function is for the SimCoupler Module for cosimulation with Matlab/Simulink
Generate Code	To generate code from the control schematic. This function is for SimCoder for automatic code generation. Please refer to *SimCoder User Manual* for more details
Open Generated Code Folder	To open the folder where the generated codes are located
Runtime Graph	To select waveforms to show in the middle of a simulation run

On the other hand, current scopes are available to elements that have current flags (such as R-L-C branches and switches). Moreover, in the free-run mode the majority of the element parameters can be changed during runtime in the middle of the simulation. This makes it possible to tune a circuit while

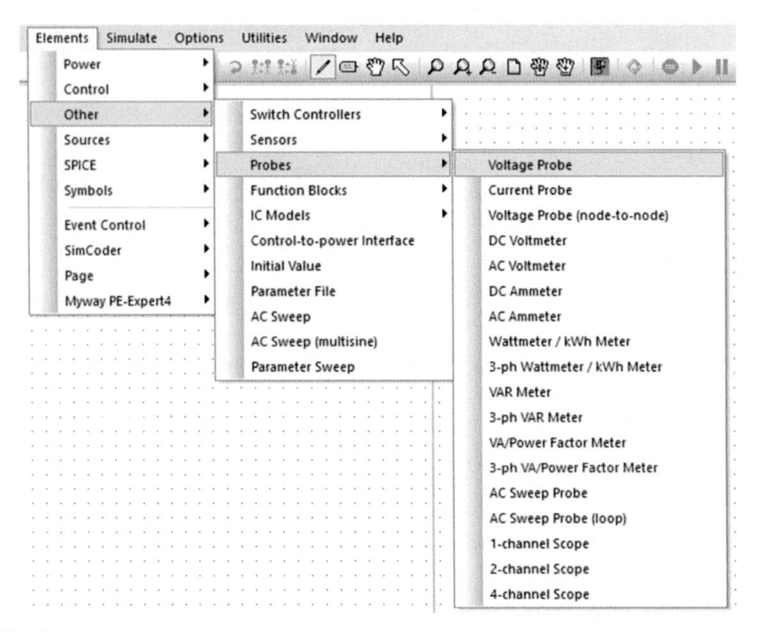

FIGURE 2.4 Probes.

inspecting key waveforms using voltage/current scopes, until the desired performance is achieved.

Simulation can also be launched with the command line option by running the program PsimCmd.exe. For example, to simulate the circuit "buck.psimsch" which is stored in the folder "c:\Powersim\examples", go to the PSIM folder, and run the following command:

```
PsimCmd -i "c:\psim\examples\buck.psimsch" -o "c:\psim\examples
\buck.smv"
```

The format of the command line is as follows:

```
PsimCmd -i "[input file]" -o "[output file]" -v
"VarName1=VarValue1" -v "VarName2=VarValue2" –t "TotalTime" -s
"TimeStep" -g
```

Note that the quotes around the parameter values must be present. The command-line parameters are:

-i: Input schematic file name
-o: Output file name (in either .txt or .smv extension)
-v: Variable name and value. This parameter can be used multiple times. For example, to define the resistance R1 as 1.5 and the inductance L1 as 0.001,
we have: -v "R1=1.5" -v "L1=0.001"

`-t`: Total time of the simulation
`-s`: Time step of the simulation
`-g`: Run SIMVIEW after the simulation is complete

Note that the number of variables that can be defined in a command line is limited to 30.

With the command-line option, one can run several circuits automatically in a batch run.

2.1.6 Simulation control

The simulation control element defines parameters and settings related to simulation. To place the simulation control in the schematic, go to the Simulate menu, and select.

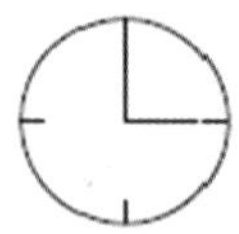

There are four tabs in the simulation control dialog:

PSIM: Define parameters for PSIM transient simulation.

SPICE: Define the analysis type and parameters for SPICE simulation.

SimCoder: Define the hardware for SimCoder simulation and automatic code generation.

Color: Define the color of the simulation control image. The default color is blue.

2.1.6.1 PSIM tab

The following parameters (Fig. 2.5) are present under the PSIM tab (Table 2.4),

2.1.6.2 SPICE tab

There are three types of analysis in SPICE: Transient, AC, and DC. Simulation parameters of different analysis types are described in Tables 2.5 and 2.6.

2.1.7 Component parameter specification and format

The parameters in the *Parameters* tab are used in the simulation. The information in the *Other Info* tab, on the other hand, is not used in the simulation. It is for reporting purposes only and will appear in the parts list in *View >> Element List* in PSIM. Information such as device rating, manufacturer, and part number can be stored under the *Other Info* tab. The component color can be set in the *Color* tab (Fig. 2.6).

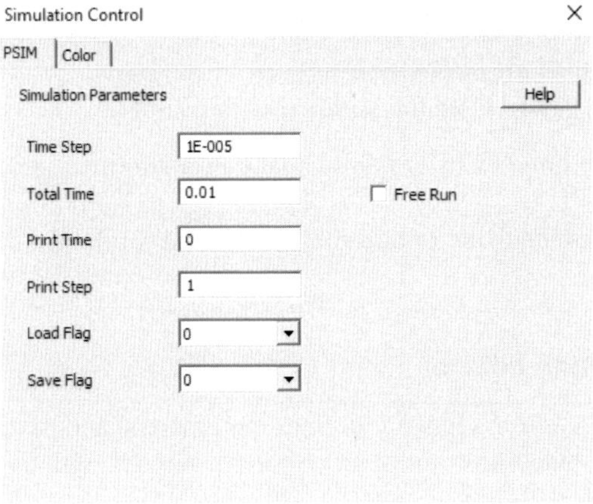

FIGURE 2.5 PSIM tab.

TABLE 2.4 PSIM tab.

Time Step	**Simulation time step, in seconds**
Total Time	Total simulation time, in seconds
Free Run checkbox	When the *Free Run* checkbox is not checked, the simulation will run up to the *Total Time* and then stop But when it is checked, the simulation will run in the free-run mode and it will keep on running until manually stopped In the free-run mode, voltage/current scopes can be used to monitor and display voltages and currents in the middle of the simulation
Print Time	Time from which simulation results are saved to the output file. No output is saved before this time
Print Step	Print step. If it is set to 1, for example, every data point will be saved to the output file. If it is set to 10, only one out of 10 data points will be saved. This helps to reduce the output file size
Load Flag	Flag for the LOAD function. If the flag is 1, the previous simulation values (saved by setting the *Save Flag*) will be loaded from a file (with the ".ssf" extension) as the initial conditions
Save Flag	Flag for the SAVE function. If the flag is 1, values at the end of the current simulation will be saved to a file with the ".ssf" extension

Power-of-ten suffix letters are allowed in PSIM (Table 2.7). The following suffix letters are supported

A mathematical expression can contain brackets and is not case sensitive (Table 2.8). The following mathematical functions are allowed:

TABLE 2.5 SPICE tab.

Transient analysis parameters	Use Initial Conditions: If this box is checked, element initial conditions will be used, and the "UIC" option will be added to the. tran command
	Time Step: Suggested computing increment and plotting increment T_{step}, in seconds. It can be left undefined (blank)
	Max Step: The maximum step size T_{max} that SPICE uses, in seconds. By default, the program uses either T_{step} or $(T_{end}-T_{start})/50$, whichever is smaller
	T_{max} is useful when one wishes to guarantee a computing interval not greater than T_{max}. It can be left undefined
	Start Time: The initial time T_{start}, in seconds. The transient analysis always begins at time zero. In the interval from zero to T_{start}, the circuit is simulated, but no outputs are stored. This parameter is equivalent to Print Time in PSIM simulation parameters. It can be left undefined
	End Time: The final simulation time T_{end}, in seconds
AC analysis parameters	Start Freq: Start frequency, in Hz
	End Freq: End frequency, in Hz
	Dec: Number of points per decade. If octive or linear variations are desired, one may use the SPICE Directive Block to write the analysis command
DC analysis parameters	Voltage/Current: Specify if the source is a voltage source or current source
	Start: Starting value, in V or A
	Step: Incremental value, in V or A
	End: End value, in V or A
Operating Point	If the Enable box is checked, SPICE simulation will determine the dc operating point of a circuit with inductors shorted and capacitors opened
Step Run Option	If the Enabled box is checked, SPICE simulation will perform parameter sweep. The parameter sweep definition is as follows:
	Parameter: The name of the parameter to be swept
	Start: Starting value
	Step: Incremental value
	End: Final value
Error Tolerance	If the Enabled box is checked, error tolerances for SPICE simulation can be changed.

(*Continued*)

TABLE 2.5 (Continued)

Option	Otherwise, default values will be used. Error tolerances are:
	RELTOL: Relative tolerance
	TRTOL: Transient tolerance
	VNTOL: Absolute voltage error tolerance
	ABSTOL: Absolute current error tolerance
	CHGTOL: Charge tolerance

2.2 Spice libraries

A PSIM library element consists of two parts: the netlist part and the image part. The netlist part comes from the netlist library, and there is only one netlist library, psim.lib. The netlist library cannot be edited.

The image part comes from an image library. There can be multiple image libraries, and all the image libraries in the PSIM directory will be automatically loaded into PSIM. The standard image library provided by PSIM is psimimage.lib. This file also cannot be edited. However, in order to facilitate users to copy images from the standard image library, the standard image library can be viewed by going to Edit >> Edit Library >> Edit library files, and choosing psimimage.lib.

Users can create their own custom image libraries. To create a new custom image library, go to Edit >> Edit Library >> Edit library files, and click on New library. Then define the library name as it appears in the PSIM Elements menu, and the library file name. This library file will be created and placed in the PSIM directory.

To edit an image library, go to Edit >> Edit Library >> Edit library files, and select the library file. The figure below shows the library editor dialog window. The dialog shows the menu tree of the library as well as various functions (Table 2.9).

To create a new element in the custom image library, click on New Element, and select the netlist that this element corresponds to from the list. For example, the netlist name of the resistor is "R." To create a new element called "My Resistor," select the netlist "R." To create the image for this resistor, click on Edit Image.

2.2.1 Creating a secondary image

It is possible that some users may find certain element images in the standard PSIM image library psimimage.lib to be different from what they are used to using. In this case, users can create their own secondary images.

TABLE 2.6 SimCoder tab.

Hardware Target	The hardware target can be one of the following:
	None: No hardware target in the circuit
	F2833x: F2833x Hardware Target for TI F2833x series DSP
	F2802x: F2802x Hardware Target for TI F2802x series DSP
	F2803x: F2803x Hardware Target for TI F2803x series DSP
	F2806x: F2806x Hardware Target for TI F2806x series DSP
	PE-Pro/F28335: PE-Pro/F28335 Hardware Target
	PE_Expert3: PE-Expert3 Hardware Target
Memory Map	Specify the memory map for compiler. For F2833x and F2803x hardware target:
Options	RAM Debug;
	RAM Release;
	Flash Release; and
	Flash RAM Release.
	For PE_Exper3 hardware target:
	PE-View9
	PE-View8
CPU Version	Specify the CPU version
	For F2833x: F28335, 28334, and 28332
	For F2803x: F28035, 28034, 28033, 28032, 28031, and 28030
	For F2802x: F28027, 28026, 28023, 28022, 28021, 28020, and 280200
	For F2806x: F28069, 28068, 28067, 28066, 28065, 28064, 28063, and 28062
InstaSPIN enabled	If the DSP is InstaSPIN enabled (for example F28069M), this box must be checked. Otherwise it should be unchecked
Default Data	This parameter is for fixed-point DSPs, such as F2803x. The default data type options are: Integer, IQ0, IQ1, ... IQ30
Type	If the box for Check Fixed-Point Range is checked, the SimCoder will check all the variables against the range and display the result
DMC Library	SimCoder has function blocks of all the functions in TI's DMC library for the following
Version	DMC versions: 4.0, 4.1, and 4.2
Comments	Comments can be entered and these comments will be inserted at the beginning of the automatically generated code

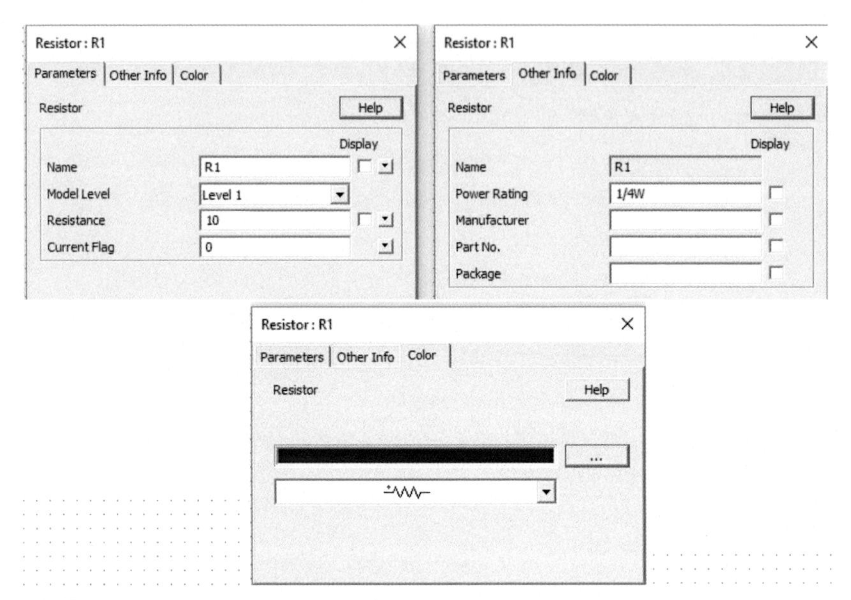

FIGURE 2.6 Component parameter specification.

TABLE 2.7 Suffix letters.

G	109
M	106
k or K	103
m	10−3
u	10−6
n	10−9
p	10−12

A secondary image can be created for an element in either standard image library psimimage.lib or the users' own custom image libraries. Secondary images are saved in a secondary image library with the .lib2 extension.

To illustrate the process, a secondary image will be created in the library "mylib.lib2" for the "Diode" element in the standard image library.

Go to *Edit >> Edit Library >> Edit secondary image library files*, and click on *New library*. In the dialog, define the secondary image library name as "mylib.lib2."

TABLE 2.8 Mathematical functions.

+	Addition
−	Subtraction
*	Multiplication
/	Division
^	To the power of
SQRT	Square-root function
SIN	Sine function
COS	Cosine function
ASIN	Sine inverse function
ACOS	Cosine inverse function
TAN	Tangent function
ATAN	Inverse tangent function
ATAN2	Inverse tangent function
SINH	Hyperbolic sine function
COSH	Hyperbolic cosine function
EXP	Exponential (base e)
LOG	Logarithmic function (base e)
LOG10	Logarithmic function (base 10)
ABS	Absolute function
SIGN	Sign function

Then select "mylib.lib2" and click on Edit selected library. The dialog window for editing secondary image library will appear.

Click on the Add button. From the PSIM library tree, navigate to Power >> Switches, and select "Diode."

The element "Diode" will appear in the list of the secondary images. The text "PSIMIMAGE" in front of the "Diode" text shows that this is for the diode element in the standard image library psimimage.lib.

Highlight "Diode," and click on the Edit button to create the image.

If this image is to be used as the default image for this element, click on the "Set as Default Image."

After the secondary image is created, this image will be available for selection in the PSIM schematic. For example, if a diode is placed on the schematic, double-click to display the property dialog of the diode, then click

TABLE 2.9 Library editor.

Up	To move an element up in the menu
Down	To move an element down in the menu
--->	To move an element one level lower in the menu
<---	To move an element one level higher in the menu
Add Separator	To add a separator between elements
Add Submenu	To add a submenu in the library
Edit	To edit the name of an element
Edit Image	To edit the image of an element
New Element	To create a new element in the image library. This element will be linked to a netlist element in the netlist library
Save Element as	To save the existing element as a new element. The new element will have the image of the existing element as the default image
Delete Element	To delete an element from the library
New Element (DLL)	To create a new element from a DLL file
New Subcircuit Element (External)	To create a new element from a subcircuit, and the subcircuit file is stored outside the library file
New Subcircuit Element (Internal)	To create a new element from a subcircuit, and the subcircuit file is stored inside the library file

on the color tab. Click on the pull-down arrow, and two images will be displayed. One from the standard image library, and the other from the custom image library mylib.lib.

If the secondary image is selected, all the images of the same element will be automatically changed to the secondary image. The selected image will also be set as the default image when a schematic is created or loaded the next time. To share the secondary images that one creates with other people, one just has to send to them the secondary image library file (with the .lib2 extension).

2.2.2 Adding a new subcircuit element into the library

There are two ways to add a custom model to the PSIM library list. One is to have the model in the form of a subcircuit, and then place the schematic file in a folder called *user defined* in the PSIM directory, or in one of the subfolders of the *user defined* folder. Any schematic files and subfolders under the *user defined* folder will appear in the PSIM library list.

Another way is to add the custom model directly to an image library. The advantage of this approach is that the custom element will have the same look and feel as the standard PSIM elements, giving it a better interface. It is also possible to associate a help file to the custom model.

There are three main steps to add a new element, modeled in a subcircuit, into the library:

- Create the subcircuit model of the new element.
- Add this element to the PSIM library.
- Create an online help file for this new element.

2.2.3 Adding a new DLL element into the library

Similar to that of a subcircuit element, there are three main steps to add a new element, modeled in a DLL, into the PSIM library:

- Create the model in the DLL file.
- Add this element to the PSIM library.
- Create an on-line help file for this new element.

To illustrate this process, an inductor is used as an example.

2.2.3.1 Creating the DLL

The first step is to create the inductance model in DLL. Please refer to the relevant section on how to create a custom DLL.

Here we assume that the DLL file, "inductor_model.dll," has already been created. It has one parameter called "Inductance," and two connecting nodes. The file is placed in the "lib" subfolder in the PSIM directory.

2.2.3.2 Adding the new element to the PSIM library

To add the DLL element into the PSIM library, follow these steps:

Go to *Edit* >> *Edit Library* >> *Edit Library Files*, and choose the library for the new element. Click on *New Library* to create a new image library, or select an existing library and click on *Edit Selected Library*.

2.2.4 Creating a symbol library

With the Image Editor in the *Edit* menu, one can easily create good component images very quickly. These images can be used as secondary images of PSIM library elements, or images of subcircuits. One can also store these images in a symbol library for the purpose of circuit wiring diagrams. Note that such a schematic is solely for display purposes, and cannot be simulated.

2.3 Rectifier PSIM model

2.3.1 Rectifier circuit structure

Half wave bridge rectifier circuit is drawn in PSIM software as follows:

2.3.1.1 AC supply section

1. AC source is taken from the following tab *Elements – Sources – Voltage – Sine* and placed in the workspace.
2. Source voltage and frequency values are edited in the parameters tab by double-clicking on the source.
3. Input voltage is measured using a voltmeter by connecting it in parallel to the input source.
4. Input current is measured using an ammeter by connecting it in series to the input source (Figs. 2.7 and 2.8).

2.3.1.2 Diode bridge section

1. Diode is taken from the following tab *Elements – Power – Switches – Diode* and placed in the workspace.

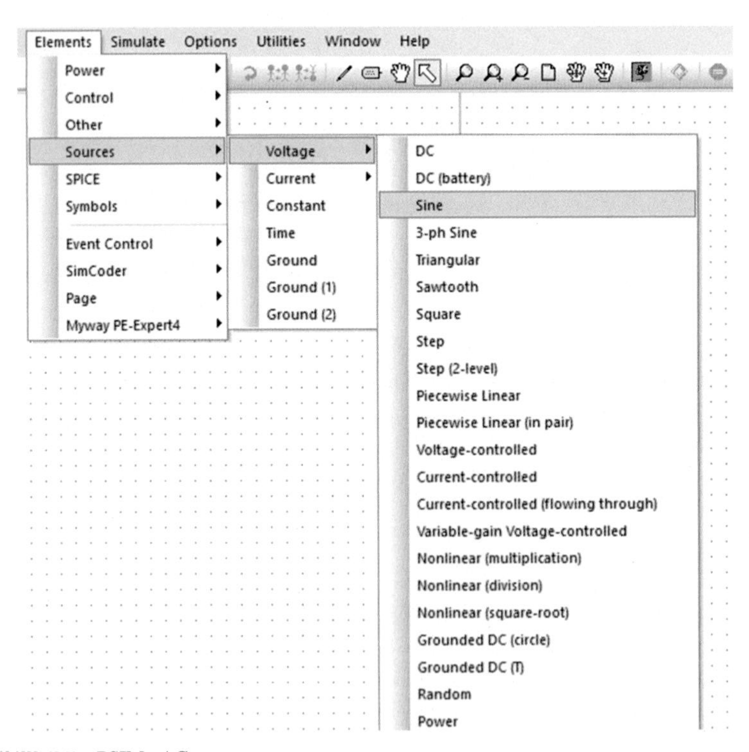

FIGURE 2.7 PSIM–AC sources.

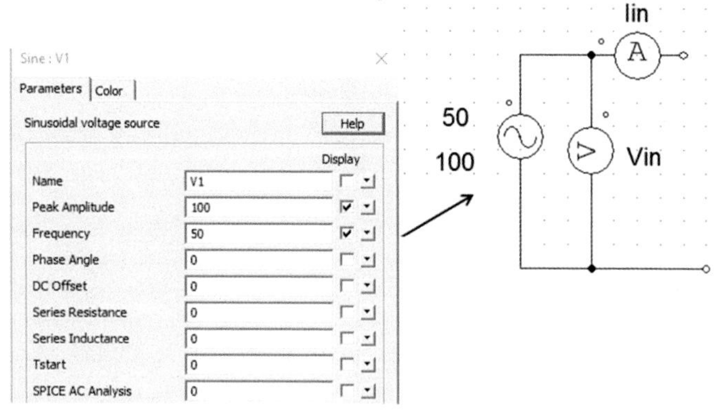

FIGURE 2.8 AC supply section.

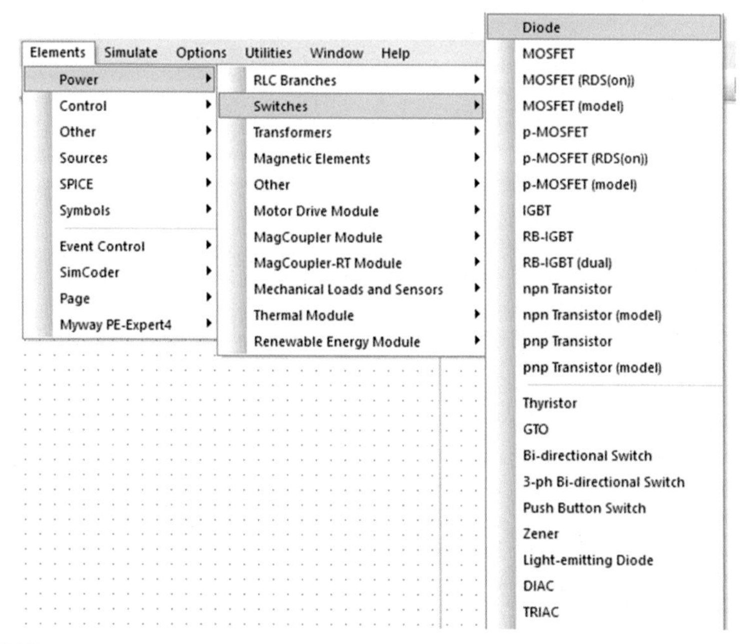

FIGURE 2.9 PSIM-switches.

2. Parameters are edited in the parameters tab by double-clicking on the diode. Color can also be changed by clicking on color tab.
3. Required number of diodes are placed in the appropriate place. Place wire option is used to connect all the diodes as per the circuit design (Figs. 2.9 and 2.10).

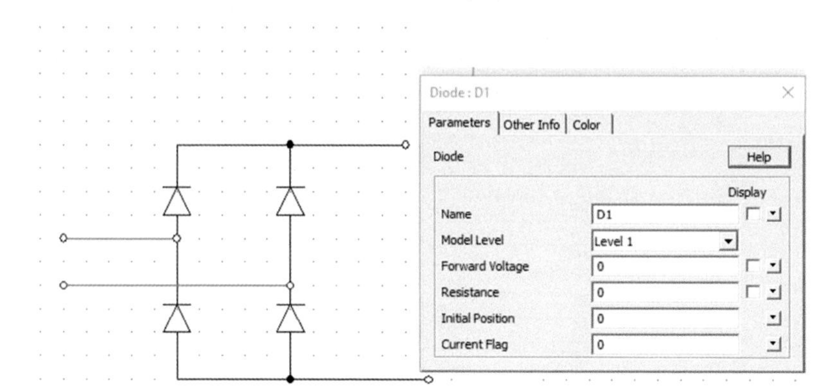

FIGURE 2.10 Diode bridge section.

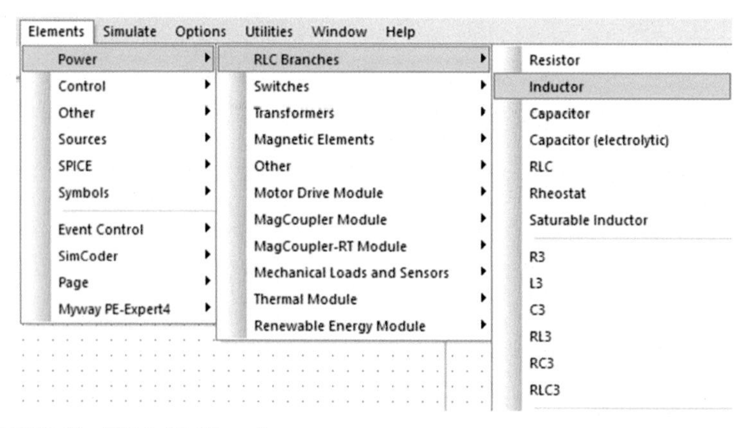

FIGURE 2.11 PSIM–RLC branches.

2.3.1.3 Filter section

1. Inductor and capacitor is taken from the following tab *Elements – Power – RLC Branches – Inductor* and placed in the workspace.
2. Inductor and capacitor values are edited in the parameters tab by double-clicking on the respective element.
3. Color can also be changed by clicking on color tab (Figs. 2.11 and 2.12).

2.3.1.4 Load and adding meters

The same steps given in the previous section are followed to add load resistance in the bridge rectifier circuit (Fig. 2.13).

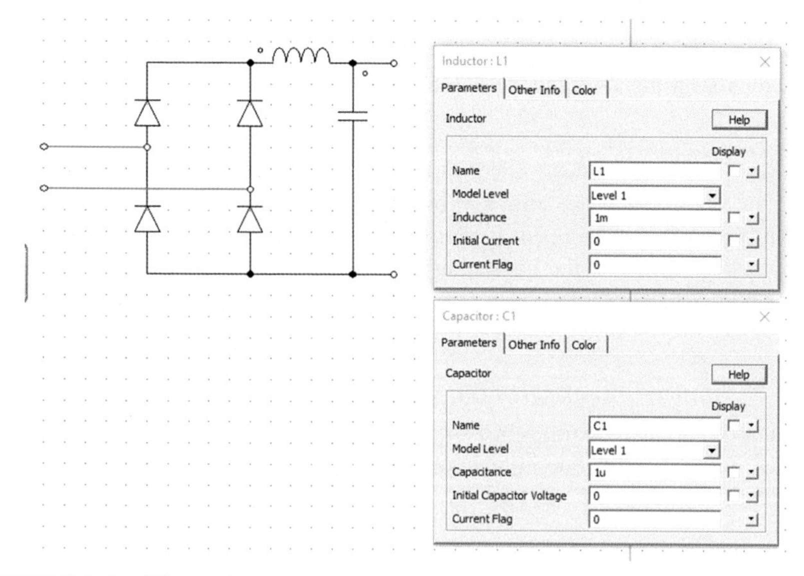

FIGURE 2.12 Filter section.

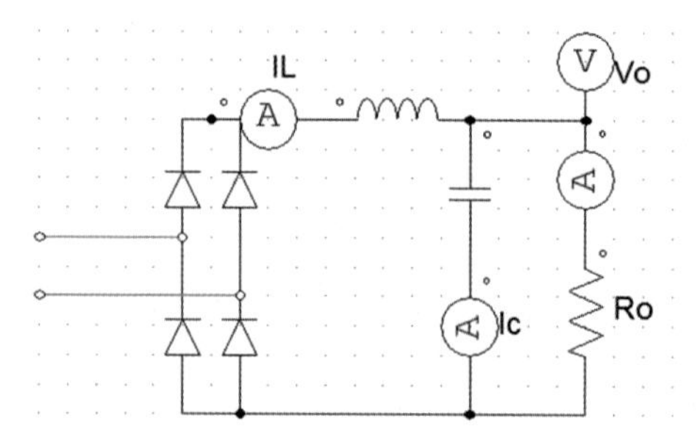

FIGURE 2.13 Load and adding meters.

2.3.2 Simulation procedure

Simulation steps are as follows,

1. Run the simulation by selecting the "Run Simulation" icon. 
2. The window switches automatically to the result screen after simulation is completed. Plot the waveforms of V_s and V_o on screen #1.

3. Add one more screen by selecting "Add Screen" icon. Plot I_s and I_L on screen #2 using "Add/delete curve" icon.

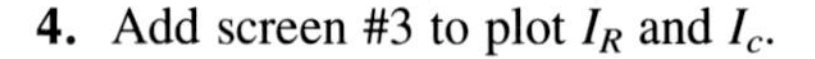

4. Add screen #3 to plot I_R and I_c.

After completing the above procedure, the window will get open as shown in Fig. 2.14. The result is plotted from zero to 0.5 seconds. In this circuit transient occurs at the beginning of simulation and reaches steady value at around 0.2 seconds.

2.3.2.1 Simulation control parameters

1. Identify the highest frequency of any sources in the circuit.
2. The maximum time step is one tenth of a period by default: for example, voltage source of 50 Hz, gives 20 ms period. Time step is $1/10 \times 20$ ms = 2 ms.
3. Estimate the time at which simulation reaches steady-state. For example 0.4 seconds. Therefore 1 seconds should give a steady-state result. So the Total time is 1 seconds.
4. Using maximum resolution, Print step is set to 1.
5. Only showing the last five cycles of the waveform by setting the Print time as:

   Print time = 1 s − (5 cycles × 20 ms) = 1000 ms − 100 ms = 900 ms
6. Increase the resolution of simulation by reducing Time step: Time step = 5 cycles/6000 points = 100 ms/6000 = 16.66 μs ~round up to 20 μs (Fig. 2.15).

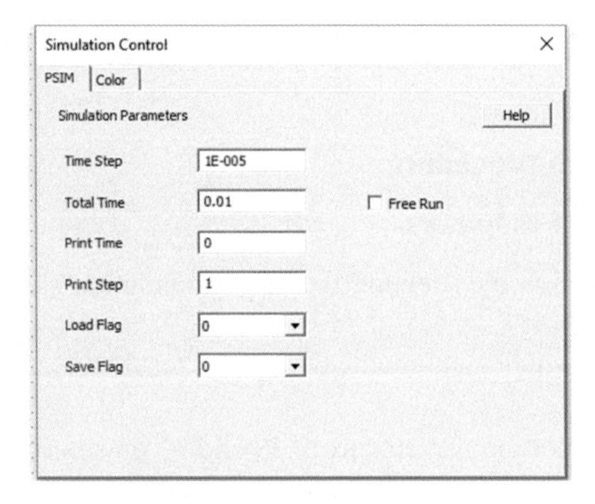

FIGURE 2.14 Simulation control parameters.

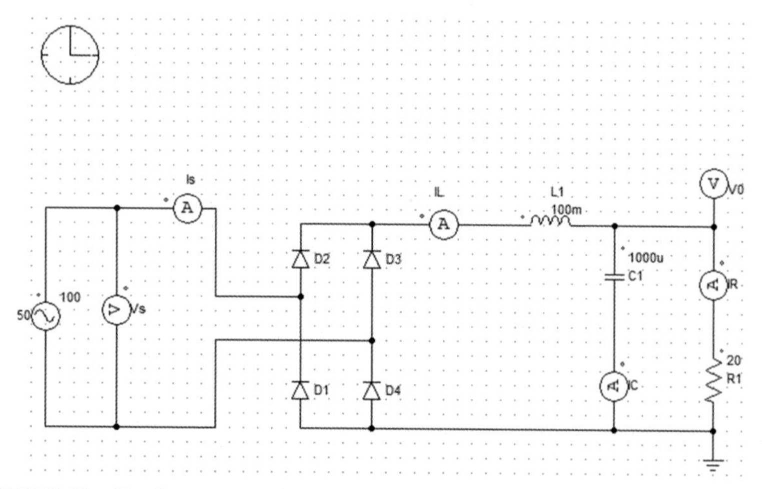

FIGURE 2.15 Circuit structure—diode bridge rectifier.

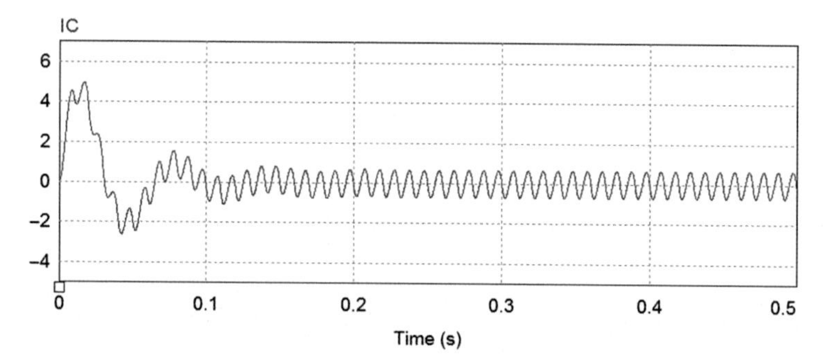

FIGURE 2.16 Simulated waveform of capacitor current of diode bridge rectifier.

2.3.3 Simulation waveforms

2.3.3.1 Capacitor current

The single phase diode rectifier simulated waveforms like capacitor current, inductor current, load current, source current, source voltage and load voltage are presented in Figs. 2.16–2.20.

2.3.3.2 Inductor current

2.3.3.3 Load current

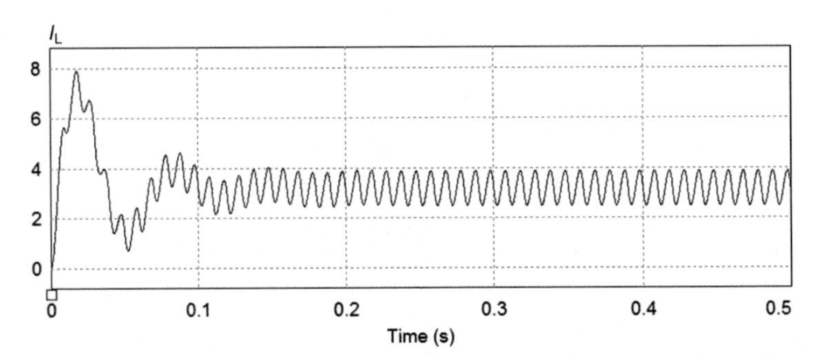

FIGURE 2.17 Simulated waveform of inductor current of diode bridge rectifier.

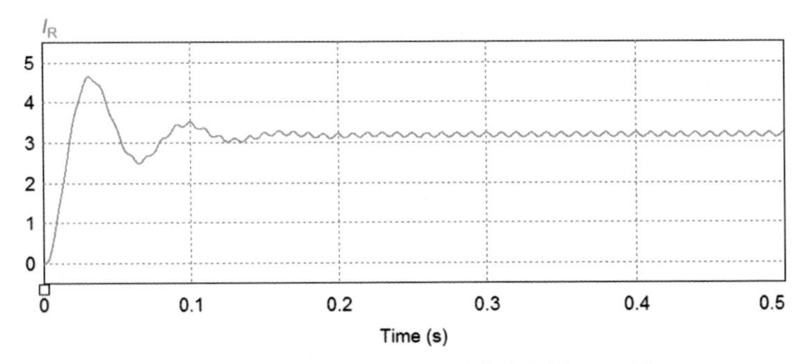

FIGURE 2.18 Simulated waveform of load current of diode bridge rectifier.

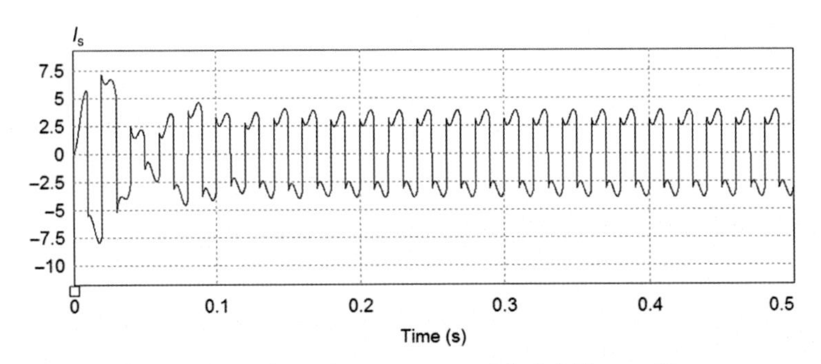

FIGURE 2.19 Simulated waveform of source current of diode bridge rectifier.

2.3.3.4 *Source current*

2.3.3.5 *Source voltage and load voltage*

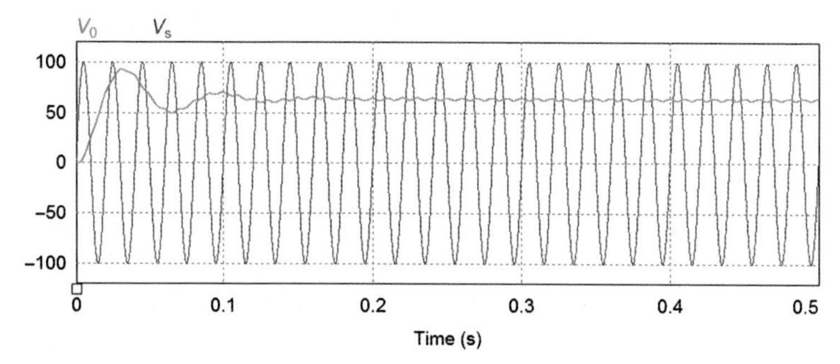

FIGURE 2.20 Simulated waveform of source and load voltage of diode bridge rectifier.

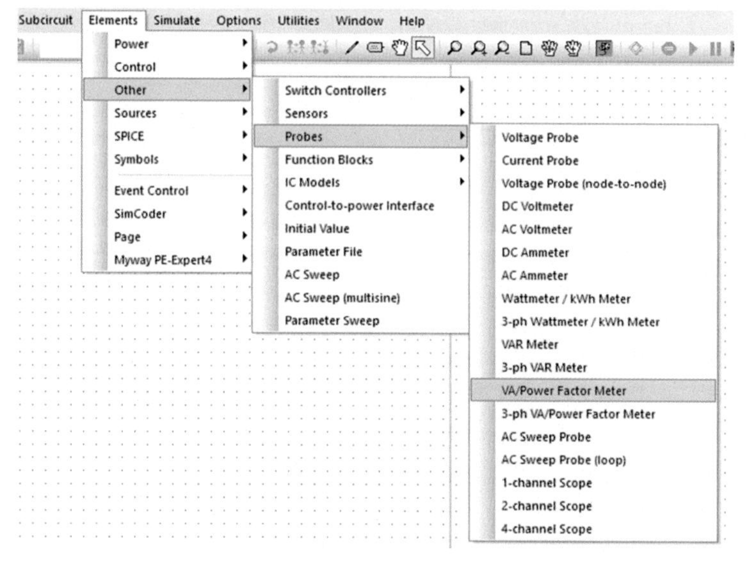

FIGURE 2.21 Power factor meter.

2.3.4 Measuring power factor

PSIM provides pf meter in it library. It can be found in *Element – Other – Probes – VAR/pf meter*. Insert this element between the AC source and rectifier (Figs. 2.21 and 2.22).

2.4 IGBT thermal model

2.4.1 IGBT device in database

An insulated gate bipolar transistor (IGBT) device has three types of packages: discrete, dual, or 6-pack. For the dual package, both the top and

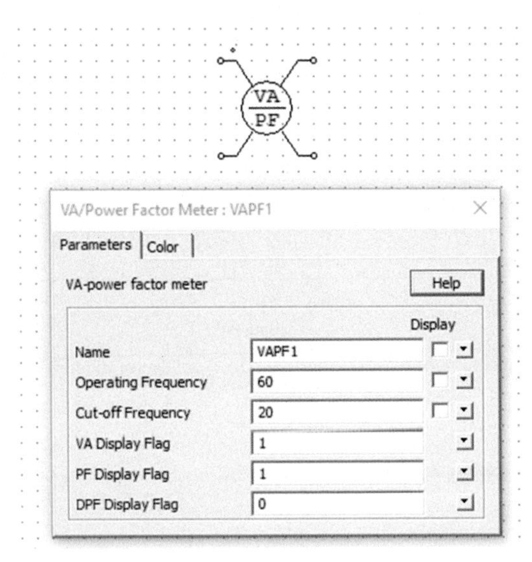

FIGURE 2.22 Power factor meter parameters.

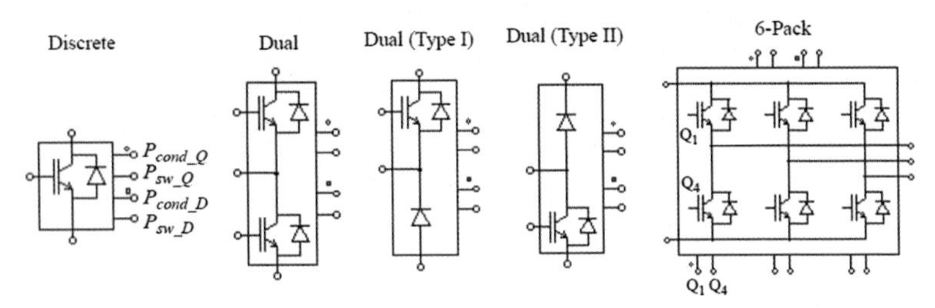

FIGURE 2.23 IGBT Device.

the bottom switches can be IGBTs (full-bridge configuration), or one of the switches is IGBT and the other is a free-wheeling diode (half-bridge configuration). For the half-bridge dual IGBT device, since the free-wheeling diode parameters can be different from these of the antiparallel diode, this type of device is referred to as the IGBT-Diode device, and is treated as a different type in the simulation. But for the convenience of discussion, both devices are referred to as the IGBT devices here. The following information is defined for an IGBT device in the database:

2.4.1.1 General information

Manufacturer: Device manufacture

Part Number: Manufacturer's part number

Package. It can be discrete, dual, or 6-pack, as shown in Fig. 2.23.

TABLE 2.10 Thermal module parameters of IGBT.

Parameters	Description
Device	The specific device selected from the device database
Frequency	Frequency in Hz under which losses are calculated
P_{cond_Q} Calibration Factor	Calibration factor K_{cond_Q} of the transistor conduction losses P_{cond_Q}
P_{sw_Q} Calibration Factor	Calibration factor K_{sw_Q} of the transistor switching losses P_{sw_Q}
P_{cond_D} Calibration Factor	Calibration factor K_{sw_D} of the diode conduction losses P_{cond_D}
P_{sw_D} Calibration Factor	Calibration factor K_{sw_D} of the transistor switching losses P_{sw_D}
Number of parallel devices	Number of identical IGBT device in parallel

2.4.2 IGBT loss calculation

An IGBT device in the database can be selected and used in the simulation for loss calculation. An IGBT device in the Thermal Module library has the following parameters (Table 2.10).

2.4.2.1 Attributes

The parameter Frequency refers to the frequency under which the losses are calculated. For example, if the device operates at the switching frequency of 10 kHz, and the parameter Frequency is also set to 10 kHz, the losses will be the values for one switching period. However, if the parameter Frequency is set to 60 Hz, then the losses will be the value for a period of 60 Hz.

The parameter P_{cond_Q} Calibration Factor is the correction factor for the transistor conduction losses. For the example, if the calculated conduction losses before the correction are $P_{cond\_Q\_cal}$, then $P_{cond\_Q} = K_{cond\_Q} * P_{cond\_Q\_cal}$

Similarly, the parameter P_{sw_Q} Calibration Factor is the correction factor for the transistor switching losses. For the example, if the calculated switching losses before the correction are $P_{sw\_Q\_cal}$, then $P_{sw\_Q} = K_{sw\_Q} * P_{sw\_Q\_cal}$. Parameters P_{cond_D} Calibration Factor and P_{sw_D} Calibration Factor work in the same way, except that they are for the diode losses.

When several identical IGBT devices are in parallel, one should have just one device in the schematic, and set the correct number of devices in the parameter input. This is because when several identical devices are in parallel in the schematic, the device currents may not be exactly equal due to small differences in the simulation. When the number of parallel devices is

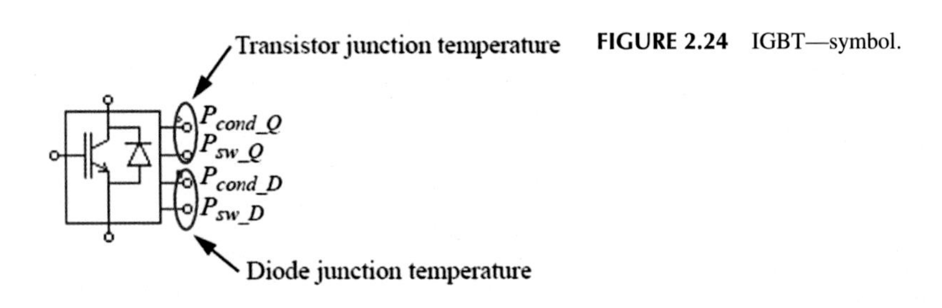

FIGURE 2.24 IGBT—symbol.

greater than one, the total current through the group of the parallel devices will be equally divided among the devices. The total losses are then obtained by multiplying the losses in each device by the number of parallel devices. The currents flowing out of Nodes P_{cond_D}, P_{sw_D}, P_{cond_Q}, and P_{sw_Q} are the total losses of all the parallel devices combined.

The voltage at the conduction losses node P_{cond_Q} or the switching losses node P_{sw_Q} represent the calculated junction temperature T_{j_Q} of the transistor, and the voltage at the conduction losses node P_{cond_D} or the switching losses node P_{sw_D} represent the calculated junction temperature T_{j_D} of the antiparallel diode (Fig. 2.24).

The calculated junction temperatures T_{j_Q} and T_{j_D} are used when the database curves are used for loss calculation. If the calculated junction temperature is between the junction temperatures of two curves, interpolation will be used. If the calculated junction temperature is lower than the lowest junction temperature or higher than the highest junction temperature in the curves, the curve corresponding to the lowest or highest junction temperature will be used. If there is only one curve, that curve is used regardless of the calculated junction temperature.

2.4.2.2 *Conduction losses*

The transistor conduction losses are calculated as:

$P_{cond_Q} = V_{ce}(sat) \times I_c$where $V_{ce}(sat)$ is the transistor collector–emitter saturation voltage, and I_c is the collector current. When the transistor is conducting periodically with an on duty cycle of D, the conduction losses are calculated as:

$P_{cond\_Q} = V_{ce}(sat) \times I_c \times D$

2.4.2.3 *Switching losses*

The transistor turn-on losses are calculated as:

$P_{sw\_Q\_on} = E_{on} \times f \times V_{cc}/V_{cc\_datasheet}$ where E_{on} is the transistor turn-on energy losses, f is the frequency as defined in the input parameter Frequency, V_{cc} is the actual dc bus voltage, and $V_{cc\_datasheet}$ is the dc bus voltage in the E_{on} and E_{off} characteristics of the datasheet, defined as "DC bus voltage (V)" in the test conditions.

The transistor turn-off losses are calculated as:

$P_{sw\_Q\_off} = E_{off} \times f \times V_{cc}/V_{cc\_datasheet}$ where E_{off} is the transistor turn-off energy losses. The loss calculation for the antiparallel diode or free-wheeling diode is the same as described in the section for the diode device.

The losses P_{cond_Q}, P_{sw_Q}, P_{cond_D}, and P_{sw_D}, in watts, are represented in the form of currents which flow out of these nodes. Therefore to measure and display the losses, an ammeter should be connected between the nodes and the ground. When they are not used, these nodes cannot be floating and must be connected to ground.

2.4.3 Curve fitting with manufacturer datasheet (SEMiX151GD066HDs)

The first step is to add the IGBT Module SEMiX151GD066HDs into PSIM's device database.

Below is the procedure to add this device into the device database.

1. In PSIM, go to Utilities >> Device Database Editor to launch the Device Database Editor.
2. One may choose to add the device to one of the existing device files that came with the PSIM software. But it is recommended that a separate device file be created. In this example, we will create a new device file called "Semikron.dev," and we will place it in the "device" subfolder in PSIM. Go to File >> New Device File, and under the "device" subfolder, create the file "Semikron.dev." This file will appear in the File Name list box at the upper left corner of the Device Database Editor.
3. Highlight the file "Semikron.dev," and go to Device >> New IGBT to create a new IGBT device. The new device will be stored in the device file "Semikron.dev" (Fig. 2.25).

Once the device is added to the device database, it can be used in PSIM for the loss calculation. To choose this device, in PSIM, go to Elements >> Power >> Thermal Module >> IGBT (database), and place the discrete IGBT element in the schematic. Double-click on the IGBT element to open the property dialog window. Click on the Browser button next to the "Device" input field, and choose the device "Semikron SEMiX151GD066HDs." The IGBT image will change to a 6-pack inverter bridge. Continue to build the rest of the circuit. The circuit below shows the completed inverter circuit using the IGBT module SEMiX151GD066HDs. The load resistances and inductances and the modulation index are selected such that the circuit operates under the specified conditions [output of 230 V_{ac}, 20 kW, 0.8 power factor (lagging)] (Fig. 2.26).

The IGBT module image shows two DC bus terminals on the left, three AC output terminals on the right, six gating signal nodes at the bottom, and four extra nodes on the top. These four nodes are for the power losses, and

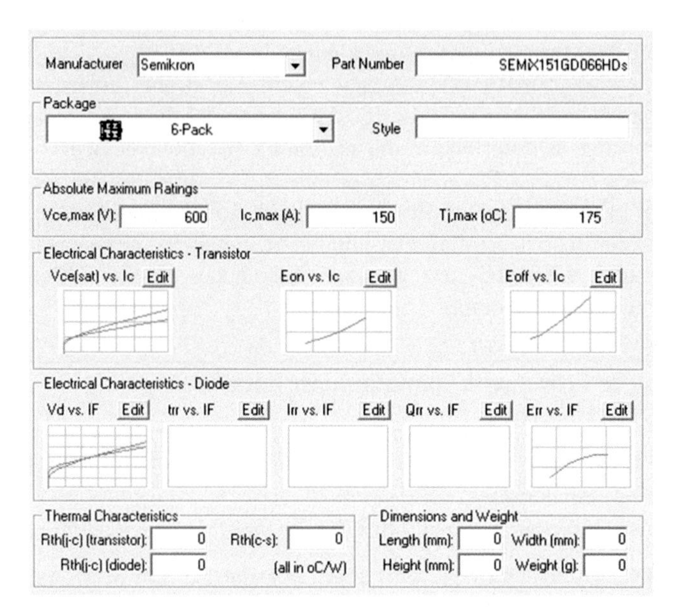

FIGURE 2.25 IGBT module SEMiX151GD066HDs.

they are (from left to right): transistor conduction losses P_{cond_Q}, transistor switching losses P_{sw_Q}, diode conduction losses P_{cond_D}, and diode switching losses P_{sw_D}. These losses are for the whole IGBT module (including all six IGBT switches). They are in the form of electric currents, and will flow out of these nodes. To measure the losses values, connect an ammeter to each node.

The parameters of the IGBT Module are defined as:

Frequency: 60

P_{cond_Q} Calibration Factor: 1

P_{sw_Q} Calibration Factor: 1

P_{cond_D} Calibration Factor: 1

P_{sw_D} Calibration Factor: 1

The frequency defines the interval under which the losses are calculated. For example, if the frequency is 60 Hz, the losses results are the average value for an interval of 16.67 ms. If the frequency is set to be the same as the switching frequency, the losses in each switching cycle are obtained. The Calibration Factors are used to scale the calculation results against experimental results. For example, for a specific device, if the calculated losses are 10 W, but the measured losses from the experiments are 12 W, the calibration factor shall then be set to 1.2 (Figs. 2.27–2.32).

The following losses results are obtained from the PSIM simulation:

Diode Conduction Losses: 44.8

Diode Switching Losses: 92.

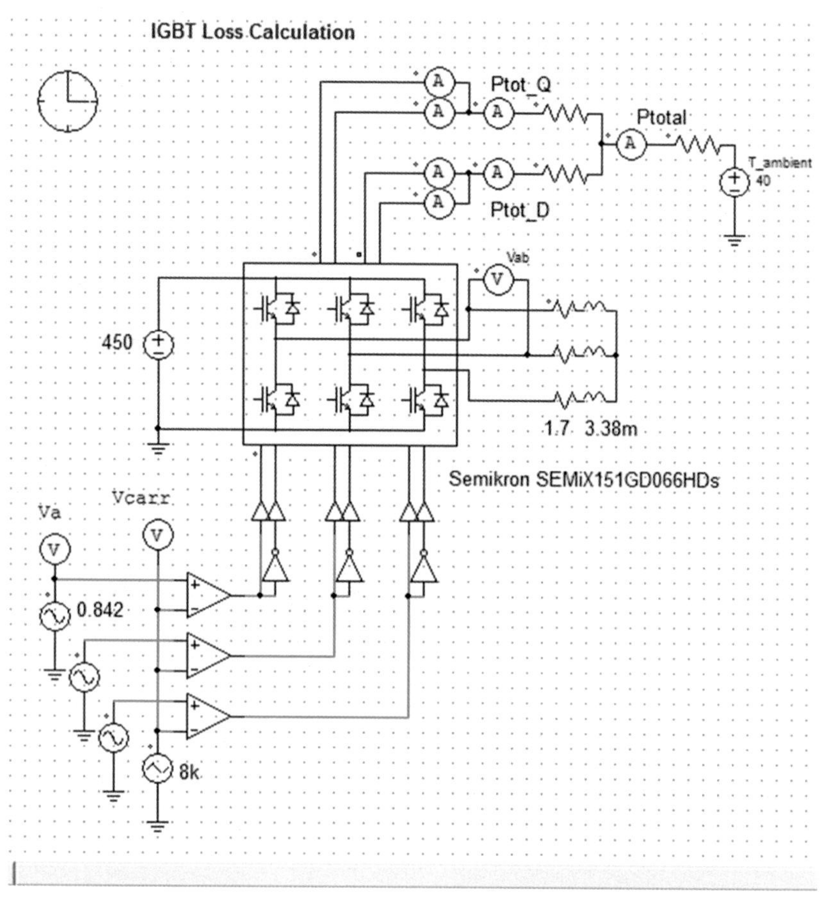

FIGURE 2.26 IGBT module.

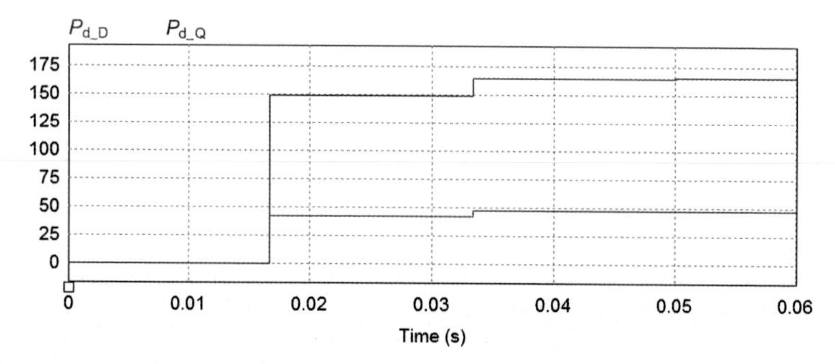

FIGURE 2.27 Waveform for conduction losses in diode and transistor.

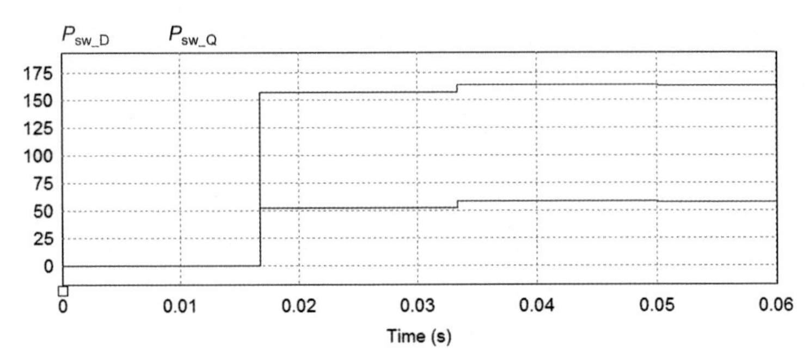

FIGURE 2.28 Waveform for switching losses in diode and transistor.

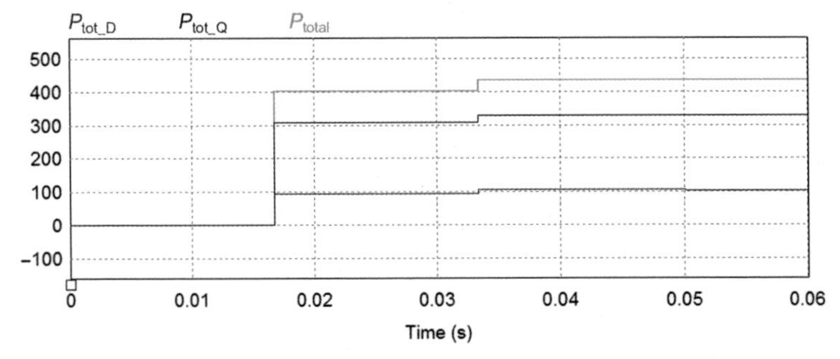

FIGURE 2.29 Waveform for total losses in diode and transistor.

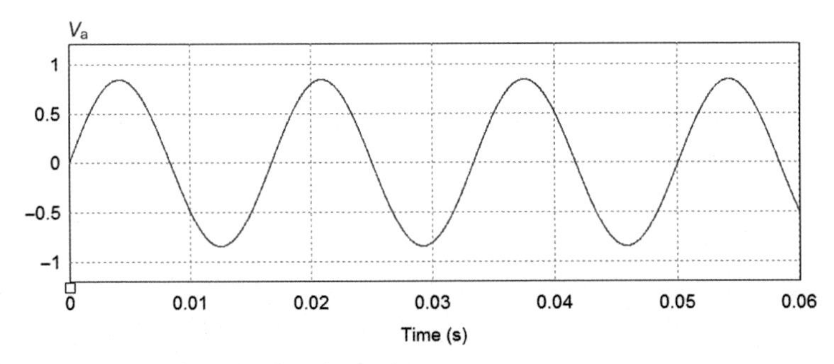

FIGURE 2.30 Waveform showing phase voltage.

Diode Total Losses: 136.8
Transistor Conduction Losses: 166.9
Transistor Switching Losses: 201.5
Transistor Total Losses: 368.4
Total Losses per Module: 505.2

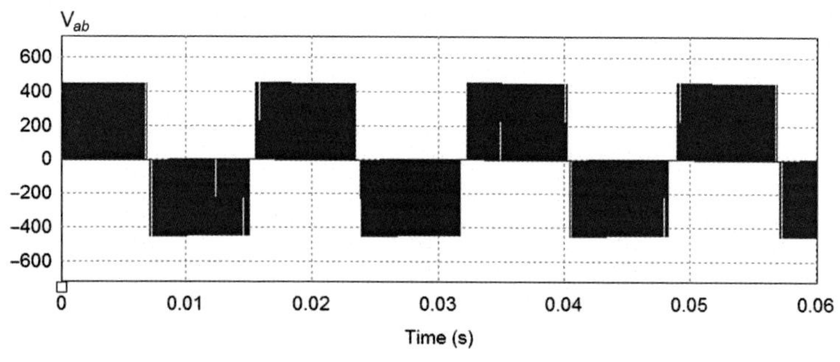

FIGURE 2.31 Waveform showing line voltage.

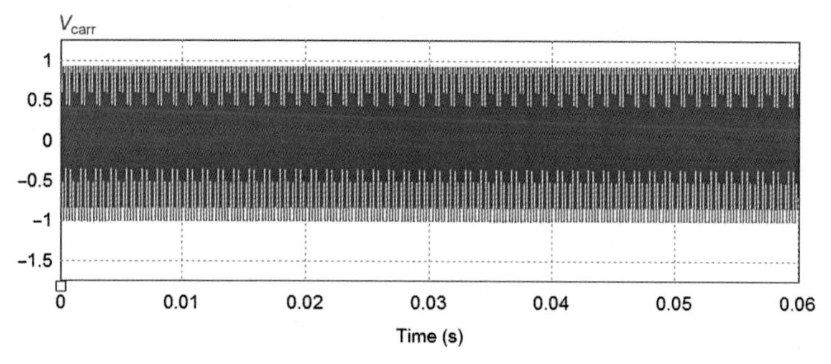

FIGURE 2.32 Waveform showing carrier wave.

The losses of the inverter are also calculated using Semikron's online thermal calculation SEMISEL.

The SEMISEL calculation is based on the following settings:

Circuit parameter:

Input voltage: 450 V

Output voltage: 230 V

Cosine phi: 0.8

Output power: 20 kW

Output current: 63 A

Switching frequency: 8 kHz

Output frequency: 60 Hz

Device: SEMiX151GD066HDs

Enter the calculation method: use typical values

Enter the correction factor of the switching losses

Transistor: 1

Diode: 1

The following losses results (for the whole Module) are obtained from the SEMISEL calculation:

Diode Conduction Losses: 46.1
Diode Switching Losses: 72
Diode Total Losses: 118.1
Transistor Conduction Losses: 162
Transistor Switching Losses: 144
Transistor Total Losses: 306
Total Losses per Module: 424.1

The results from the Thermal Module are close to the results from SEMISEL.

2.5 Renewable energy module

The Renewable Energy Module library contains the following elements:

Solar modules: physical model, functional model, cSi model, and thin-film model;

Wind turbine; Lithium-ion battery; and Supercapacitor.

2.5.1 Solar module—physical model

The physical model of the solar module simulates the behavior of the solar module with more accuracy because it takes into account the light intensity and temperature variation (Fig. 2.33).

In the image, the nodes with the " + " and " − " signs are the positive and negative terminals. The node with the letter "S" refers to the light intensity input (in W/m^2), and the node with the letter "T" refers to the ambient temperature input (in °C). The node on the top is theoretical power (in W) given the operating conditions. While the positive and negative terminal nodes are power circuit nodes, the other nodes are all control circuit nodes.

The attributes are given in Fig. 2.34. (1) Number of cells Ns of the solar module. A solar module consists of Ns solar cells in series. (2) Light intensity S_0 under the standard test conditions, in W/m^2. The value is normally 1000 W/m^2 in manufacturer datasheet. (3) Temperature T_{ref} under the standard test conditions, in °C. (4) R_{se} and R_{sh} are series and shunt resistances of each solar cell in ohms. (5) Short circuit current I_{sc0} of each solar cell at the reference temperature T_{ref}, in A. (6) Diode saturation current I_{s0} of each solar cell at the reference temperature T_{ref}, in A. (7) Band energy of each solar cell, in eV. It is around 1.12 for crystalline silicon, and around 1.75 for amorphous silicon. (8) Ideality factor A of each solar cell, also called emission coefficient. It is around 2 for crystalline silicon, and is less than 2 for amorphous silicon. (9) Temperature coefficient C_t, in A/°C or A/°K. (10) Coefficient ks that defines how light intensity affects the solar cell temperature (Fig. 2.34).

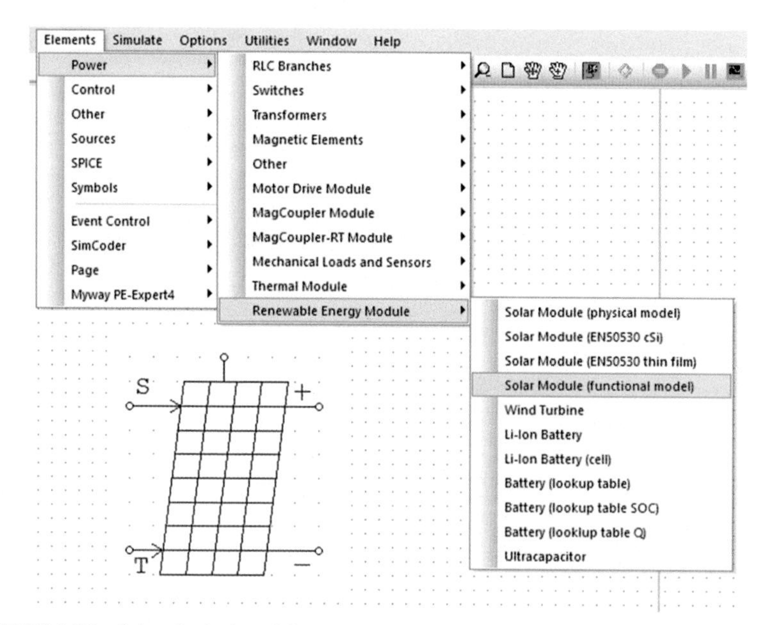

FIGURE 2.33 Solar physical model.

Solar Module (physical model) : SCP1

Parameters | Color

Solar module (physical model) — Help

		Display
Name	SCP1	
Number of Cells Ns	36	
Standard Light Intensity S0	1000	
Ref. Temperature Tref	25	
Series Resistance Rs	0.008	
Shunt Resistance Rsh	1000	
Short Circuit Current Isc0	3.8	
Saturation Current Is0	2.16e-8	
Band Energy Eg	1.12	
Ideality Factor A	1.2	
Temperature Coefficient Ct	0.0024	
Coefficient Ks	0	

FIGURE 2.34 Solar physical model attributes.

The solar cell test setup is shown in Fig. 2.35. The outputs are presented in Figs. 2.36–2.39. The manufacturer datasheet data are entered in the utility tool dialog window. The *I*–*V* and *P*–*V* curves, and the maximum power point are calculated. The datasheet and experimental data for different

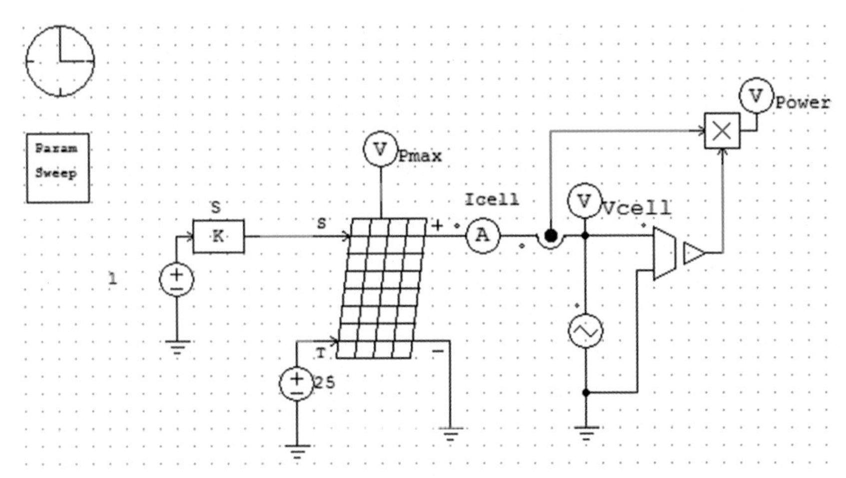

FIGURE 2.35 Solar cell test setup.

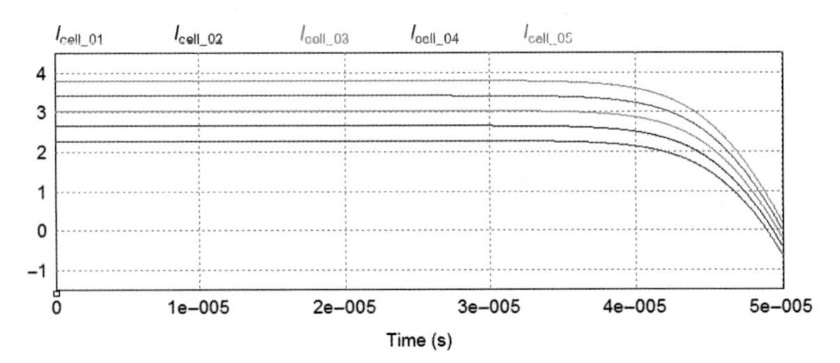

FIGURE 2.36 Solar cell current.

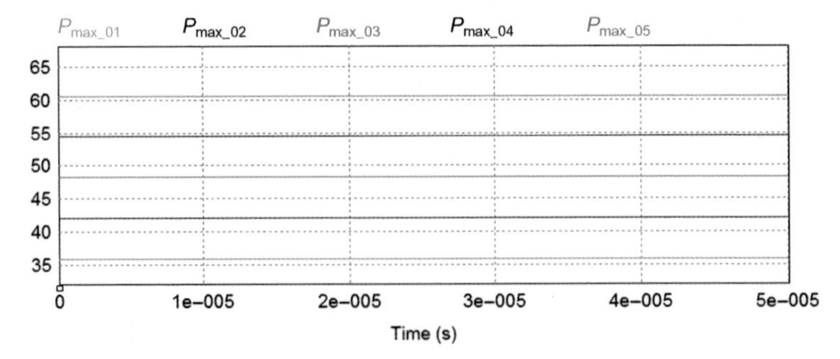

FIGURE 2.37 maximum solar Power.

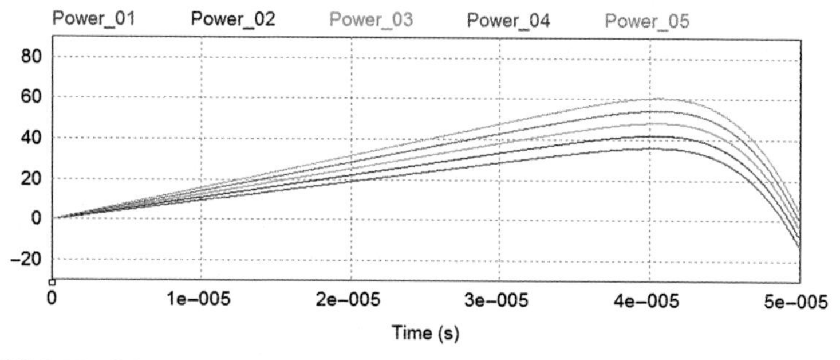

FIGURE 2.38 Solar cell power.

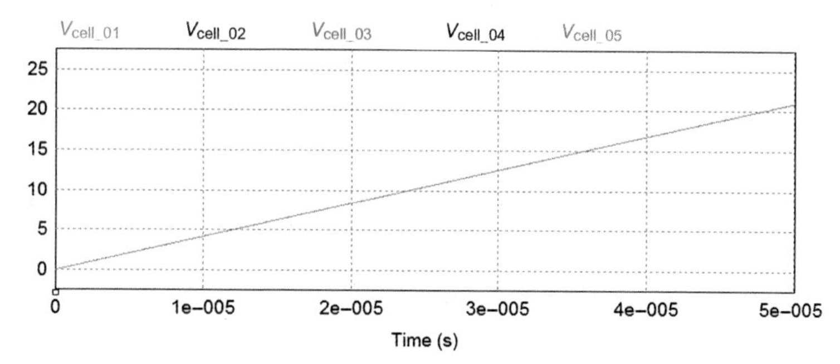

FIGURE 2.39 Solar cell voltage.

operating conditions are compared, and the parameters are fine-tuned. Many iterations and much trial and error may be needed to obtain a good fit to the datasheet or experimental data. After the parameters are finalized, click on the Copy PSIM Parameters button to copy the model parameters to the PSIM schematic. To save the datasheet and parameter values to a text file to later use, click on the Save button, and save it to a file (for example "Solarex MSX-60.txt"). To load the data of a specific solar module back, click on the Load button.

2.6 Summary

At the end of this chapter, students should be able to identify the tools available in PSIM and the role of PSIM in the simulation software field. The users will be able to understand how to design the power circuit depending on the applications. The detailed converter design procedure will effectively enable the learners to design and to simulate their own circuits. Also the users will have learned about the analysis of any power electronic devices and circuits.

2.7 Review questions

1. Draw the structure of PSIM simulation software.
2. Describe how to install the PSIM software and explain the procedure in detail.
3. Describe the step-by-step simulation running procedure in PSIM software.
4. What are the options available under simulation control in PSIM software? Explain the same with suitable diagrams.
5. Explain about the usage of SPICE libraries and their importance in PSIM software.
6. Write short notes on secondary image creation in the PSIM environment.
7. Draw the diode rectifier circuit using PSIM software and measure the electrical parameters across each element present in the circuit.
8. How do you fit some manufacturer data sheet parameters with device characteristics available in PSIM software?
9. Calculate the following losses in the IGBT module using PSIM software:
   a. Conduction loss
   b. Switching loss
10. Model the solar cell and plot the following characteristics:
    a. Solar cell current
    b. Solar cell voltage
    c. Solar cell power

Chapter 3

Basics of PSpice Simulation Tool

Chapter Outline

3.1 Introduction to PSpice

OrCAD PSpice is a technology program that models and simulates the activity of a circuit comprising analog equipment. Applied with OrCAD Acquiring for design launching, it could be thought of PSpice as a software-dependent breadboard of the circuit that could be applied to check the design prior to forming the physical hardware.

PSpice can perform the following tasks:

1. DC, AC, and transient study, hence testing the outputs of circuits to various signals.
2. Parametric, Monte Carlo, and sensitivity/worst-case study, hence testing the outputs of circuits to various component values.

3.1.1 Nomenclature, File structure

1. In the Windows Start menu, select the OrCAD Release 9 program folder and then the Acquire Shortcut to start Capture.
2. With Project Manager, in the File menu, select New and choose Project (Fig. 3.1).
3. Choose Analogue Circuit Wizard (Fig. 3.2).
4. Within the Name text box, enter the name of the project (CLIPPER).
5. Click OK, and Finish.

Software Tools for the Simulation of Electrical Systems. DOI: https://doi.org/10.1016/B978-0-12-819416-4.00003-X

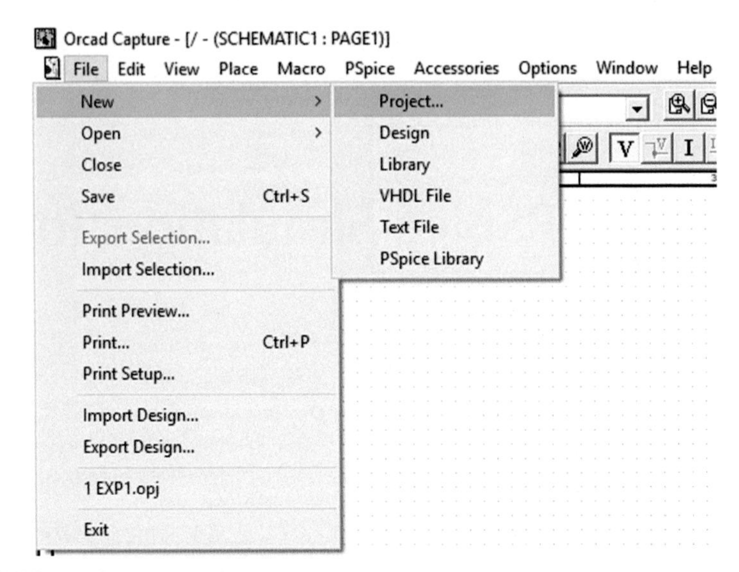

FIGURE 3.1 Open New Project.

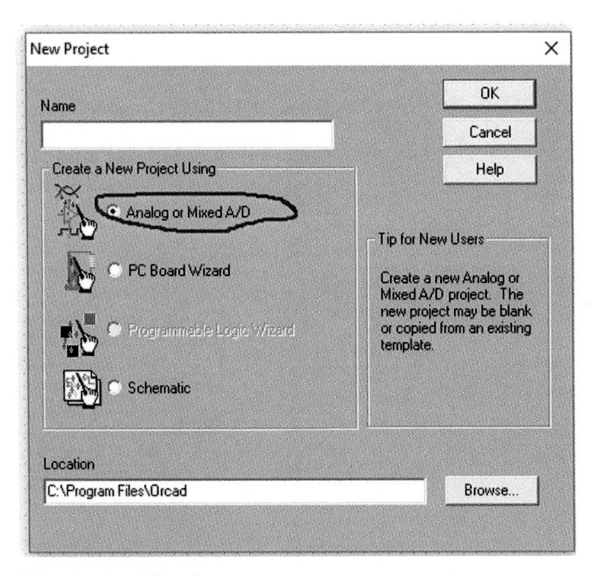

FIGURE 3.2 Select Analog Wizard.

No specific libraries need to be organized now. A fresh page will be shown in Capture and a fresh project will be organized in Project Manager.

To place the voltage sources

1. Within Capture, go to the schematic page editor (Fig. 3.3)
2. From the Place menu, prefer Part to show the Place Part dialog box.

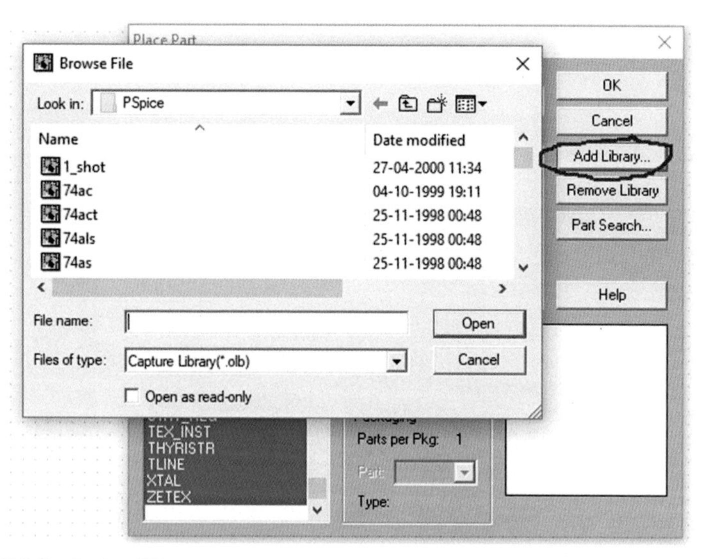

FIGURE 3.3 Assign File name.

3. Join the library for the parts that are needed to be placed: Click the Add Library button. Then, choose SOURCE.OLB (in PSpice library) and click Open.
4. Within Part text box, type VDC (Fig. 3.4).
5. Select OK.
6. Adjust the pointer to the exact point on the diagram page and click to choose the basic part.
7. Adjust the cursor and click once again to keep on the second part.
8. Right-click and prefer End Mode to finish placing parts (Fig. 3.5).

To locate the diodes

1. In the Place menu, select Part to exhibit the Place Part dialog box.
2. Join the library for the parts which are needed to be placed: Add Library button. Then, select DIODE.OLB (from the PSpice library) and click Open.
3. Within the Part text box, type D1N39 to enumerate the diodes (Fig. 3.6).
4. Choose D1N3940 and click OK.
5. Press R to turn the diode to the proper positioning (Fig. 3.7).
6. Place the initial diode (D1), then click to position the second diode (D2) (Fig. 3.8).
7. Right-click and take End Mode to stop placing parts.

To locate another part

1. From the Place menu, select Part to show the Place Part dialog box.

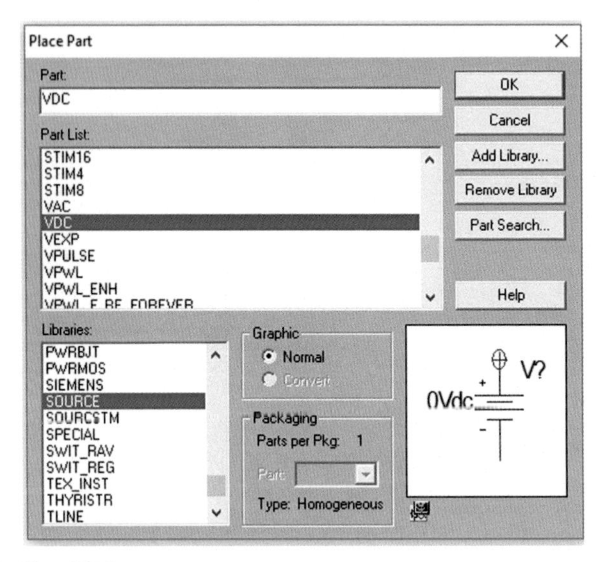

FIGURE 3.4 Place VDC.

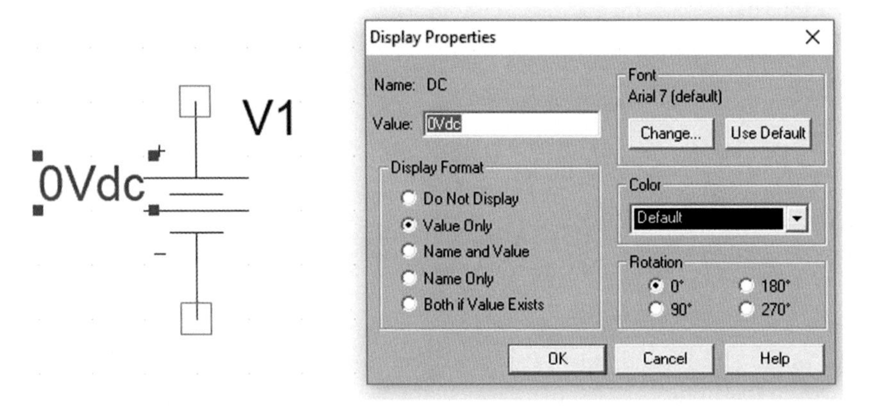

FIGURE 3.5 Properties of VDC.

2. Join the library for the parts to be placed: Add Library button. Select ANALOG.OLB (from PSpice library) and Open it.
3. Do the corresponding course as represented for the diodes to keep the parts cataloged as under, accordant to designing. The part names needed to be typed within text box of the Part box, are exhibited in parentheses: resistor (R) and capacitor (C) (Figs. 3.9 and 3.10).
4. To spot the off-page connective devices (OFFPAGELEFT-R), select the Off-Page Connector button on the tool board.

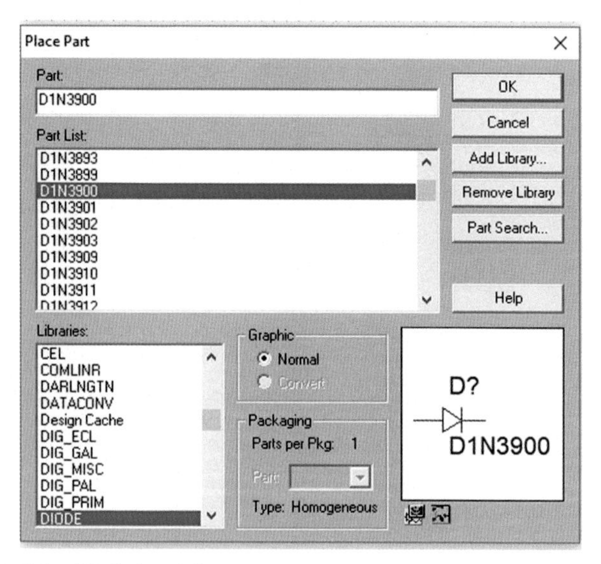

FIGURE 3.6 Select Diode from Library.

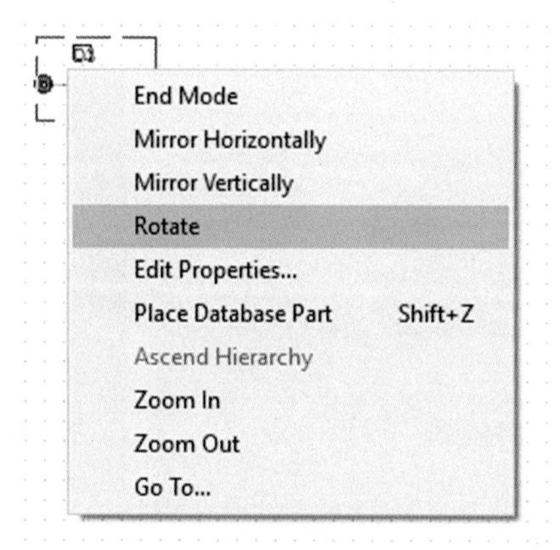

FIGURE 3.7 Rotate Option.

5. Join the library for the devices needed to be located: (1) Join with Library button. (2) Choose CAPSYM.OLB (from the Capture library) and Open it.
6. Position the off-page connective devices according to schematic design.
7. To place the ground device (0), select GND button in tool palette (Fig. 3.11).
8. Adjoin the library for the devices which are required to be placed: (1) Join with the Library button. (2) Choose SOURCE.OLB (in PSpice library) and Open it.
9. Place the grounding devices parts as per Fig. 3.12.

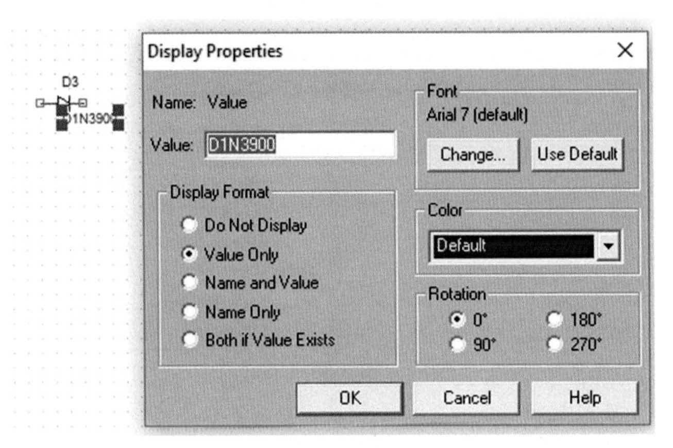

FIGURE 3.8 Diode Properties Edit.

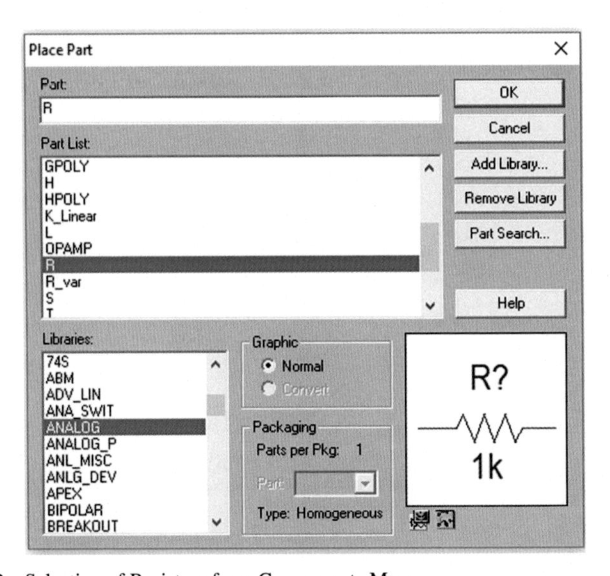

FIGURE 3.9 Selection of Resistors from Components Menu.

To link the devices

1. In the Place menu, to start with choose wire parts. The pointer effect, to a crosshair.
2. Choose link element (the very end) of the pin in off-page connector, in the input of circuit.
3. Choose closest link point of the input resistor R1.
4. Link the far end of R1 with output condenser.

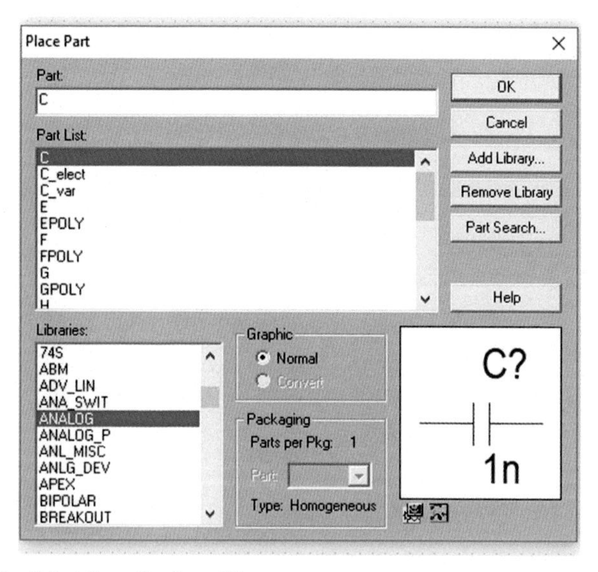

FIGURE 3.10 Select Capacitor from Library.

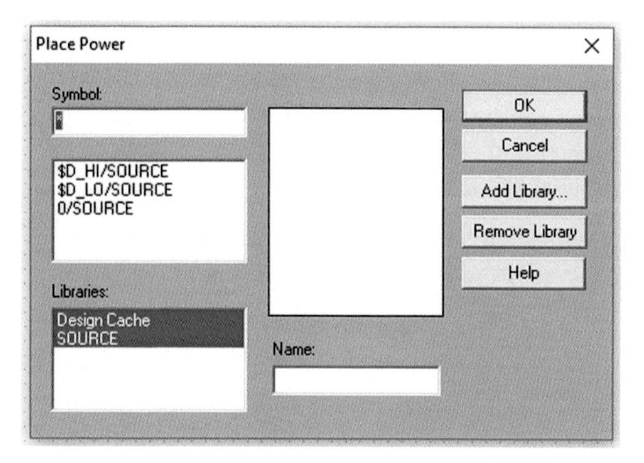

FIGURE 3.11 Select Ground from Library.

5. Link the diodes with one another and wire in between them:
   a. Connect point of cathode with corresponding down end diode.
   b. Alter the cursor, to straight up and to wire between the diodes. Now wire termination, and the conjunction of the wire section become visualization.
   c. In the conjunction to continue to do circuit.
   d. Select terminal of top side diode's anode pin.
6. Proceed linking devices until the circuitry is connected as per diagram blueprint.

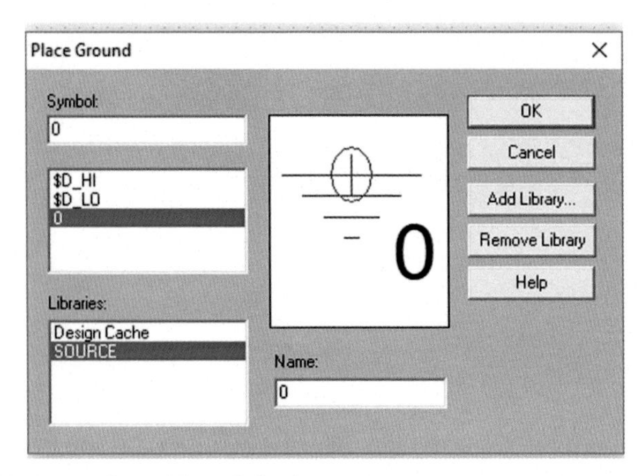

FIGURE 3.12 Locate Ground from Collection.

To assign names (labels) to the nets

1. In the Place menu, pick out Net Alias to show the Spot Net Alias dialog box.
2. Within the Name text box, give as "Mid."
3. Select OK.
4. Put the alias on any section of the circuit that links R1, R2, R3, the diodes, and the condenser. The low left nook of net alias makes necessarily contact the circuit.
5. Select (right-click) End Mode to leave the Net Alias utility.

To assign names (labels) to the off-page Terminals

Spread out the off-page termination according to Fig. 3.12.

1. Choose name in off-page (double-click) linkage to display the Property dialog box.
2. Provide the new name.
3. Snap OK.
4. Choose and move the new name as desired.

To assign names to the parts

1. Choose second VDC part (double-click) to show the Devices spreadsheet.
2. Select the initial cell subordinate to the Reference column.
3. Give new figure "Vin."
4. Click Employ to update the alteration to the device, then close the spreadsheet.
5. Proceed, forgiving the name left out of the devices until the schematic drawing appears like the primary designing.

To change the values of the parts

1. The voltage label (0 V) is selected (double-click) on V1 to show the Exhibit Properties dialog box.
2. Within Value text box, type 5 V.
3. Give OK.
4. Proceed forgiving name left out devices until schematic drawing visage like primary designing with all names of devices.

To save the design

1. In File menu, select Save.

3.1.2 Model Libraries

A model explains the electrical activity of a device. On a diagram page, this compatibility is given by a device execution property, which is allotted with the framework name. Relying on the device kind that it describes, a model is defined as one of the following:

- a model factor set
- a subcircuit net list

Either methods of explaining a design are written in the document as a special regulation of syntax.

Device kind and subcircuit distinctness are arranged into the examples collection in the library. Possible libraries are textual matter files that bear one or higher number of model explanation. The specific model library bears the name of LIB extension. Almost all these libraries incorporate examples of an analogous type. For vendor-made examples, libraries are divided by maker. To learn more about the model library, see the remark in the file header.

3.1.2.1 Model library configuration

PSpice probes model libraries for names specified by the MODEL execution for parts in the design, and they explanation that PSpice exercise to simulate the required circuit (Fig. 3.13).

For PSpice to see for the required model account, the libraries should be configured, which should say:

- Stipulate the directory route or paths to its model libraries.
- Giving a name to every library that PSpice will look up and list those in the required lookup command.
- Distribution of designing reach to the model library.

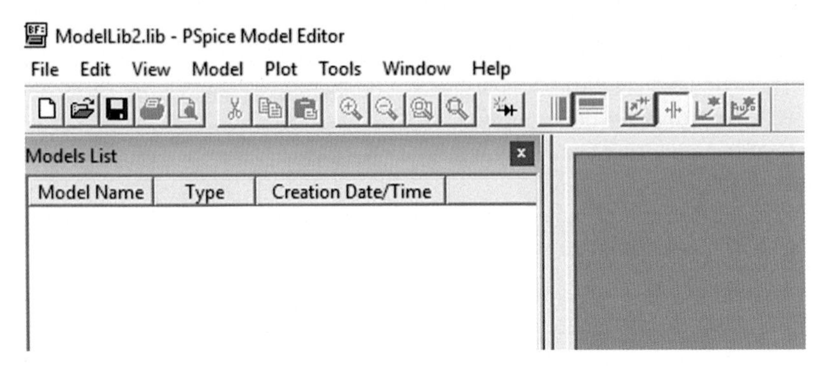

FIGURE 3.13 Open Model List.

OrCAD-provided models

The model libraries at the start of installation with OrCAD programs are cataloged in NOM.LIB. This file presents how could be nested mention with other libraries and models.

If the Libraries tab is chosen in the Simulation Set dialog box straightaway after installing, the NOM.LIB* entry in the Library Files database will be seen. The asterisk denotes that the model library, and any other mentioned, comprises global model explanation.

Tools to make and modify models

There are three tools that can be applied to make and modify model definitions:

- Model Editor when it is needed:
  - obtain models from data sheet curves given by supplier, or
  - change the activity of Model Editor-supported model.
- Alter the PSpice, command syntax (text) for .MODEL and .SUBCKT explanation.
- Give Subcircuit command within diagram page editor when there is a hierarchical position in the designing, which is wanted to be done as the same part with activity represented as a subcircuit net list (.SUBCKT syntax).

The Model Editor is applied in five ways:

- *To specify a fresh model, and to automatically make a part.* a few new model and part are by design accessible to every design.
- *To describe a new model only (no part).* You can optionally switch off the device formation characteristic for fresh models. The model description is obtainable to any design, for instance, by altering the model execution for a part example.
- *To change a model meaning for a part example in drawing*: This means you need to start the Model Editor from the drawing sheet editor

TABLE 3.1 Different model types.

Part type	Use this definition form	Name prefix
Diode	.MODEL	D
Bipolar transistor	.MODEL	Q
Bipolar transistor, Darlington model	.SUBCKT	X
IGBT	.MODEL	Z
JFET	.MODEL	J
Power MOSFET	.MODEL	M
Operational amplifier	.SUBCKT	X
Voltage Comparator	.SUBCKT	X
Nonlinear magnetic core	.MODEL	K
Voltage Regulator	.SUBCKT	X
Voltage Reference	.SUBCKT	X

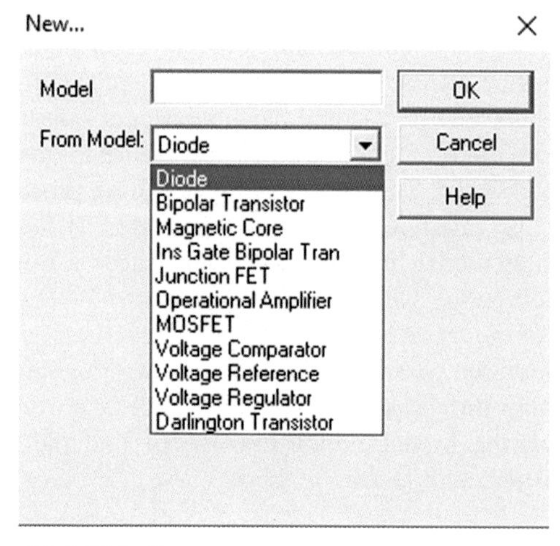

FIGURE 3.14 New Model Selection.

subsequent to choosing a device case in drawing. The schematic editor by design encloses the fresh model functioning (that Model Editor create) to the particular part occasion.

- *To inspect electrical character of a model exclusive of applying PSpice.* It could be used in the Model Editor on its own to ensure characteristics of a model rapidly, with given model limitation standards, or evaluate characteristic curves to data sheet values or calculated data (Table 3.1 and Fig. 3.14).

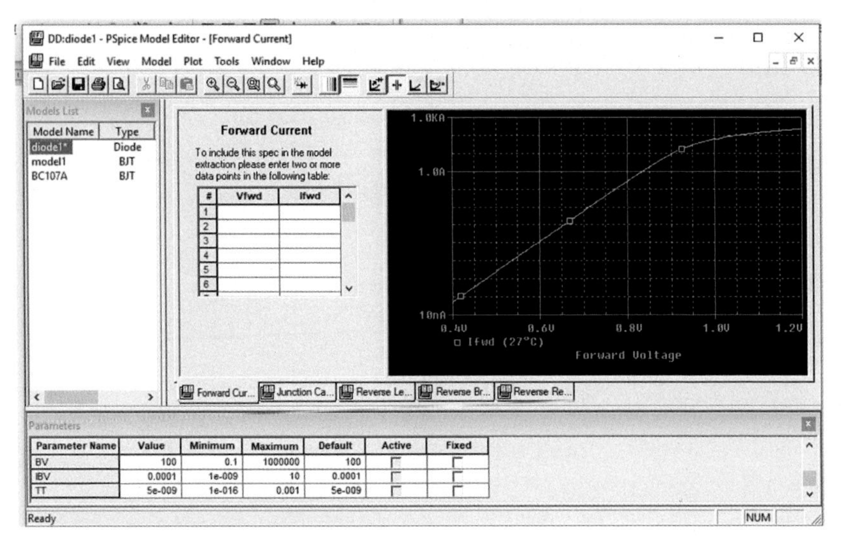

FIGURE 3.15 Diode Model Editor.

Making models with data sheet value

The major spread for characterizing models is to input the data sheet value for every device attribute. After it has been fulfilled with the performance of every characteristic, it could have the Model Editor rough calculate (or extract) the respective model attributes and create a graph presetting the performance of the feature. This is named as the fitting process. This process is iterated, and while satisfying results are processed, it may be saved; the Model Editor creates model libraries, which envisages a suitable model and subcircuit definition.

Evolving the consequence of model parameter on device features

It could be edited the parameters of model in a straightaway method and observe how altering their value influence a device feature. As model parameters are altered, the Model Editor reworks to find the behavior of the device characteristics and shows a fresh curve for every affected ones (Fig. 3.15).

To fit the model

1. For every gadget feature that is needed to set up: Within the Spec Entry frame, choose tab of the apparatus characteristic. Then, the device data from the data sheet is entered.
2. In Tools menu, choose to dig out Parameters to take out all pertinent model values for the current condition.

   A check mark pops up in the Active column of the parameter casing for each extracted model parameter.
3. Do steps 1–2 until the model meets our objective behavior.

Executing only the Model Editor

Execute the Model Editor only to carry out of the underlined:

- to produce a model and apply the model in any design;
- to generate a model and its definition accessible to any design (without producing a part); or
- to inspect characteristics of a specified model exclusive of using PSpice.

Operations from only the Model Editor denotes the model that is being created or examined and is not at present attached to a part occurrence in the drawing sheet or to a part editing portion.

Initial the Model Editor

1. In Start menu, select the OrCAD program folder, and select Model Editor.
2. In File menu, prefer New or Open, and select existing or fresh model library name.
3. In Part menu, select New, Copy From, or Import to load a model.

To create parts for fresh models by itself

1. In Tools menu, prefer Options (Fig. 3.16).
2. If not previously tested, opt for Always Create Part to allow automatic part formation.

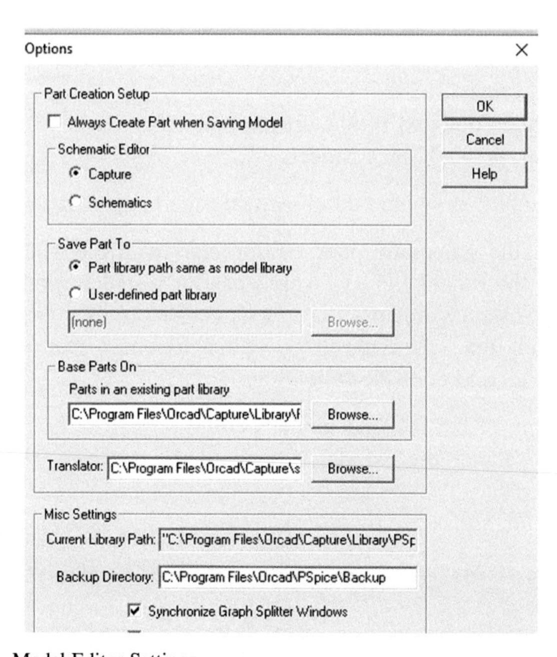

FIGURE 3.16 Model Editor Settings.

3. Beneath Save Part to give name of the part library for a fresh part, prefer either:
   - Part Library Path Same as Model Library: To build or to release the *.OLB file that has the identical name prefix as the presently open model library (*.LIB).
   - User-Defined Part Library: Select the respective file name in the Part Library text box.

Global model (and parts) storing

While storing the changes, the Model Editor carries the functions of:

Stores the model description to the model library that was initially opened.

If the auto mode of part formation preference is enabled, it stores part definition to MODEL_LIBRARY_NAME.OLB.

Operation of the Model Editor from the schematic edit page

If it is required to:

- check performance variation on an element, or
- filter a model in previous of assembling it to be available for every design

then execute the Model Editor in schematic page editor in Capture.

This implicates editing models for part performance in the schematic page. When it is selected as a part occurrence and corrects its model, the editor by itself generates a case model that could then be changed.

To begin editing an instance model

- In Capture, pick one part on the drawing sheet.
- In Edit menu, prefer PSpice Model.

For the schematic drawing edit choice, see the model libraries example.

- If observed, the schematic page edition start with the Model Editor, by itself, open the model library, which has an instance model and brings forward the instance model.
- If not found, the schematic page editor assumes that this is a new instance model and does the following:

Makes a copy of the original model definition, names its original model name-Xn, and starts the Model Editor with the new model loaded (Fig. 3.17).

Generating the half-wave rectifier system

1. In the Project Manager, File menu go to New and then to Project (Fig. 3.18).
2. Create fresh task with its name (RECTFR) and create it.
3. In Capture's Place menu, go to part.

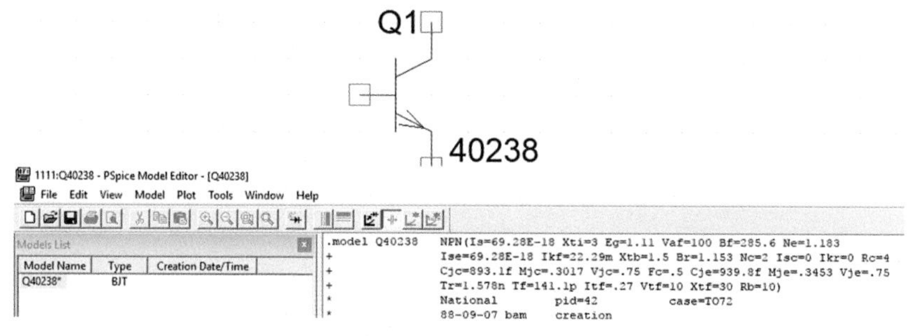

FIGURE 3.17 Model Editor.

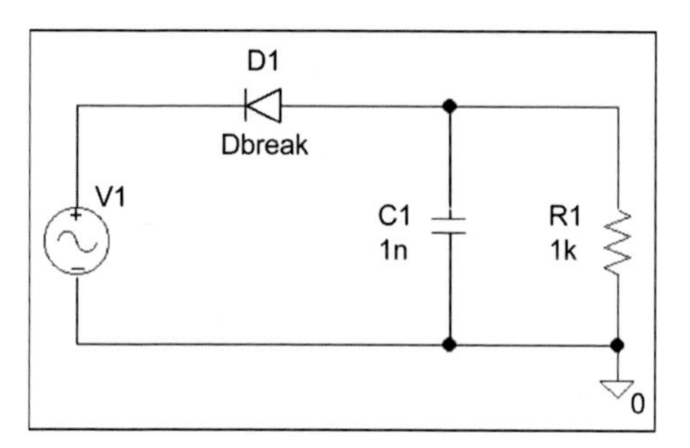

FIGURE 3.18 Circuit for Instance Model.

4. Position any of the parts as listed under (reference designator exhibited in parentheses) as in Fig. 3.30:
   - Dbreak (D1 diode) · C (C1 capacitor) · R (R1 resistor) · VSIN (V1 sine wave source)
5. Select Ground button and locate the analogue ground.
6. In Place menu, choose Wires, and finish the connections of the devices as per Fig. 3.30.
7. Save it in file.

3.1.3 Way to scrutiny

3.1.3.1 DC Sweep

The DC scrutiny valuates circuit operation with respect to a DC source.

The following table gives the PSpice calculations for every DC analysis type.

DC Sweep	Steady-state voltages and currents during the time of sweeping a source, a model Factors, or temperatures with the extent of values
Bias point detail	Bias point data in addition to what is automatically computed in any simulation
DC sensitivity	Sensitivity of a net or part voltage as a function of bias point
Small-signal	Small-signal DC gain, i/p & o/p resistances, as a function of bias point

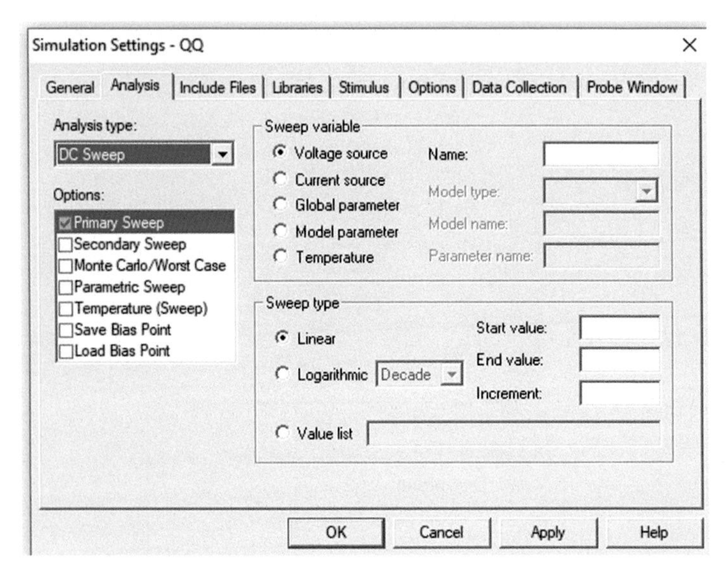

FIGURE 3.19 DC Sweep window.

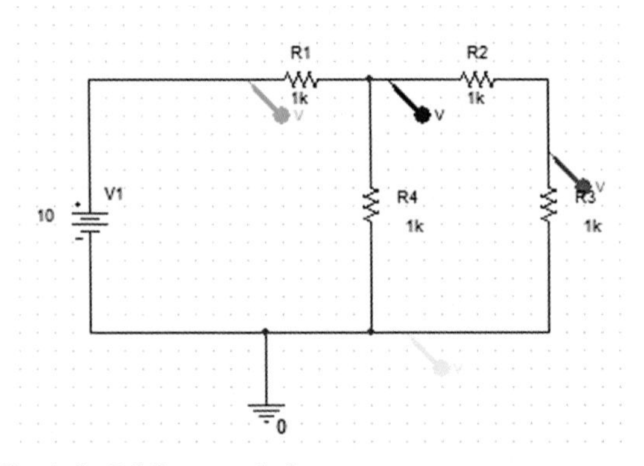

FIGURE 3.20 Circuit for DC Sweep analysis.

DC transfer

In Capture, opt for fresh Simulation design or correct the Simulation setting in PSpice menu. (If in case of fresh simulation, give the name of the profile and then OK.)

1. The Simulation options dialog box looks like the under given figure.
2. In Analysis type, opt for DC Sweep (Fig. 3.19).
3. For initial Sweep choice, give the data of required values and opt for respective check boxes to finish the analysis specifications.

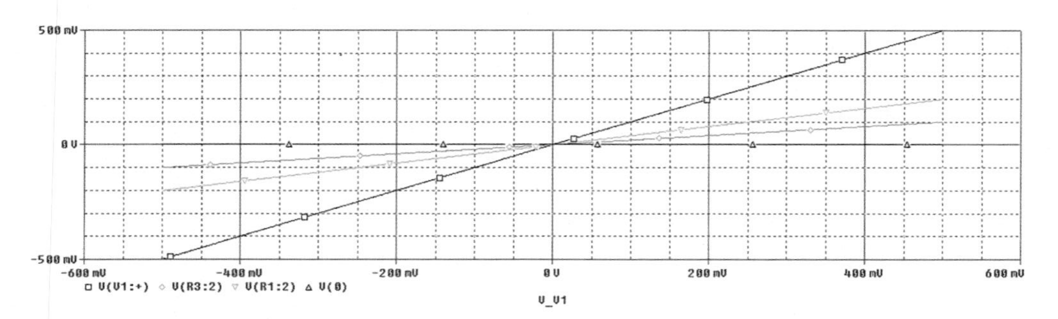

FIGURE 3.21 DC Sweep analysis Waveforms.

4. Save in simulation profile.
5. Opt for Run under the PSpice menu to begin the simulation.

To compute the DC action of an analog circuit, PSpice does not take time from the circuit. This is carried out by operating all condensers as open, all inductors as short circuited, and applying just the DC magnitude of voltage and current sources. To resolve the circuit equations, PSpice applies a repeating algorithm. For analogue elements, the equations are uninterrupted (Figs. 3.20 and 3.21).

3.1.3.2 *Transient analysis*

1. Commence a fresh work.
2. Design the circuitry as per Fig. 3.22.
3. Within the Place menu available with schematic editor prefer Part to be shown in the Place Part dialog box.
4. Adjoin the library for the devices which are needed to be placed:
   a. Add the Library button.
   b. opt for SOURCESTM.OLB (in PSpice library) and Open it.
5. In the Part text box, give the command as VSTIM and position the device in the drawing editor.
6. When the circuitry of Fig. 3.23 is completed, choose Save to save the design.

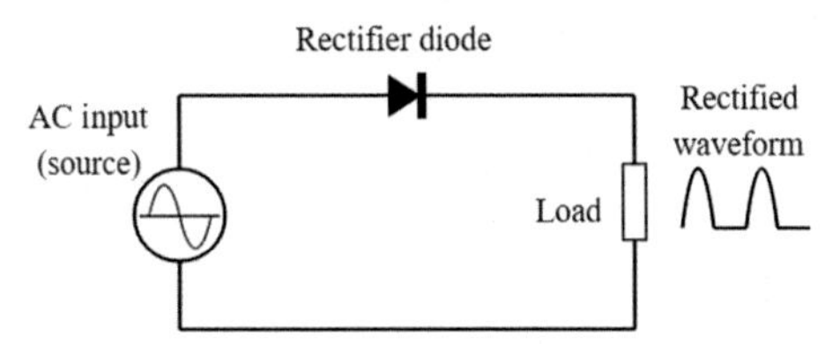

FIGURE 3.22 Rectifier Circuit.

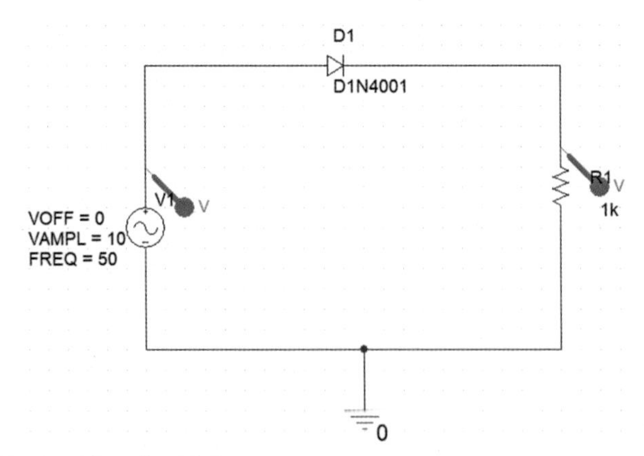

FIGURE 3.23 Rectifier using PSpice.

To bring the stimulus

1. Opt for the VSTIM device (V2).
2. Within Edit menu, opt PSpice Stimulus and its fresh dialog box is shown.
3. In the fresh Stimulus dialog box, give command as INPUTLAB21.
4. Choose Pulse, and give OK.
5. Within the Pulse dialog box, set the data as follows which contains:
   a. Voltage level = 1 V
   b. Pulse width = 1 ms
   c. Periodicity = 2 ms
   d. Rising duration = 2 ns
   e. Falling duration = 2 ns
   f. delaying period = 0 s
6. Apply the values, and observe the resultant output square waveform in display, in the stimulus editor.
7. Proceed with OK.
8. In File menu, pick save option for stimulus information. Opt out "Yes" to revise drawing.
9. In File menu, prefer Exit with Stimulus Editor.

For building and to execute the transient analysis

1. In Capture's PSpice menu, prefer fresh project Profile, and the new Simulation dialog box pops up.
2. Give the name as "Lab21TimeResponse."
3. From the Inherit in listing, go for None, then "Create" which brings up "Simulation Settings dialog box."
4. Go for Analysis tab.
5. In that list, pick Time Domain (Transient) and preferred time interval is given over which the computer will work out the time response:
   a. TSTOP = 6 ms
   b. Start saving data after = 0 s (Fig. 3.24).
6. Give OK to complete the Simulation setting dialog box.
7. Within PSpice menu, opt to Run and carry out the analysis. PSpice sets by itself for internal time steps for calculation, which is attuned according to the needs of the transient analysis as it progresses. PSpice save data into the waveform data file for every inner time step (Figs. 3.25 and 3.26).

3.1.3.3 AC Sweep analysis

The AC sweep analysis option in PSpice is a linear (or small-signal) frequency domain study that could be observed for frequency responses of several circuits at its bias point.

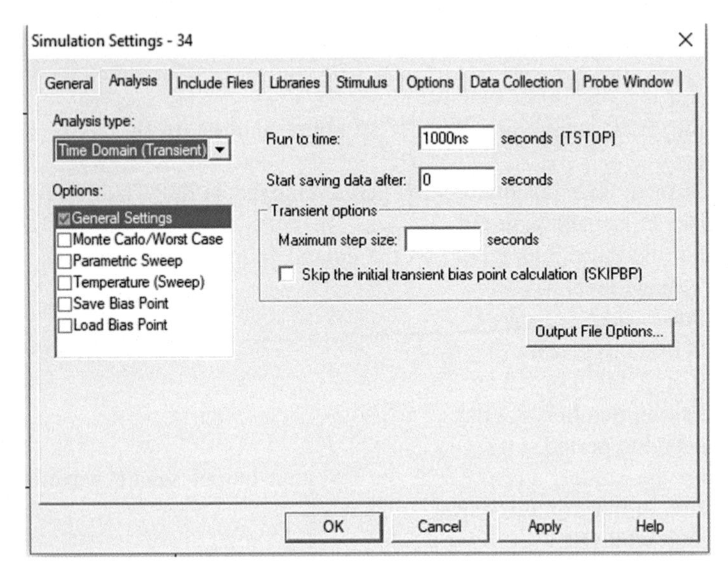

FIGURE 3.24 Transient Analysis Window.

Designing and executing an AC Sweep analysis
Put up the AC voltage source over the drawing folio, as exhibited in diagram.
To display the devices over spreadsheet (double-click) choose over VAC part (0 V).
Modify the Referring cell to V in and vary the ACMAG cell to 1 V.
Apply it to renew the alterations and complete the spreadsheet (Figs. 3.27 and 3.28).

For building and to execute the AC Sweep

1. In Capture's PSpice menu, opt for new Simulation Project.
2. Give the name as "AC Sweep," and create it. The Simulation options dialog box pops up.
3. Opt for Analysis tab.
4. In Analysis type listing, prefer AC Sweep/Noise and make the settings as per Fig. 3.29.
5. Complete the Simulation Settings dialog box (OK).
6. Within the PSpice list of options, choose Run to start the simulation. PSpice executes the AC analysis.

To adjoin markers for waveform analysis

1. In Capture's PSpice Opt for Markers/Advanced, then prefer db scale of Voltage.

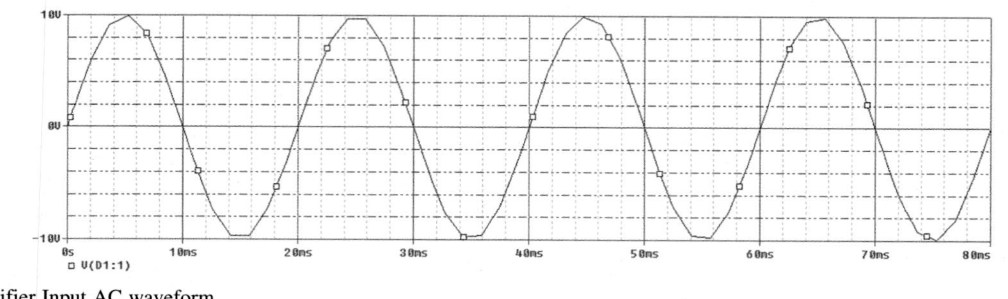

FIGURE 3.25 Rectifier Input AC waveform.

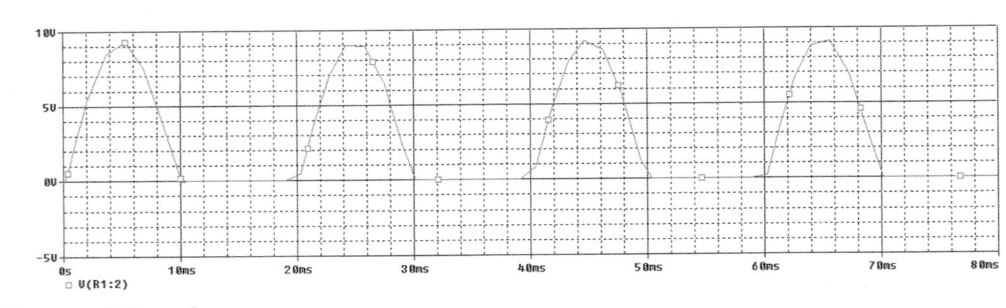

FIGURE 3.26 Rectifier output DC waveform.

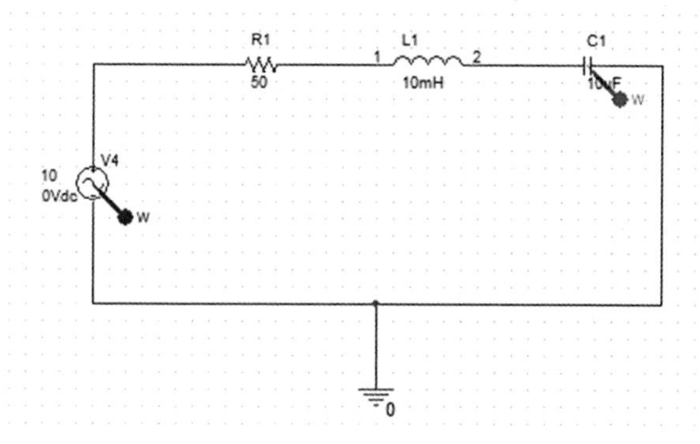

FIGURE 3.27 Series RLC Circuit.

2. Position Vdb markers over the Out net, then position one more in the mid net.
3. Save the design.

To exhibit a Bode plot of the output signal voltage, consider phase

1. Captures PSpice list, prefer in order of markers/Advanced/Phase of Voltage
2. Remove the Vdb marker on Mid.
3. Substitute to PSpice.
4. Prefer trace name VP (Out) to take the trace.
5. In Edit menu, pick out, Cut.
6. In Plot menu, adjoin Y Axis.
7. In Edit menu, Paste and Bode plot come out.

Parametric Analysis (Fig. 3.30)

1. Alter the time value of R1 in expression {Rval}.
2. Orient the PARAM device to state the factor Rval.
3. Generate and run a constant synthesis to get the numerical quantity of R1 applying Rval.

for alteration of measurement of R1 to the equation of {Rval}
Modification the measure of R1 to the expression {Rval}

1. Within Capture, open file.
2. Take the value (1k) of part R1 (double-click) to exhibit the Presentation Attributes dialog box.
3. Within Value text option, substitute 1k with {Rval} and choose OK.

To add a PARAM part to declare the parameter Rval

1. In Capture's Place menu, opt for Part.
2. Within Part text box, give as PARAM (available in PSpice library SPECIAL.OLB), then OK.

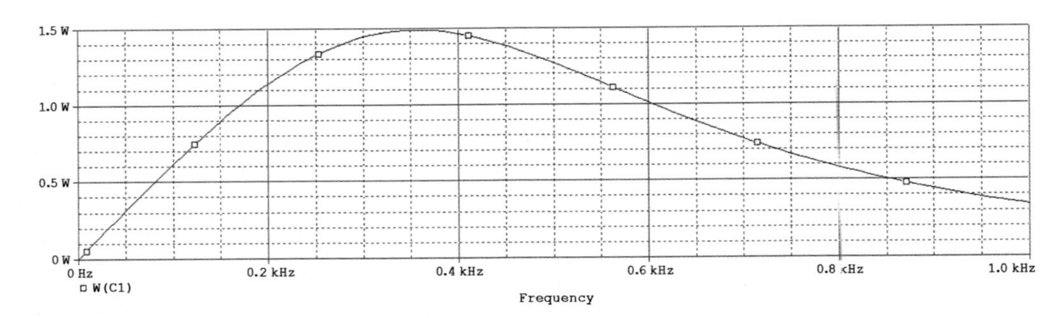

FIGURE 3.28 Frequency Response of Series RLC Circuit.

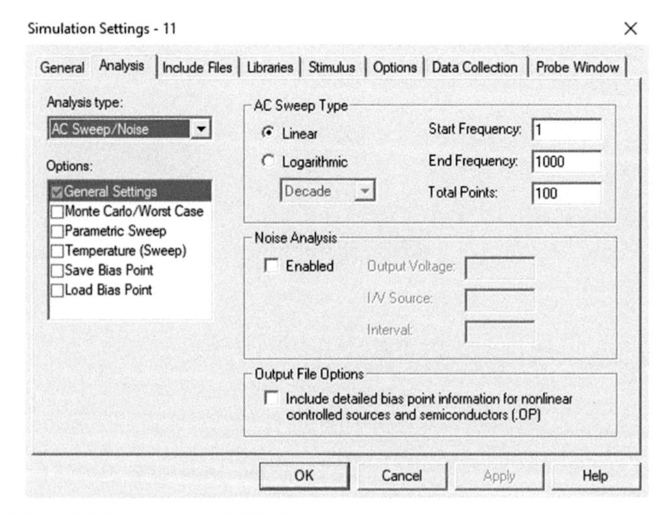

FIGURE 3.29 AC Sweep Analysis Window.

3. Position one PARAM device in free space in the diagram page.
4. Opt for the PARAM part to exhibit the device spreadsheet, and the New.
5. Within Property text box, opt Rval (no curly braces), then OK which generates a fresh attribute for the PARAM part, as exhibited by the fresh column tagged as Rval within the spreadsheet.
6. Opt the cell beneath the Rval file and enter 1k as the first quantity of the parametric range.
7. During this cell is selected, opt for Display and choose Name and Numerical quantity, and OK.
8. Within Display Information frame,
9. Opt to apply to update all the modification to the PARAM part.
10. Finish the Parts spreadsheet.
11. Take the VP marker and go for D to withdraw the marker in drawing page.
12. In File list, opt to save (Fig. 3.30).

To Design and process a constant quantity analysis to step the numerical quantity of R1 Applying Rval

1. In Capture's PSpice listing, opt Fresh Simulation Project, its dialog box pops up.
2. In the Name option give Parametric.
3. With Inherited From listing, choose AC Range, and Create. The Simulation Settings dialog box come out.
4. Opt for Analysis tab.

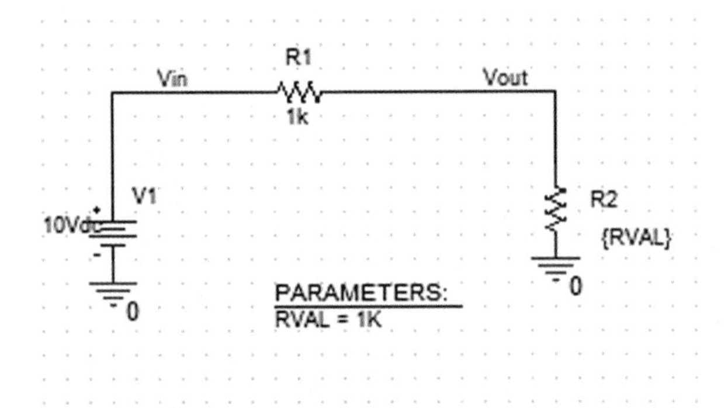

FIGURE 3.30 Circuit for Parametric Analysis.

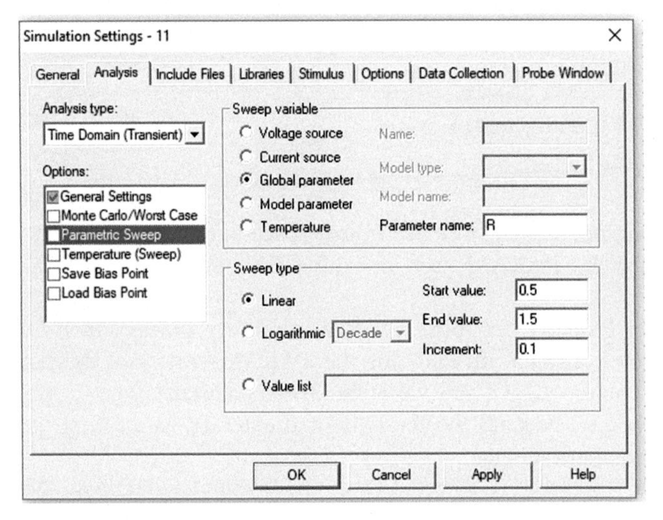

FIGURE 3.31 Parametric Window.

5. In Options, select Parametric Sweep, further carry as per settings as depicted in Fig. 3.31 and 3.32.

3.2 Designing and simulation of power IGBTs

The IGBT transistor adopts the better qualities of two variants of transistors, the higher input impedance with switching rate of a MOSFET with the minimum saturation voltage of a bipolar transistor, and amalgamates them jointly to give out one more variant of transistor switching equipment which can handle sizable collector-emitter currents with no gate current drive.

The IGBT amalgamates the insulated gate technology of the MOSFET with o/p operation characteristics of a customary bipolar transistor. The effect of this hybrid process is that the "IGBT Transistor" has the o/p

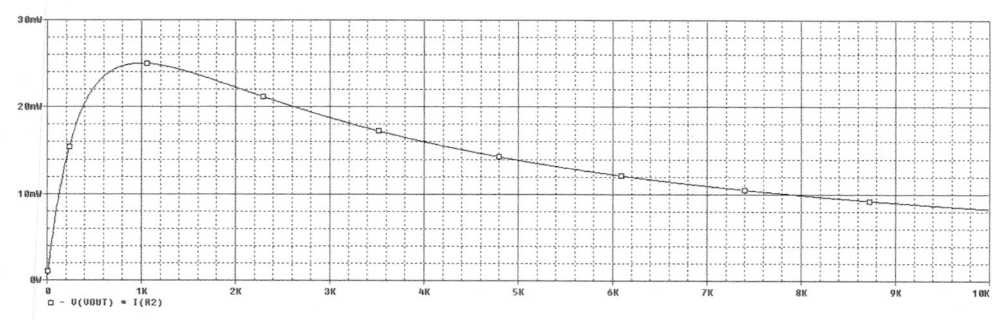

FIGURE 3.32 Output waveform-Parametric Analysis.

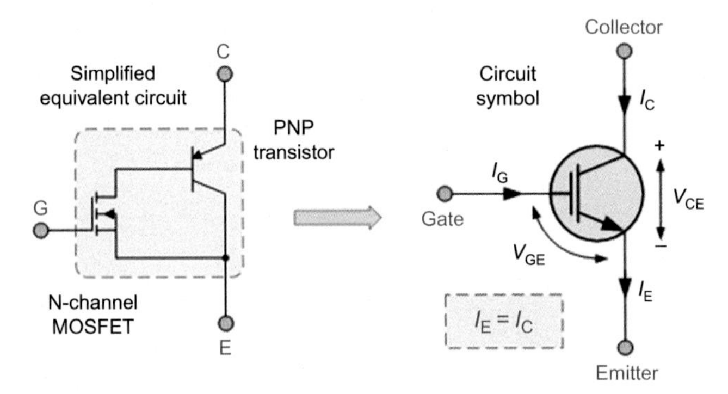

FIGURE 3.33 IGBT symbol.

switching and conduction characters of a bipolar transistor but the voltage-control is similar to MOSFET.

IGBTs are applied in inverters, converters, and power supplies, where the need of the solid-state switching devices is not fully met by power bipolar/MOSFETs. High-current/voltage bipolar is obtainable, with low switching speeds, but power MOSFETs have high switching speeds; high-voltage/current equipment are high-priced and challenging to achieve.

The benefit achieved by the IGBT against a BJT or MOSFET is that it gives higher power vantage compared to standard bipolar type transistor added with the high-voltage working capacity and reduces input losses of the MOSFET. In the result, it is an FET integrated with a bipolar transistor in a design of Darlington type design as exhibited (Fig. 3.33).

3.2.1 IGBT models

Even if the area of improving power semiconductor design for circuit simulator has won extra attention emanates from research group and software vendor in latest years, still there is break with growth of real semiconductors and the growth of actual accessible simulator design of the equipment. During designing of IGBTs, there is bargaining in between correctness and ease of the design. High exactness is frequently able to be achieved barely at the price of process speed, the exchanges within various circuit simulators, or difficult factor mining.

There are various kinds of methods in designing of IGBTs, so every design has an inherent kind and tries to follow the realism. Hence it is necessary for designers to aim for the reason of the simulations, and depending on this, jointly with existing materials, it builds on the suitable options of design. Also, in spite of design, the data taken out it is of great significance and several of data and the mining process are taken care previous to selecting design.

3.2.2 Characteristics of IGBT

Static $I-V$ Characteristics of IGBT

Fig. 3.34 depicts static $I-V$ characteristics of an n-channel IGBT beside a circuit drawing showing details, such as the $I-V$ characteristics of an n-channel IGBT. This is comparable with that of a BJT excepting that the factors that keeps stable for a plot is VGE since IGBT is a voltage dependent device unlike BJT which is a current forced device.

Transfer Characteristics of IGBT

During device is in OFF condition (VCE is positive and VGE < VGET) the reverse voltage is choked-up by J2 while it is reverse biased (Fig. 3.35).

3.2.3 PSpice model of IGBT

1. In menu file, opt for New, and in that it can designed required circuit drawing on it.
2. Save schematic design as a file, then in menu File a popup window appears. Opt for the directory created. E.g., c:\name in C: drive. Choose

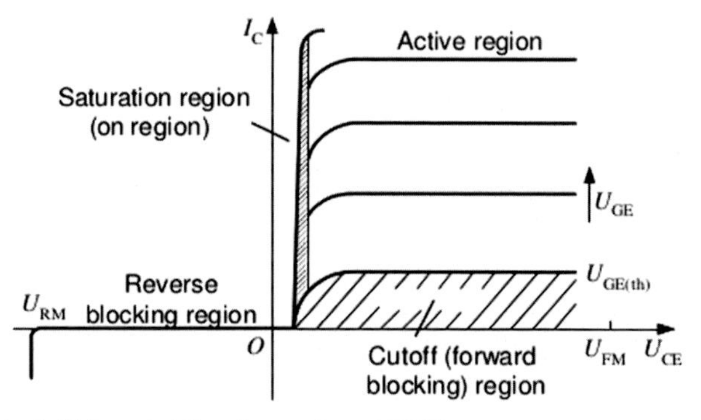

FIGURE 3.34 $I-V$ characteristics of an n-channel IGBT.

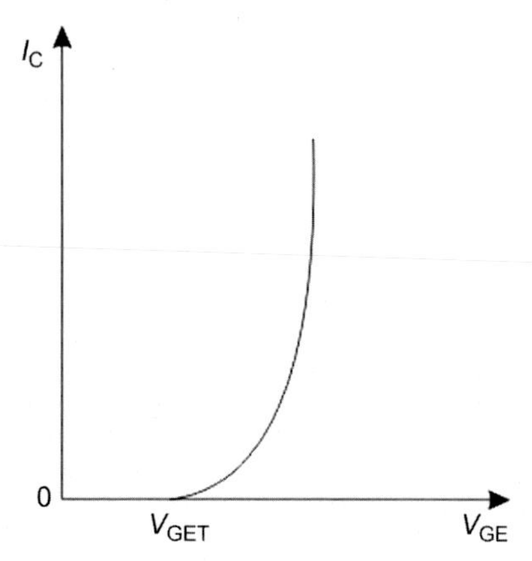

FIGURE 3.35 Transfer characteristics of IGBT.

that folder in "Save in." Describe a name for the file. Opt for the name in "File name" row and "Save" it.

3. To design the circuit in schematics, the primary thing is locating the devices in the circuit. Which has, DC supply, analogue earth, resistor, and IGBT. All these devices are available in libraries.

   In the "Part Name:" give the input as "Vdc," which fetches DC voltage source. Each device is opted and placed suitably and opt for "Place and Close" which positions the devices in appropriate positions.

   During this step, it can be observed that DC voltage source symbol is dragged along with the mouse. Click the mouse at two different places, which gives out two DC voltage sources in drawing. To remove the DC source opt out "ESC" button in the keyboard and it will exit the DC source insertion option. The DC source value is not concerned as off now (which are 0 V now) and the values of the parts are assigned afterwards. Choose circuit device and its color will turn to red. It can be dragged to anywhere in schematic, and can be deleted with Keyboard
4. Iterate the course of action as step 5 to include a resistor and analogue earth. Apply the "R" resistor, and name it "GND_ANALOG" to take from analogue. In the case of IGBT, pick out the device as in Fig. 3.36.
5. Once the devices are positioned, the devices could be rotated when it is necessary. For instance, in the circuit if R1 to be rotated, the resistor is taken so its color is converted to red. After that click over menu "Edit—Rotate" and R1 will be twisted by 90 degrees. Again, with R1 sorted out, click menu "Edit—Flip." This will flip R1 upside down. This is to check the R1is having exact current polarity so as to flow from node 1 to 2, and not from node 2 to 1.
6. Now connect all the devices using wire (Fig. 3.37).
7. Now set the analysis in PSpice—the type of simulation. In the menu "Analysis—Setup," a fresh popup window comes out in which it could be selected for DC Sweep analysis.
8. PSpice is about to simulate the project. Choose menu "Analysis—Simulate." PSpice will process the simulation and exhibits an "PSpice A/D Student Demo" window, as in Fig. 3.38. Rarely the window may not be available, in such cases iterate the same step (click "Analysis—Simulate") and the desired results could be got.
9. The simulation is processed, and it could be observed for the analysis of results. In PSpice A/D window, menu opt for "Trace—Add Traces," which gives a popup window as in Fig. 3.39. This lists every existing voltage and current signals. For instance, to observe current flow in R1, opt "I(R1)," and I(R1) will be exhibited on "Trace Expression" line.
10. PSpice also processes mathematical data in multidimensional signals and draws the output. If the voltage in between drain and source VDS of transistor IGBT in a new plot is needed to be observed in PSpicc, solve it. It should be taken care that V = V (V1) − V (V2); V1 and V2

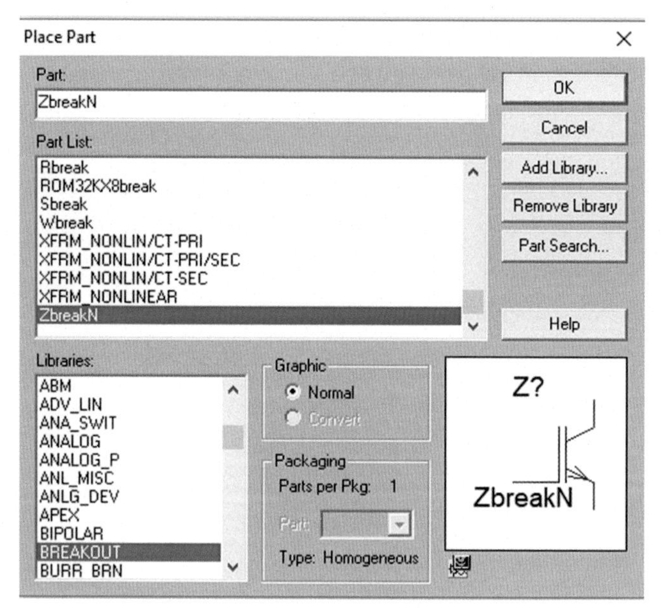

FIGURE 3.36 Place IGBT from Library.

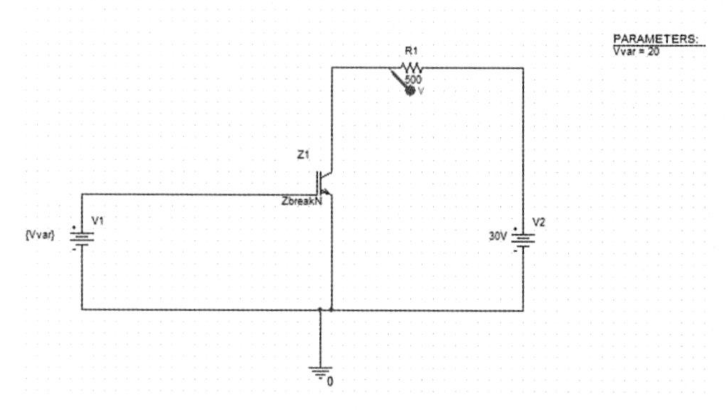

FIGURE 3.37 Simulation Circuit: Input Characteristics.

are the node description that are pronounced earlier. Choose "Plot—Add Plot to Window," and a fresh empty plot will come out in the window. The "SEL>>" sign under the empty plot shows the recently added signal trace. Now opt in to the menu "Trace—Add trace"; in the popup option, choose "V(V1)" in left column, and then opt "-" sign in the right column (Analogue Operations and Functions column), after that once more opt V(V2) in the left column, it is observed expression "V(V1) − V(V2)" exhibits the "Trace Expression" line. The characteristics are presented from Figs. 3.40–3.42.

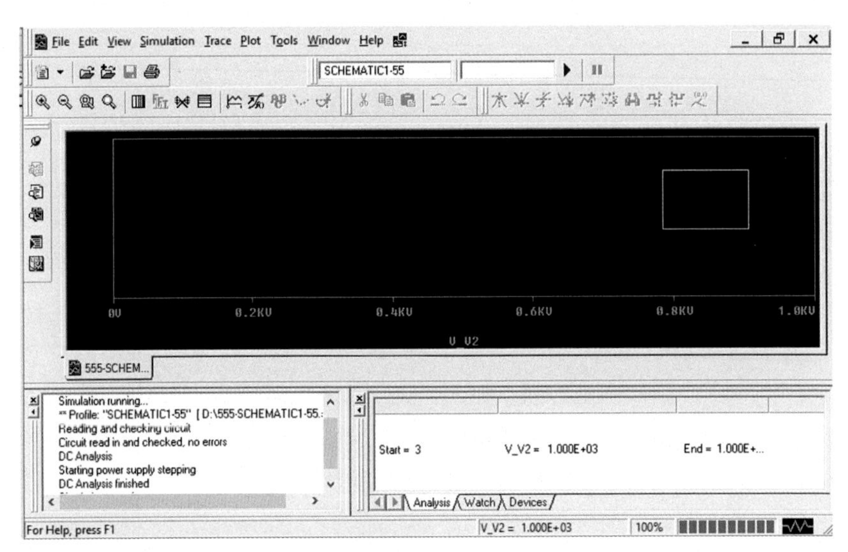

FIGURE 3.38 PSpice A/D Student Demo.

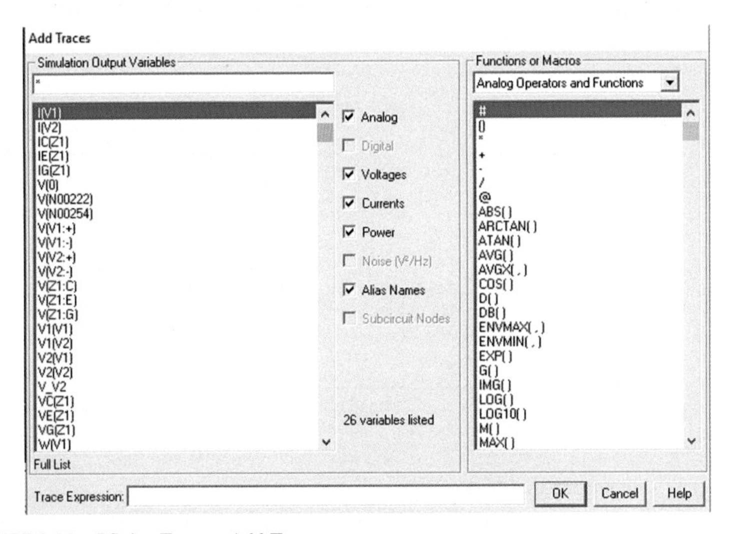

FIGURE 3.39 PSpice Trace—Add Traces.

3.3 Design and simulation of TRIAC

The TRIAC is a power electronics device applied for the AC switching equipment. This can control the flow of current over both halves of an alternating cycle. Only the thyristor can alter in one-half of a cycle. In the leftover half no conduction takes place and consequently only half of the waveform can be applied (Fig. 3.43).

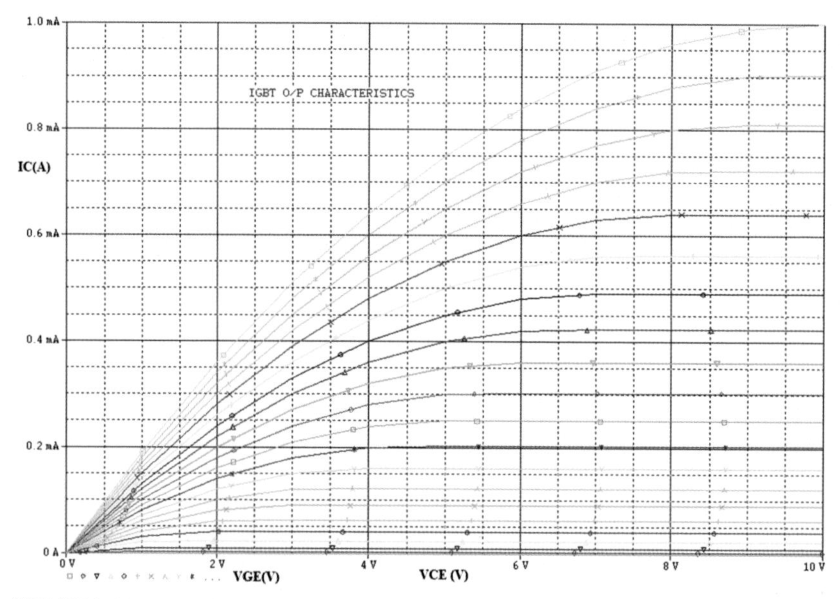

FIGURE 3.40 Input characteristics of IGBT.

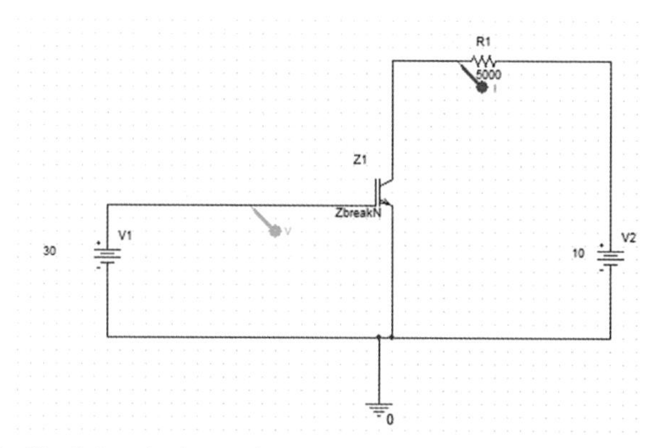

FIGURE 3.41 Simulation circuit: transfer characteristics.

3.3.1 Introduction to TRIAC

The symbolization has three terminals, which are Gate, Anode, or Main terminal. The two terminals are named as anode1, anode2, main terminal MT1, or MT2. In external orientation, the symbol is observed as back-to-back thyristors, which is depicted in the symbol (Fig. 3.44).

3.3.2 *I*–*V* characteristics of TRIAC

The figure under depicts the representative of TRIAC characteristics. The Triode for AC current has ON and OFF state characteristics and are related

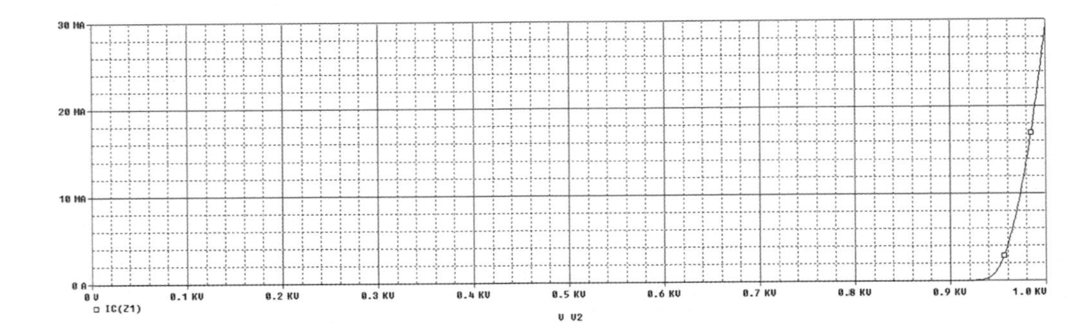

FIGURE 3.42 Transfer characteristics of IGBT.

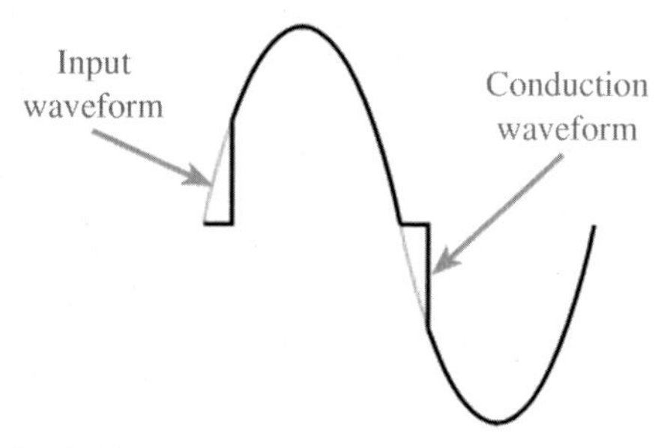

FIGURE 3.43 TRIAC switching operation.

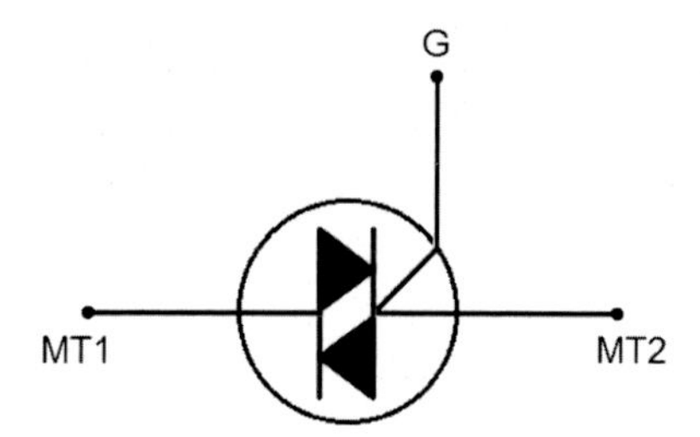

FIGURE 3.44 TRIAC symbol.

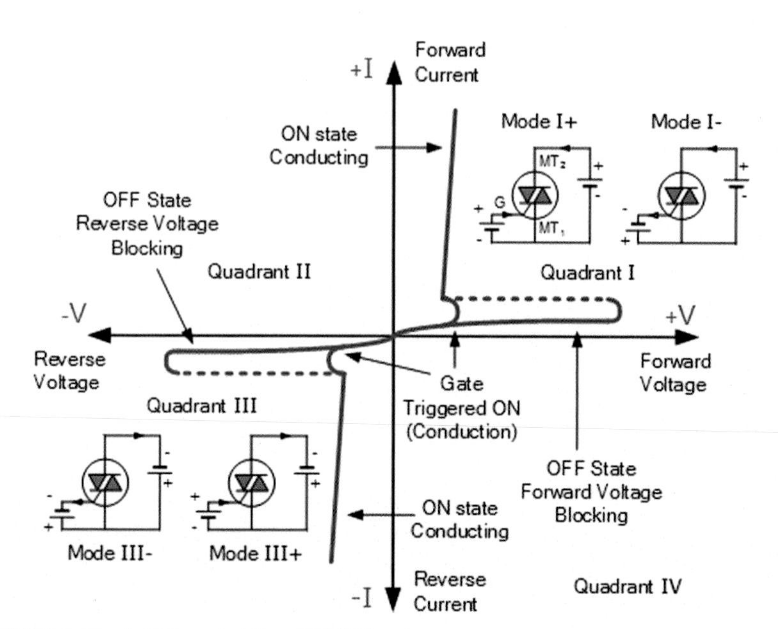

FIGURE 3.45 $I-V$ characteristics of TRIAC.

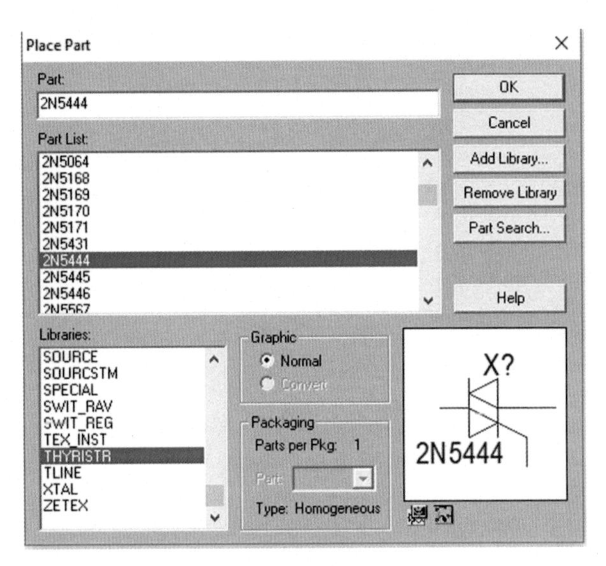

FIGURE 3.46 TRIAC in library.

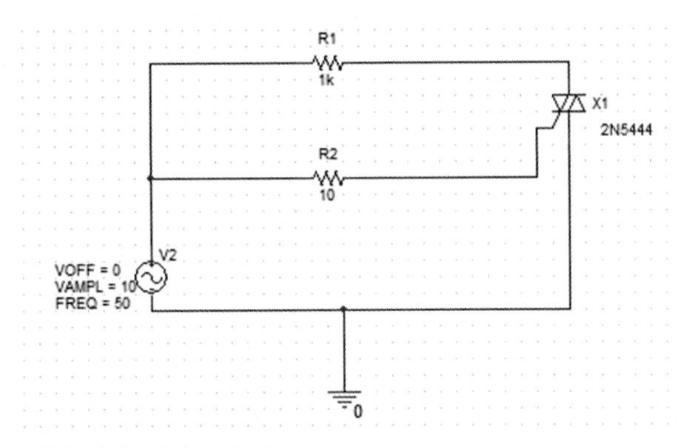

FIGURE 3.47 TRIAC simulation circuit.

to the SCR, which are relevant to both the voltages of both polarities (Fig. 3.45).

This is predictable because the Triode for AC has two SCRs, which are adjoining in collateral and in diametric directions. In the first quadrant the MT2 is positive, which corresponds with MTX, and in the fourth quadrant, it is in the negative.

The gate pulse may be triggered in one of four quadrant condition of operations. If it is in "ON" the conduction authorize a vast level of current in it. This current is limited by the resistance; if not, the device may break down. In TRIAC, the gate is a control end and the correct trigger is given with gate, hence the firing angle is controlled.

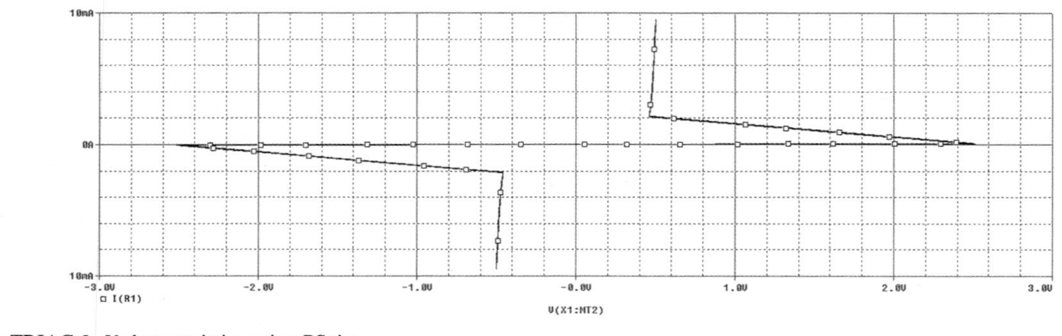

FIGURE 3.48 TRIAC $I-V$ characteristics using PSpice.

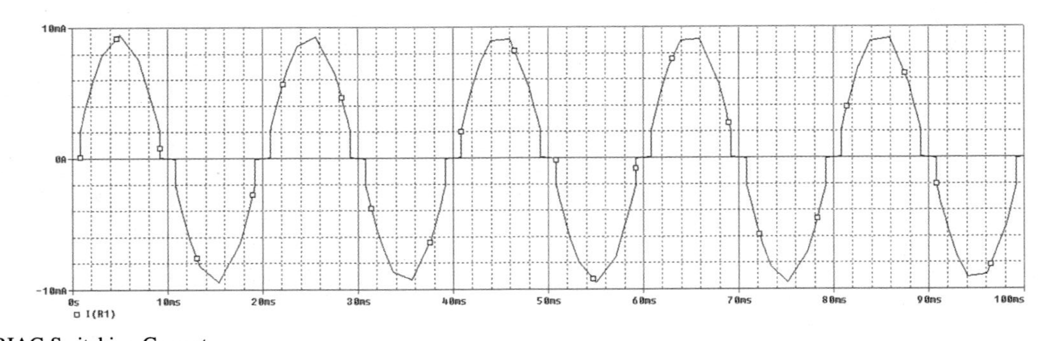

FIGURE 3.49 TRIAC Switching Current.

The gate for trigger the block is called the gate triggering circuit, and or the TRIAC it is same as to the SCRs. Mostly, the triggering circuits will render the firing pulse of the devices and have enough property and time, so that the firing of the device is secure. To keep up the firing, a time period of 35 μs is needed.

3.3.3 *I*–*V* characteristics of TRIAC using PSpice

1. Ensue the corresponding steps presented in the early section.
2. *VAC*: The AC small-signal part of PSpice calculates the ac O/P variables depending on frequency. The program initially calculates the dc operational location of the circuit and finds linear zed, small-signal models for the nonlinear devices of circuit. The subsequent linear circuit is then studied for, in required limits of frequencies. The desired output of an AC small-signal synthesis is normally a transfer function (voltage gain, Tran impedance, etc.). If the circuit has an AC input, it is accessible to archive that input to unity and zero phases, like that the output multivariate have the unique value as the transfer function of resultant variable when compared with input.
3. TRIAC is derived from library as in Figs. 3.46–3.49.
4. Link up the devices according with circuit presented as under,
5. Simulate this circuit and do transient synthesis. The waveforms are obtained after the simulation process.

3.4 Summary

This chapter explains many uses of the PSpice simulation software, from the fundamentals to the harder problems. The reader can effortlessly comprehend the designing means of a range of circuits applying PSpice and will be able to excel in all levels of design in PSpice. The simulation of the electric circuit is explained with every procedure that starts with positioning devices until the output plot of waveforms, which facilitates good perception.

3.5 Review questions

1. What are the ways to include components that are drawn from the library in PSpice software?
2. Narrate the means to design and simulate any circuit applying PSpice.
3. In brief, explain the Model Libraries in PSpice.
4. List the ways to apply the Model Editor with PSpice software.
5. Illustrate the methods of studying the impact of model data on device characteristics.
6. Explain the instance model.
7. Describe the DC Sweep study with an example.

8. Describe the how the transient analysis is done out in PSpice software, and explain with a simple example.
9. Plot the frequency response of series RLC circuit.
10. When is the "PARAM" applied in PSpice software "PARAM"?
11. Explain the step-by-step method used to obtain the V–I characteristics of IGBT.
12. Draw transfer characteristics of IGBT.
13. Design and simulate the V–I characteristics of TRIAC.

Chapter 4

Multisim

Chapter Outline

4.1 Multisim introduction

Multisim is the systematic catch and reenactment use of National Instrument Circuit Design Suite, which is a retinue of electronic design automation apparatuses that help you in completing the significant strides in the circuit configuration stream. Multisim is intended for schematic passage, simulating, and encouraging to downstage advances, for example, printed circuit-board design.

4.1.1 Menu bars

The menu bars listed underneath are accessible in Multisim, and they are

- Standard Menu bar
- Main Menu bar
- Simulation Menu bar
- View Menu bar
- Components Menu bar
- Virtual Menu bar
- Graphic Annotation Menu bar
- Instruments Menu bar

Software Tools for the Simulation of Electrical Systems. DOI: https://doi.org/10.1016/B978-0-12-819416-4.00004-1

4.1.1.1 Standard toolbar

Sl. no.	Button symbol	Description
1		This creates the new file for circuit
2		This opens the subsisting circuit file
3		This opens a file holding unit and getting commenced files
4		This saves the simulated circuit
5		This prints the simulated circuit
6		This analyzes the circuit since it is going to be printed out
7		This exits the selected components and put them over the windows notepad
8		This reproduces the selected components and puts them over the windows notepad
9		This includes the content of the windows notepad at the cursor position
10		This overturns the latest performed activity
11		Redoes the most recently performed undo

4.1.1.2 *Main toolbar*

Sl. no.	Button symbol	Description
1		This shifts the Design Menu bar on or off
2		Switches Spreadsheet View on or off
3		Starts the Database Manager pop-up box
4		Launches the Element Wizard
5		Displays the graph
6		Displays the Postprocessor dialog box
7		Verifies that the electronic rules set up for the wirework of the circuitry have ensued
8	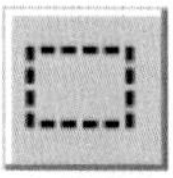	Captures Screen Area
9		This back Expound via Ultiboard
10		Forward Expound
11	 ?	Launches the support file

4.1.1.3 *Simulation toolbar*

Sl. no.	Button symbol	Description
1		Pauses at following MCU instruction border
2		Stepping into
3		Stepping over
4		Stepping out
5	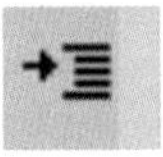	Running to the cursor
6		Toggle breakpoint
7		Remove all breakpoints

4.1.1.4 *View toolbar*

Sl. no.	Button symbol	Description
1		Represents just the workspace, without menu bards or components
2		Magnifies the simulated circuitry
3		It decreases the enlargement of the simulated circuitry
4		Drags the cursor to choose a location on the workspace to enlarge

(*Continued*)

(Continued)

Sl. no.	Button symbol	Description
5		Shows an entire circuit in the workspace

4.1.1.5 Elements menu bar

	Description
	Chooses the *Source* elements group within the browser
	Chooses the *Basic* elements group within the browser
	Chooses the *Diode* elements group within the browser
	Chooses the *Transistor* elements group within the browser
	Chooses the *Analog* elements group within the browser
	Chooses the *TTL* elements group within the browser
	Chooses the *CMOS* elements group within the browser
	Chooses the *Miscellaneous Digital* elements group within the browser
	Chooses the *Mixed* elements group within the browser
	Chooses the *Power* elements group within the browser
	Chooses *the Indicator* elements group within the browser
MISC	Chooses the *Miscellaneous* elements group within the browser
	Chooses the *Electromechanical* elements group within the browser
	Chooses the *RF* elements group within the browser
	Chooses the *Advanced Peripherals* elements group within the browser
	Chooses the *MCU* module elements group within the browser

4.1.1.6 Graphic annotation toolbar

Press on this toggle to position a *picture* over the workspace

Press on this toggle to design a *polygon*

	Press on this toggle to design an *arc*
	Press on this toggle to design an *ellipse*
	Press on this toggle to design a *rectangle*
	Press on this toggle to design a *multiline*
	Press on this toggle to design a line
A	Place the text toggle
	Place the comment toggle

4.1.1.7 *Instruments toolbar*

	Positions a multimeter over the workspace
	Positions a function creator over the workspace
	Positions a wattmeter over the workspace
	Positions a scope over the workspace
	Positions a four-channel oscilloscope over the workspace
	Positions a Bode plotter over the workspace
	Positions a frequency counter over the workspace
	Positions a word generator over the workspace
	Positions a logic analyzer over the workspace
	Positions a logic level converter over the workspace
	Positions an IV Analyzer over the workspace
	Positions a distortion observer over the workspace

4.1.2 Building blocks

1. *Open or create systematic*: A clear systematic circuit 1 is made automatically. To make another systematic, tap on File, then New, and then Schematic Capture. To save the systematic tap on File or Save As.

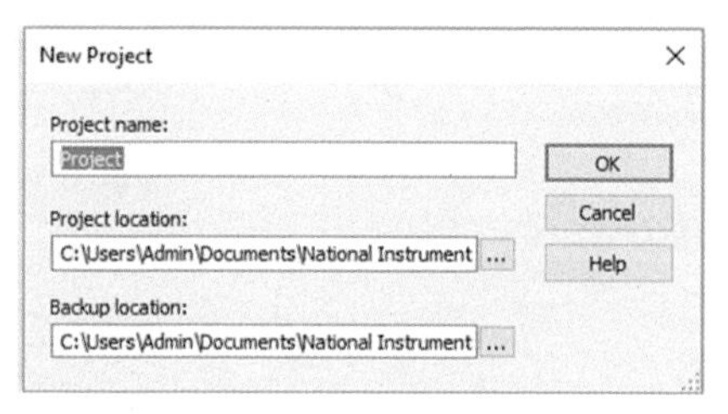

FIGURE 4.1 Open new project.

Now, to open a current document tap on File or Open in the menu bar (Fig. 4.1).

2. *Place elements*: To position Elements tap on Place or Components. Over the Select Element Window tap on Group to choose the elements required for the circuitry. Tap OK to put the element on the systematic (Fig. 4.2).
3. *Virtual elements*: Elements can likewise be placed on the circuitry utilizing Virtual segments. Tap on View, then Toolbars and choose the menu bar required for the circuitry.
4. *Rotate components*: To pivot the parts right tap on the Resistor to turn the segment on 90-degree Clockwise or press Ctrl + R and 90-degree Counter-Clockwise or Ctrl + Shift + R.
5. *Position wire/connect elements*: To interface resistors tap on Place/Wire dragging and spot the wires. Parts can likewise be associated by tapping the mouse on the terminal edges of one segment and drag to the edge of some other segment.
6. *Change component value*: To sift segment value double tap on the part this raises a window that shows the attributes of the segment (Fig. 4.3).
7. *Grounding*: Every circuit should be grounded prior to the circuitry can be reproduced. Tap on Ground in the menu bar to ground the circuitry. In the event that the circuitry is not grounded, Multisim will not operate the simulating process.
8. *Simulation*: For simulating the finished circuitry tap on Simulate or Run or even F5. This element can likewise be accessed from the menu bar.
9. *Analysis*: Select the kind of analysis that you need to activate by tapping Simulate > > then Analyses >>. Example select a transient analysis (Fig. 4.4).
10. Choose Simulate > > then Analyses > > and then Transient Analysis and tap over the Output tab. Include I or v1 to the right-side section by first tapping on I or v1 in the left-side segment and afterward clicking on Add.
11. Likewise, select the time (using transient Frequency parameter).
12. *View output*: Tap on Simulate. The output pop-up shows up which comprises of the tab for cathode yield just as transient corrugates. Distinctive color waves may be seen by picking the color of particular

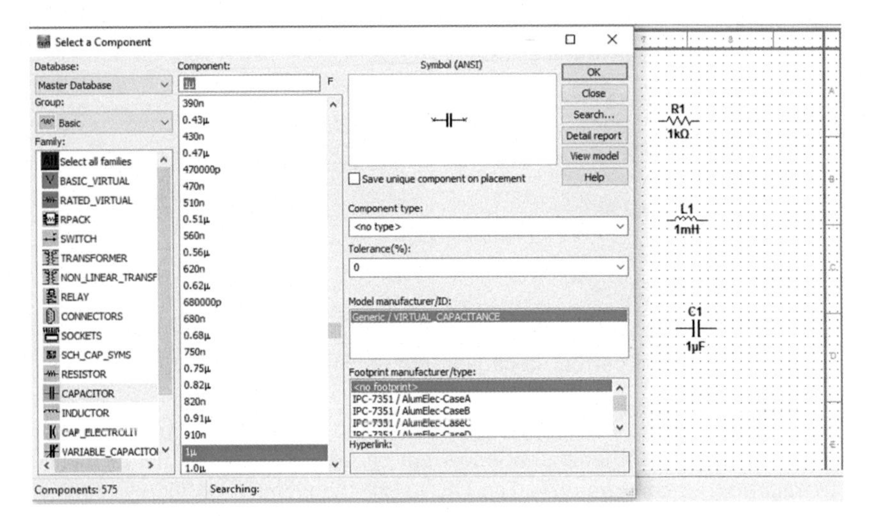

FIGURE 4.2 Placing component.

Capacitor

Label Display Value Fault Pins Variant User fields

Capacitance (C): 1µ F

Tolerance: 0 %

Component type:

Hyperlink:

Additional SPICE simulation parameters

Initial conditions: 0 V

Layout settings

Footprint: Edit footprint...

Manufacturer:

Replace OK Cancel Info Help

FIGURE 4.3 Component value changing.

wires of the electronic amount (voltage and flow). Right tap on the wire in the circuitry and afterward tap on Color fragment . . . to pick the color of wires and therefore the waveform colors (after reproduction).

4.1.3 Electrical Rule Check

The Electrical Rule Check (ERC) makes and shows a report itemizing association blunders (e.g., a yield pin associated with the power pin) and

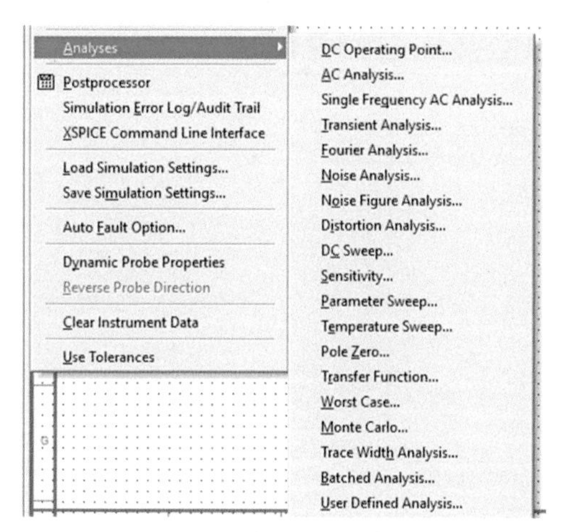

FIGURE 4.4 Types of analysis.

detached pins. When the circuitry is wired, verify the associations for accuracy dependent on the standards established in the Electrical Rule Check box.

Contingent upon your circuitry, you might wish to contain admonitions given if a few sorts of associations are available, error notifications for other association types, and no alerts or blunders for different associations. You handle the kind of associations that are accounted for once ERC is finished by establishing the guidelines inside the grid discovered within the ERC Rule of the Electrical Rule Check box.

ERC might be operated over a whole plan, or just over specific zones of a structure. At that point, once an ERC is operated, any inconsistencies are accounted for into the outcomes sheet at the base of the screen, whereas the circuitry is interpreted on with roundabout blunder markers. Tapping on a blunder will fixate and magnify the mistake area.

The ERC Options along with ERC Rule are utilized to design the ERC.

To operate the electronic guidelines check:

1. Choose Tools or Electrical Rule Check to show the Electrical Rule Check box (Fig. 4.5).
2. Establish the reporting course of actions utilizing the ERC Options (Fig. 4.6).
3. Establish the rules utilizing the ERC Options (Fig. 4.7).
4. Tap OK. The outcomes are shown in the configuration chosen within the Output in the ERC Options.

4.1.4 Running simulating process

To see the aftereffects of your simulating process, you should utilize the virtual instruments or run the examination to show the reenactment yield.

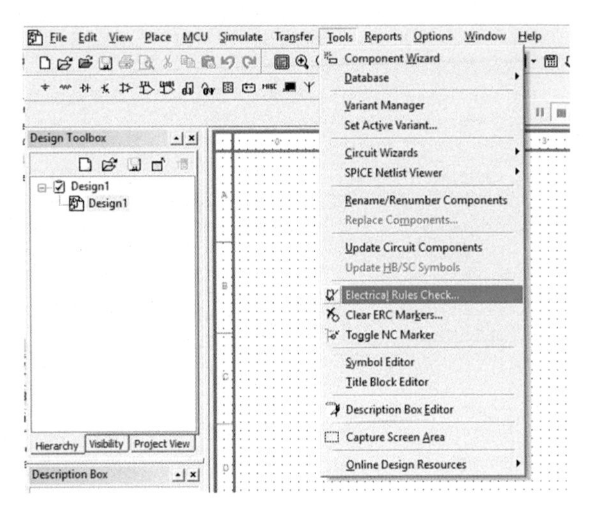

FIGURE 4.5 Electrical rule checking tab.

FIGURE 4.6 Electrical rule checking options tab.

This yield will incorporate the joined consequences of all Multisim recreation motors.

At that point, when you utilize an intelligent simulating process within Multisim (by tapping on the Run or Resume Simulation), you notice the recreation results in a flash by reviewing virtual instrument, for example, the Oscilloscope. Likewise, you can see the impact of reproduction on parts, such as LED and 7-portion digital presentations.

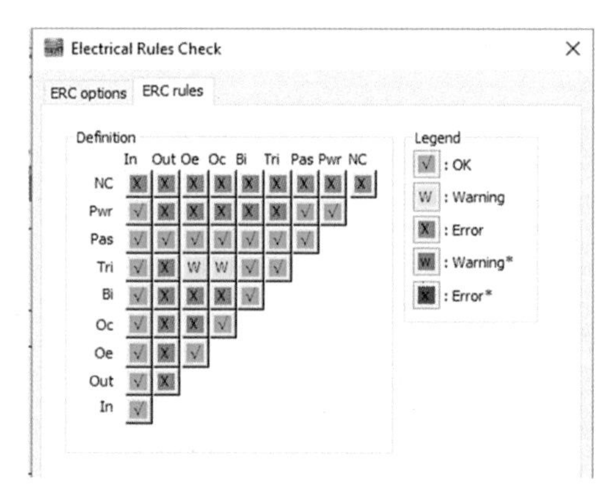

FIGURE 4.7 ERC rules. *ERC*, Electrical Rule Check.

Through intuitive recreation you can operate various analyses over your circuitries. Aftereffects of analyses are shown within the Grapher and can likewise be put something aside for later control within the Postprocessor.

All simulating processes need a referencing net regarding every voltage is given. Within SPICE, that is constantly net 0. In this manner, someplace in the circuitry a net is known as "0" should be characterized. You can either call this net 0, or that will occur if the ground segment is wired inside a circuitry.

4.1.4.1 Intuitive elements

Intuitive elements' value can be modified by tapping a particular key (established in the segment's Value tab) over your keyword. So you can modify the estimation of intuitive elements during recreation and view the impact right away.

On the off chance that you tap on a key, the estimation of every single intuitive gadget on the system that is plotted to the key within their Value tab will modify. On the off chance that you wish to modify the estimation of a particular segment, in the event that you move the cursor on that segment to show a control component. For instance, on the off chance that you move the cursor on a potentiometer, the slider bar creates the impression that you can move to increase or decrease the setting of the potentiometer.

4.1.4.2 Components tolerance

Multisim allows you to add changes for your simulation that are presented because of segment resiliences. For instance, 1 kω circuit breaker with the 10% resilience could differ give or take 100 Ω, and that would thusly influence the consequences of its circuit's reproduction. Parts that have client-settable resiliences are circuit breakers, inductors, capacitor, and a few sources. It is preposterous to expect to set the resilience for all circuit breakers, each and every capacitor, and so on from one area. Resistances must be fixed separately by following the systems beneath.

1. To establish the resistance for a set part:
   a. Double tap on the segment and choose the Value tab.
   b. Choose or insert the ideal value within the Tolerance box and tap on OK.
2. To establish resistances within the Spreadsheet View, choose the ideal segment within the Components
   a. The tab of the Spreadsheet View and modify the values within the Tolerance box.

   Tip: Whether you need to change the resistance to a similar value for various sets parts, choose them all within the Spreadsheet View utilizing Shift along with Ctrl key, and modify the values within any Tolerance box.
3. To utilize part resistances during reproduction, choose Simulate or Use Tolerances. The checkmark shows up close to the menu.

4.1.4.3 *Start, Pause, or Stop Simulation*

Symbol	Description
	To reproduce a circuitry, tap on the Run/continue button. Multisim starts to recreate the circuit's conduct. You can likewise choose Simulate or Run
	For pausing the recreation while it is operating, choose Simulate or Pause. To continue the simulating process from a similar point when you stopped, choose Simulate or Run
	For stopping the reproduction, press the Stop Simulation or choose Simulate or Stop. On the off chance that you restart the reproduction in the wake of halting it, it will reactivate from the earliest starting point (in contrast to Pause, which permits you to restart from where you stopped)

4.1.4.4 *Simulation Run Indicator*

To demonstrate that recreation is operating, the Simulation Run Indicator shows up within the status bar shown in the model beneath. This marker flashes until you stop the recreation process. This is particularly valuable when seeing an instrument which has arrived at a relentless state, for example, the IV Analyzer.

4.1.4.5 *Speed of the simulation*

There are numerous parameters influencing reenactment speed and combination. These are available in the Interactive Simulation Setting box. Probably the most significant settings have appeared on the primary tab. The significant reenactment setting administering the rate of reproduction is TMax. So TMax is the greatest time period that the system is permitted to take. So as to create results, the test system may make littler time periods at its circumspection, anyway, it will not make a bigger period than that predetermined by TMax. The littler TMax is, the exact reenactment results will be. Anyway, it will consume more time to arrive at the random recreation results.

4.1.5 Plotting

4.1.5.1 Grapher

For having the Grapher show up, choose View or Grapher. Actually, the Grapher is a display apparatus with multipurpose facilities that allow you to see, modify, save, and share diagrams and outlines. It is used to show the consequences of every Multisim analysis in diagrams and graphs. A diagram of Fig. 4.8 for certain instruments (for instance the consequences of the postprocessor, scope, and lastly the Bode plot).

The showcase shows the diagrams as well as graphs. In a diagram, information is shown as at least one traces along upright and level axes. In an outline, content information is shown in lines and sections. The window is comprised of a few selected pages, contingent upon the number of analyses, and so forth have been operated. Every page has two potential dynamic regions and will be shown by the red bolt: the whole page demonstrated with the bolt in the left edge close to the page title of the outline/diagram showed with the bolt in the left edge close to the dynamic outline/chart. A few capacities, for example, cut/duplicate/paste, influence just the dynamic region, so be certain you have chosen the ideal territory before playing out an activity.

4.1.5.2 Working with graphs-legends along with the grids

To implement a grid to a diagram:

1. Choose a diagram by tapping anywhere on that.
2. Tap the Show or Hide Grid option. To eliminate the grid, tap the button once more. (Or)

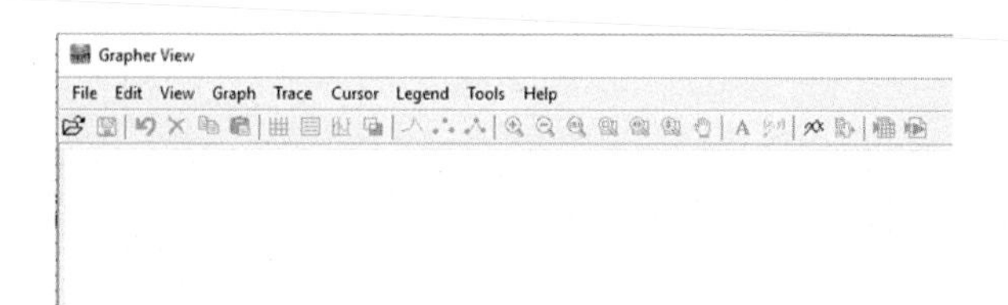

FIGURE 4.8 Graph view.

Choose a diagram by tapping anywhere on that.

1. Tap the Properties option. The Graph Properties box shows up.
2. Tap the General tab.
3. Authorize the Grid On button. If wanted, modify the grid's pen dimension and colors.

To implement a legend to a diagram:

1. Choose a diagram by tapping anywhere on that.
2. Tap the Show or Hide Legend option.

To eliminate the legend, tap the button once more (Or) (Fig. 4.9).

4.1.5.3 Cursors

Once you initiate the point of the cursor, two upright cursors show up on the chosen diagram. Simultaneously, a window appears, presenting a listing of data for a single or each of the traces (Fig. 4.10).

The cursor data comprise:

- x1, y1 or (x, y) correlates to the left-side cursor
- x2, y2 or (x, y) correlates to the right-side cursor
- dx or x-axis Δ among both cursors
- dy or y-axis Δ among both cursors
- 1/dx or correlative of the x-axis Δ
- 1/dy or correlative of the y-axis Δ
- minima x, minima y—x along with y min in the graph extents
- maxima x, maxima y—x along with y max in the graph extents

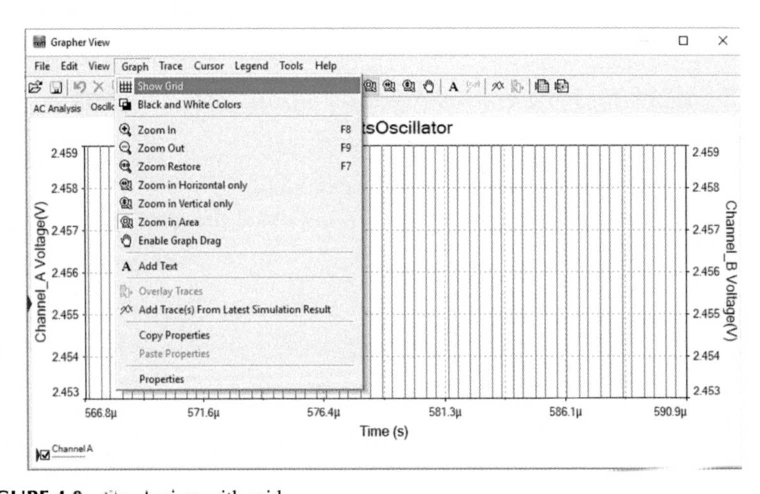

FIGURE 4.9 Graph view with grid.

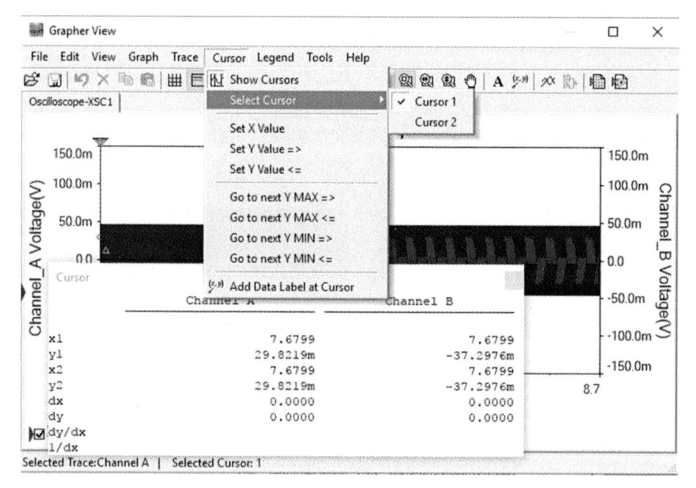

FIGURE 4.10 Graph view cursors.

4.1.5.4 Zoom and restore

1. Choose a diagram by tapping anywhere on that.
2. Tap the Properties option. The Graph Properties box shows up.
3. Choose the General tab.
4. Allow the Legend On button.

Zoom along with Return
To enlarge on any element of a diagram:

1. Choose a diagram by tapping anywhere on that.
2. Press and move the cursor until the spotted amplification box coats the area of the diagram that you need to enlarge.
3. Leave the mouse cursor. The axes are mounted and the diagram is redrawn on the basis of the zoom in box.

Or

1. Choose a diagram by tapping anywhere on that.
2. Tap the Properties option. The Graph Properties box shows up.
3. Tap the axis tab to enlarge along the axis. For instance, select the Bottom Axis to enlarge along with the parallel sizes. (Review the Traces to observe which axle is utilized for the extent you need to zoom.)
4. Insert a new least and highest.
5. To return a diagram to its initial measure, tap the Zoom Restore option.

4.1.5.5 Traces

Left tap to choose a trace. Trace choose marks show up on the chosen trace (tiny triangles on the trace) (Fig. 4.11).

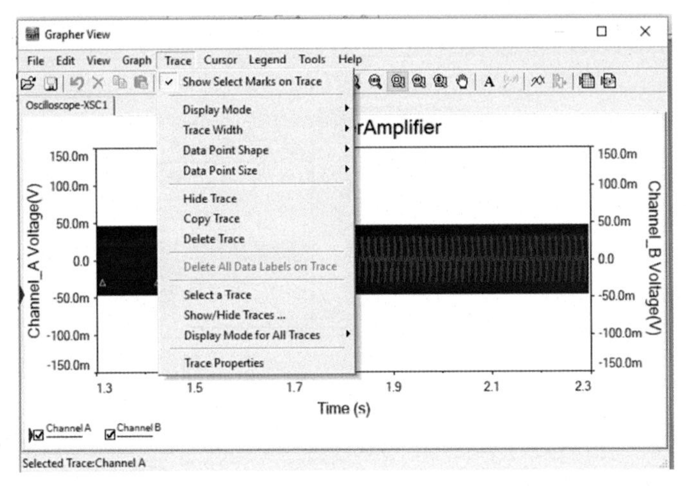

FIGURE 4.11 Graph view traces.

Graph Properties
General Traces Left axis Bottom axis Right axis Top axis
Trace label: Channel A
Trace ID: 1 Show
Sample
Color: Red
Show trace lines
Width: 1
Show data points
Shape:
Size: 4
X-horizontal axis
Bottom axis
Top axis
Y-vertical axis
Left axis
Right axis
Offsets
X offset: 0
Y offset: 0
Auto-separate
Restore OK Cancel Apply Help

FIGURE 4.12 Graph properties.

To replace other trace attributes:

1. Right tap over the trace to show the pop-up window.
2. Choose the desired element from the pop-up window.

You can modify numerous features of every trace in a diagram from the Traces within the Graph Properties box (Fig. 4.12).

To modify the properties of a trace:

1. Choose a diagram by tapping anywhere on that.

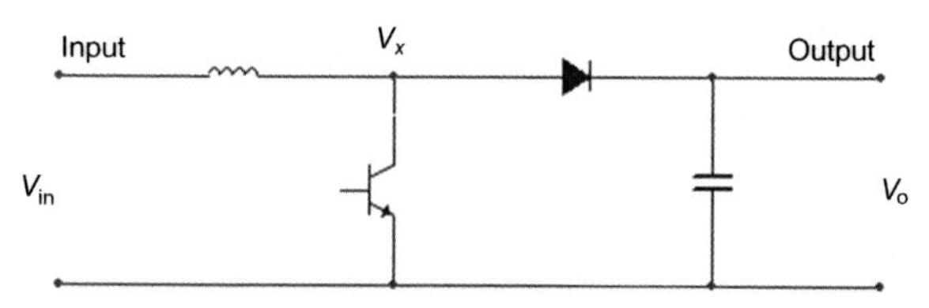

FIGURE 4.13 Step-up inverter.

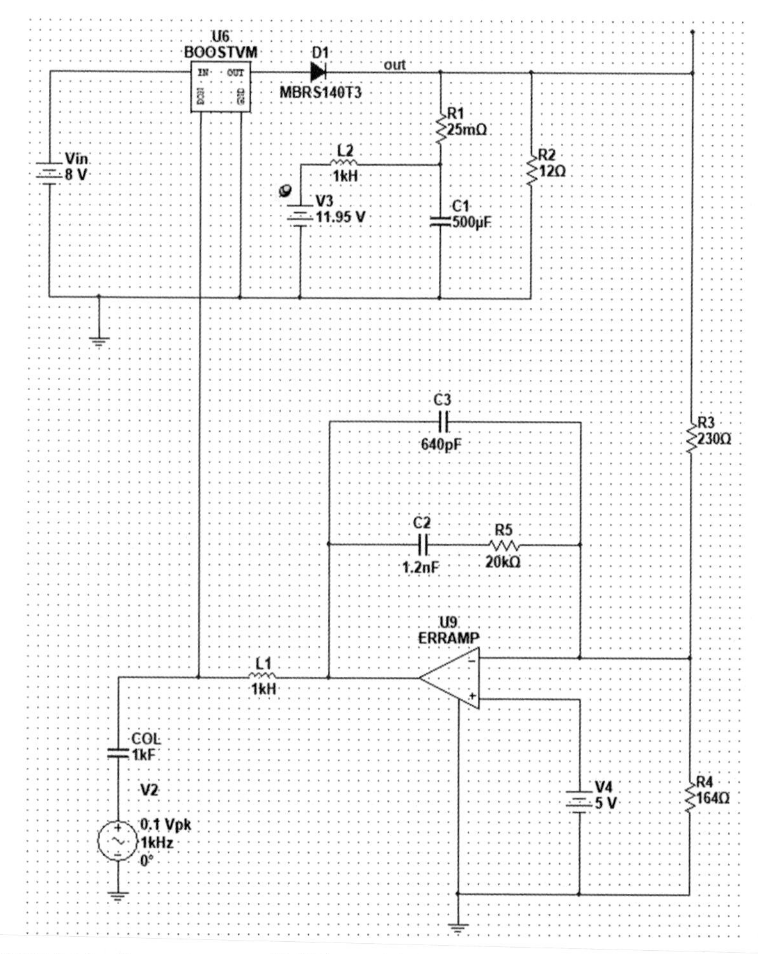

FIGURE 4.14 Step-up converter Multisim model.

2. Tap the Properties option. The Graph Properties box shows up.
3. Tap the Traces.
4. Choose a trace.
5. Modify one of the trace's properties, utilizing the following areas
   a. Trace—Specifies the trace whose properties are being affected.

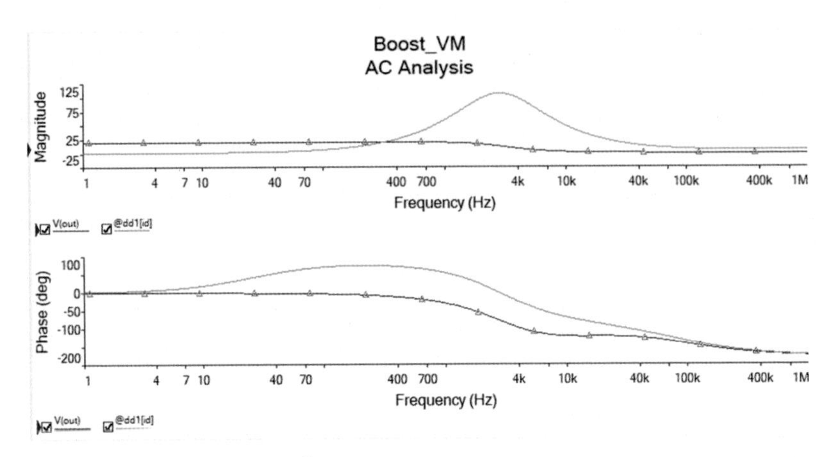

FIGURE 4.15 Step-up converter frequency response.

b. Label—Specifies a label for the trace. Appears in legend.
c. Pen size—Controls the thickness of the trace.
d. Color—Controls the color of the trace. The Sample box shows a preview.
e. Bottom axis/Top axis—Controls the X range of the trace.
f. Left axis/Right axis—Controls the Y range of the trace.
g. X offset/Y offset—Value to offset the trace from its original coordinates.

4.1.6 Converters-using Multisim Model Maker

4.1.6.1 Boost converter design

Boost Converter—Voltage Mode pulse width modulation (PWM) Control (Average Analysis). A step-up converter is a boost DC to DC inverter with a direct DC power higher compared to its indirect DC power. In this example AC Analysis is utilized to view the recurrence reaction to a duty cycle disturbance (Figs. 4.13–4.15).

1. Via the Model of the Component Properties box, tap Add or Edit. Choose a Model box that shows up.
2. Tap Start Model Creator. The Choose Model Creator box shows up.
3. In the Model Creator List, choose Boost Inverter and tap Accept. (Tap Cancel to revert to the Model.) The Step-up Converter Model box shows up.
4. Insert wanted quantities in the Step-up Converter Model box.
5. At the point when all quantities are inserted, tap OK to finish the design, or tap Cancel to quit.

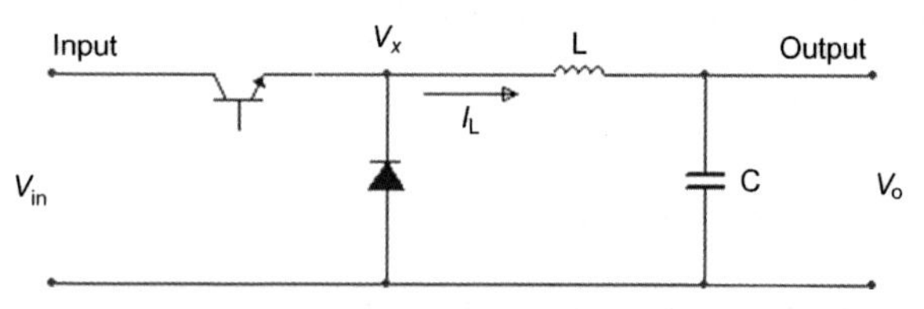

FIGURE 4.16 Step-down converter.

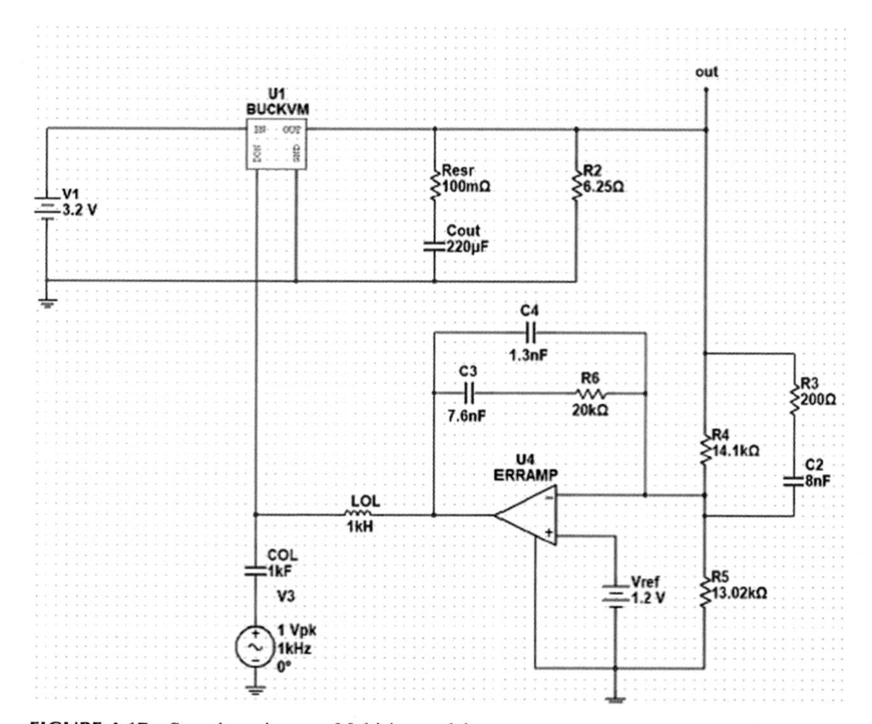

FIGURE 4.17 Step-down inverter Multisim model.

4.1.6.2 Buck converter: voltage mode PWM control

A buck inverter is a step-down DC to DC inverter with the outputs DC voltage power compared to its input DC power. In this example AC analysis is utilized to view the recurrence reaction to a disturbance within the controlling signal (Figs. 4.16–4.18).

1. In the Model of the Component Properties box, tap Add or Edit. Choose a model box that shows up.
2. Tap Start Model Creator. The Choose Model Maker box shows up.
3. In the Model Maker Listing, choose Buck Converter and then tap Accept. (Tap Cancel to revert to the Model.) The Step-down Converter Model box shows up.

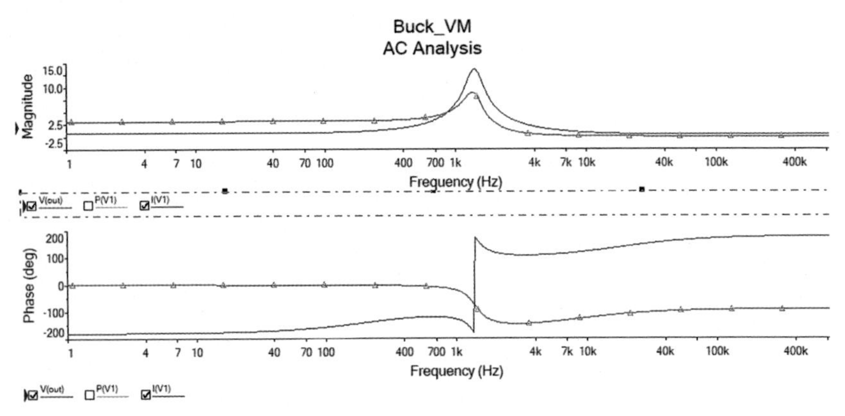

FIGURE 4.18 Step-down inverter frequency response.

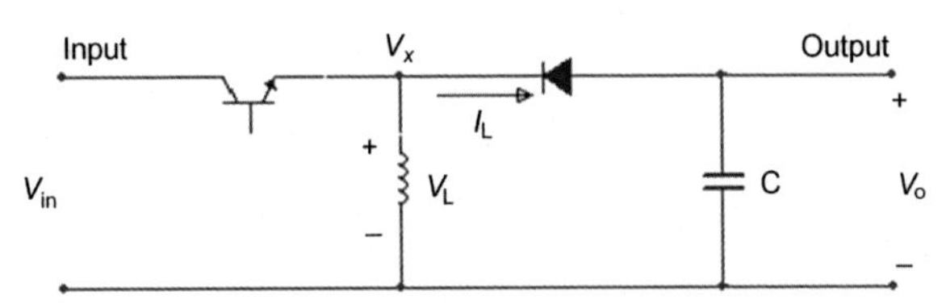

FIGURE 4.19 Step-down step-up converter.

4. Insert desired quantities in the Step-down Converter Model box.
5. Once all quantities are inserted, tap OK to finish the model or tap Cancel to quit.

4.1.6.3 Buck-boost converter: current mode PWM control

The Step-down Step-up inverter is a kind of DC–DC inverter that contains an output power that can be higher than or lesser compared to the input power. In the example below, AC analysis is used to view the open-loop frequency response to a disturbance in the input voltage (Figs. 4.19–4.21).

1. In the Model of the Component Properties box, tap Add or Edit. Choose a Model box that shows up.
2. Tap Start Model Creator. The Select Model Creator box shows up.
3. In the Model Maker Listing, choose Step-down Step-up Converter and press Accept. (Tick Cancel to revert to the Model.) The Step-down Step-up Converter Model box shows up.
4. Insert desired quantities in the Step-down Step-up Converter Model box.
5. Once all quantities are inserted, tap OK to finish the model or tap Cancel to quit.

4.1.6.4 Flyback converter: voltage mode PWM control

The flyback inverter is a DC–DC inverter with electric isolation along the input source and the output source. During the first half of the switching

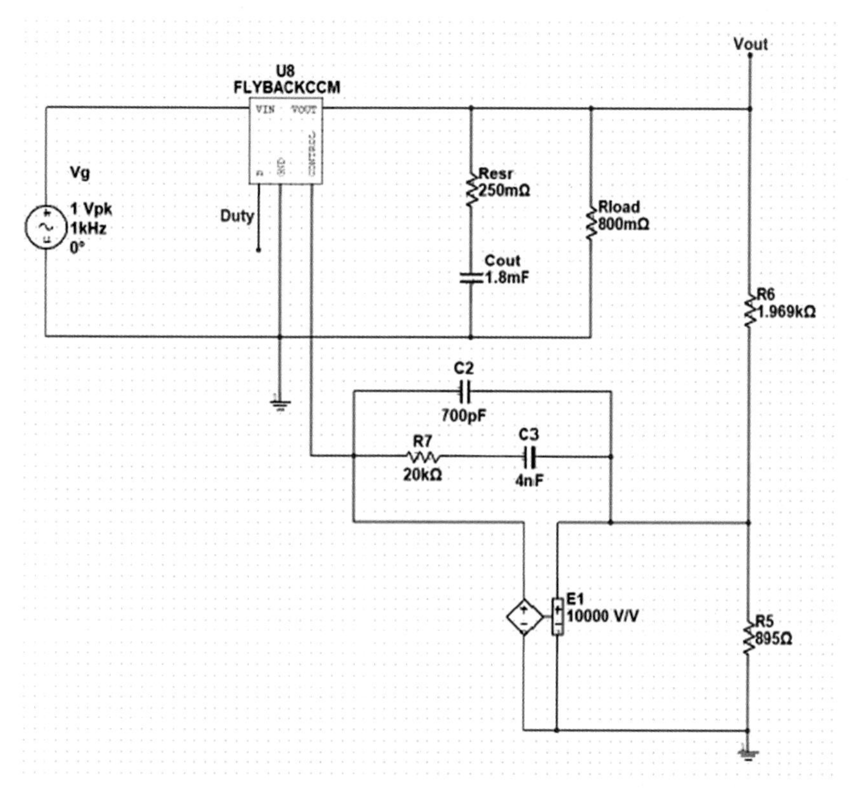

FIGURE 4.20 Step-down step-up converter Multisim model.

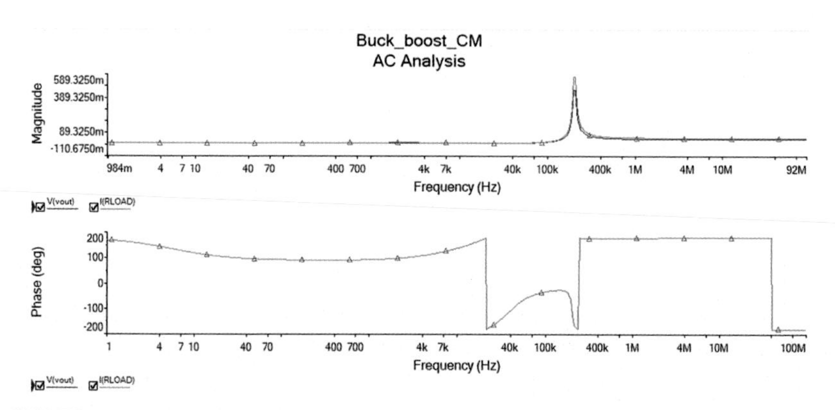

FIGURE 4.21 Step-down step-up converter frequency response.

period energy is saved in a converter primary. Throughout the other half of this period, the energy is moved to the converter secondary and to the charge. In this instance, AC Analysis is utilized to view the frequency reaction to a disturbance within the controlling signal (Figs. 4.22 and 4.23).

4.1.7 Clipper and clamper design

4.1.7.1 Clamper applications circuit

The circuits from this example clamp the peak of a waveform to a specific DC level. The diodes in these circuits operate primarily in reverse bias. In circuit A, the capacitor initially charges on the negative peak, which briefly forward biases the diode. The capacitor cannot discharge as there is no resistive component to the circuit. The capacitor charges to 10.3 V [(−20) to (9.7)] which keeps the diode in reverse bias; −9.7 V is the voltage necessary to forward bias the diode. At this point, imagine a 10.3 V battery in place of the capacitor. The output waveform will be clamped at 10.3 V above zero and the output will follow the input (Figs. 4.24 and 4.25).

4.1.7.2 Clamper applications circuit

The clipper application where circuit input is limited to TTL levels (0.5 V). The simple circuit which produces rectangular output from sinusoidal input (Figs. 4.26 and 4.27).

4.1.7.3 Precision clipper

This circuit demonstrates the electronic clipper, it is constructed by including Rc resistor to the bipolar-output source dead-zone circuitry. It clips off all signals above the positive reference voltage and below a negative reference voltage. By removing resistor Rc the output looks similar to the output from a bipolar-output source dead-zone circuitry (Figs. 4.28 and 4.29).

4.1.8 Filter design

4.1.8.1 First-order low-pass filter

This circuit demonstrates the operation and characteristics of a first-order low-pass activated strainer. The straining is performed by an RC network, along with that, the op-amp is utilized as unity-gain amplifiers. It passes all signals with frequencies below its cutoff frequency (F_c) (Figs. 4.30–4.32).

4.1.8.2 Active bandpass filter

This circuit demonstrates the operation and characteristics of an activated bandpass strainer. It enables all input signal frequencies in a given scope (called bandwidth) to go along, while opposing those outside the range (Figs. 4.33 and 4.34).

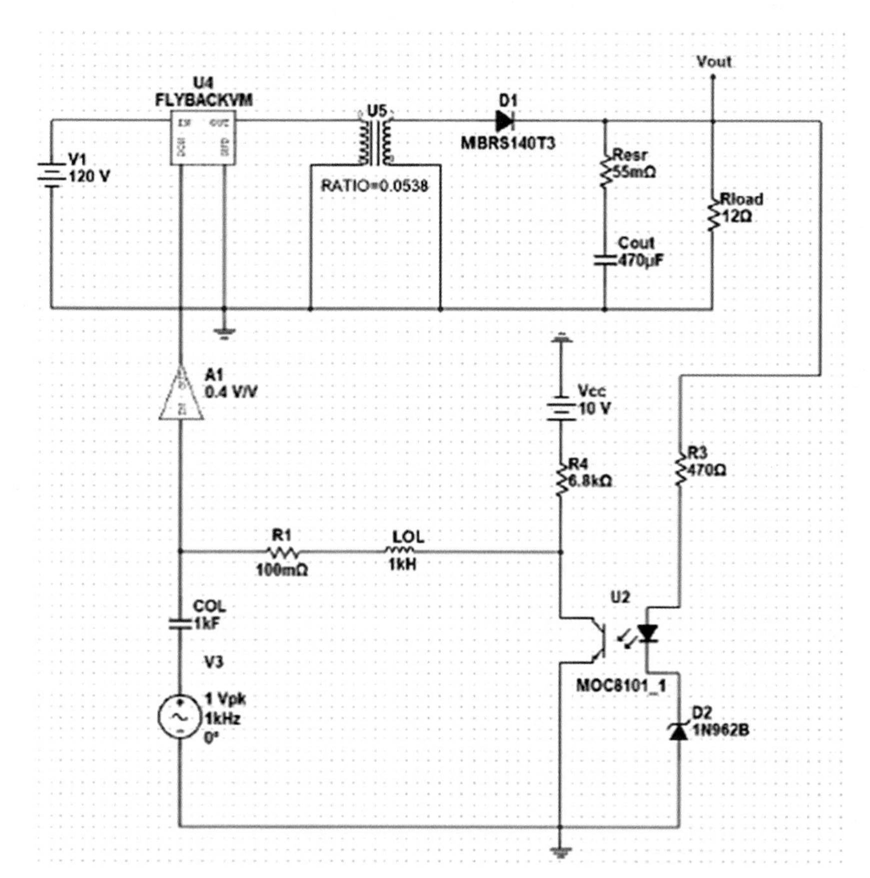

FIGURE 4.22 Flyback converter.

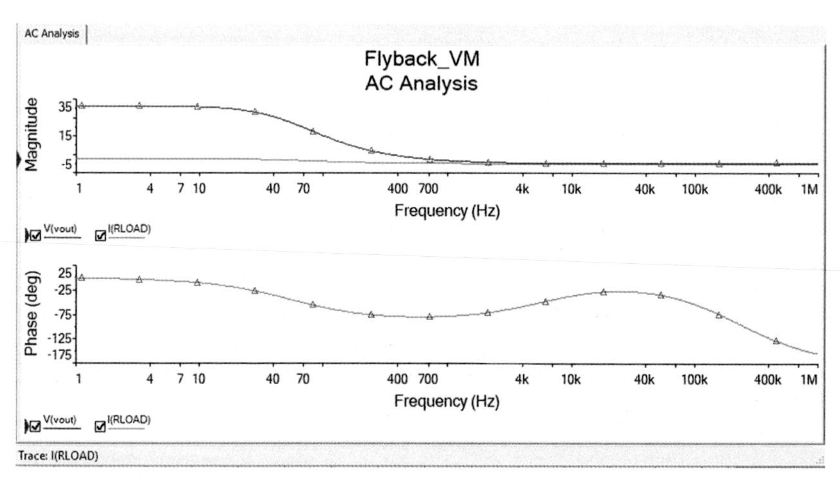

FIGURE 4.23 Flyback converter waveforms.

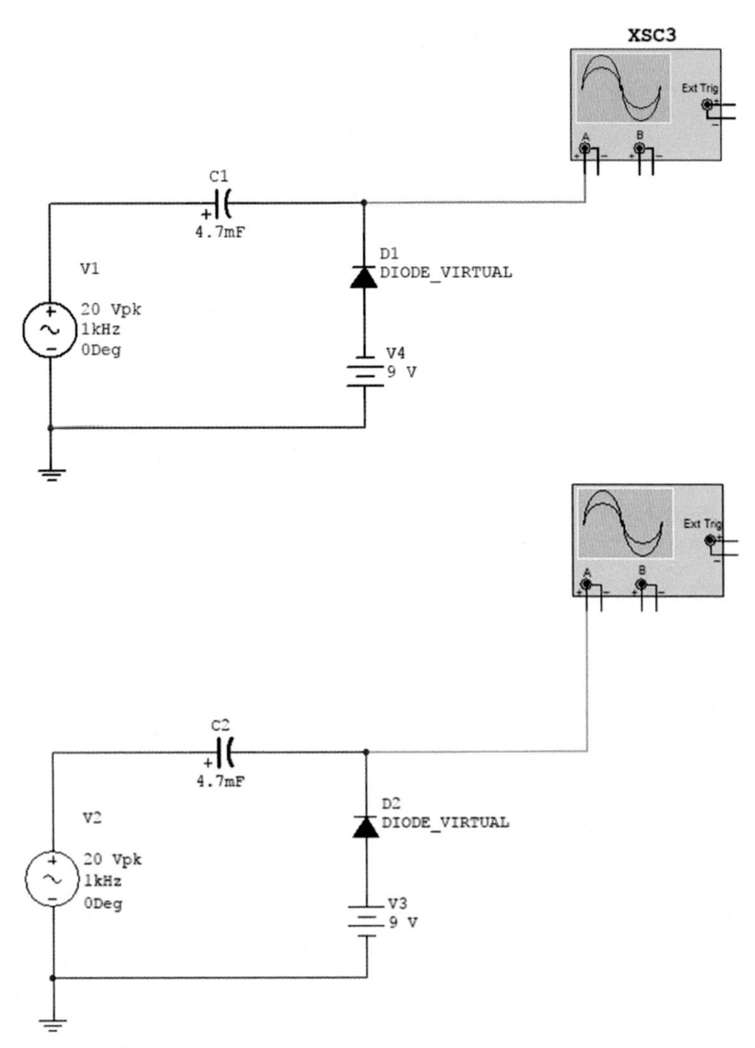

FIGURE 4.24 Clamper circuits.

4.1.8.3 High-pass active filter

This circuit demonstrates the operation and characteristics of a second-order high-pass active filter, which was designed for a cutoff frequency of 100 Hz. It rejects every signal underneath the particular cutoff frequency while passing every signal whose frequencies are over the cutoff frequencies. It creates a reel-off of 40 dB per decade (Figs. 4.35 and 4.36).

4.1.8.4 Basic differential amplifier

This circuit illustrates a basic differential amplifier. It is used to magnify tiny signals buried in much greater ones. The output voltage of the circuit is

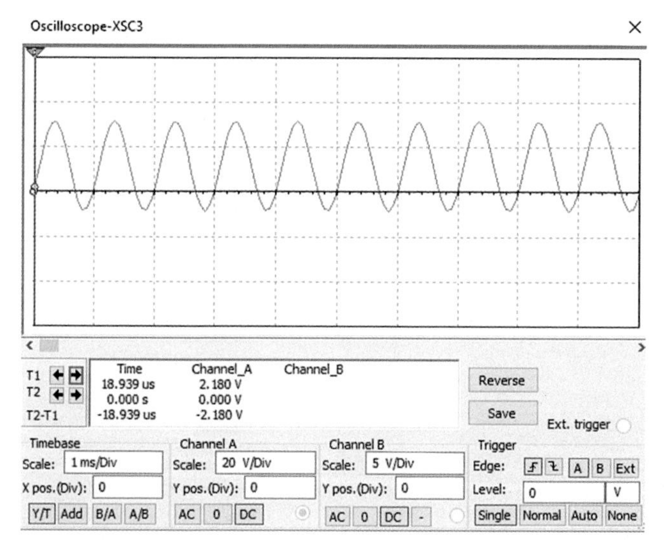

FIGURE 4.25 Clamper circuit waveform.

direct to this difference in the input voltages connected to the positive along with the negative input of the op-amp (Figs. 4.37 and 4.38).

Vout = m(E1−E2),where m is known as the distinct gain and it is incubated by the ratios of the resistors, m = mR/R.

4.2 Circuits design using Multisim Model Maker

4.2.1 Amplifier design

4.2.1.1 Class-A amp

This circuitry illustrates a Class-A normal-emitter power amp and is recommended for conducting an analysis of the Class-A amp. For the highest possible output signals, the Q-point should be centered. The noncentered Q-point restricts the output swings (Figs. 4.39 and 4.40).

4.2.1.2 Class-AB amp

The circuitry exhibits the structure and activity of the class-AB amp. The class-AB amp is accessed from the class-B push-pull amp that has the hybrid contortion impact wiped out. The hybrid bending impact is disposed of by biasing the two transistors somewhat over their cutoff points. By interfacing both silicon diodes known as D1 and D2 to the bottom ends of the transistor, the biasing power volt given to the transistors is equivalent to the forward volt dropping of the diodes. Both diodes are for the most part known as Biasing Diodes and Compensating Diodes so are picked to coordinate the qualities of the coordinating transistors (Figs. 4.41 and 4.42).

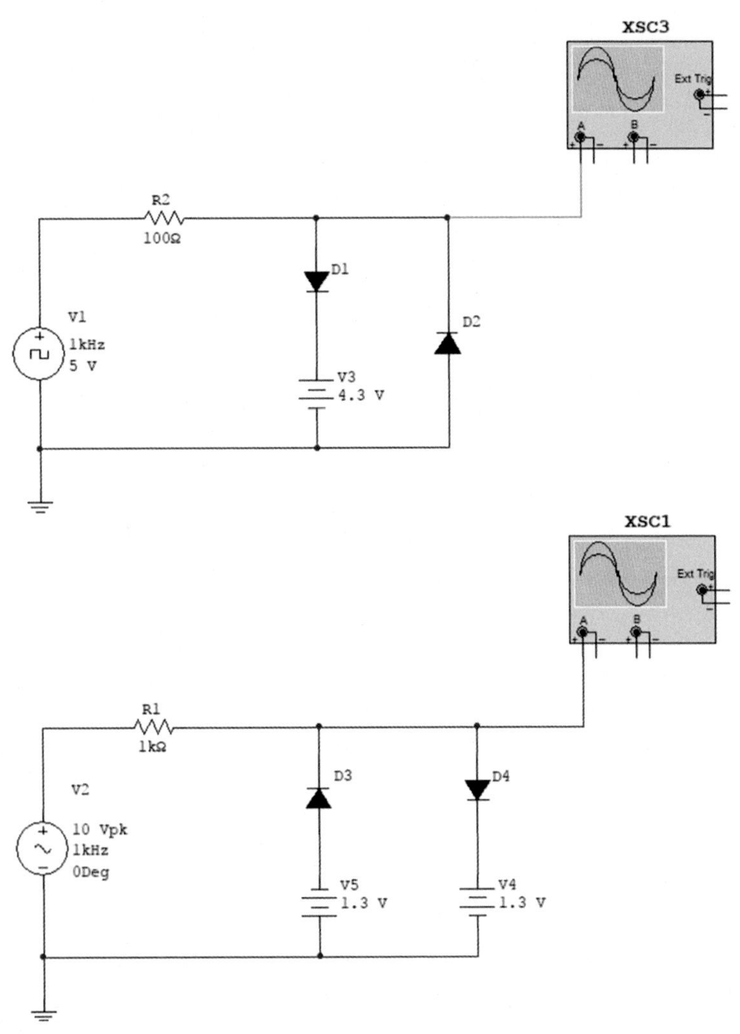

FIGURE 4.26 Clipper circuits.

4.2.1.3 Darlington pair

This circuit illustrates a Darlington pair, one of the ways to boost input impedance. The Darlington resistor is a composite design comprising of two bipolar pairs associated so that the current enhanced by the principal transistor is intensified further with the second one. The producer of the input source transistor is associated legitimately to the end of the second. The two collectors are associated together. Thusly the end current from the primary transistor goes into the bottom of the second. This arrangement provides a

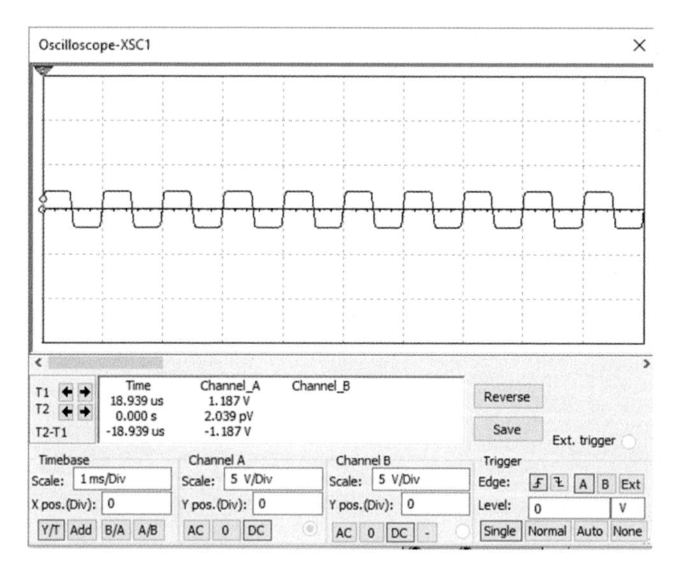

FIGURE 4.27 Clipper circuits.

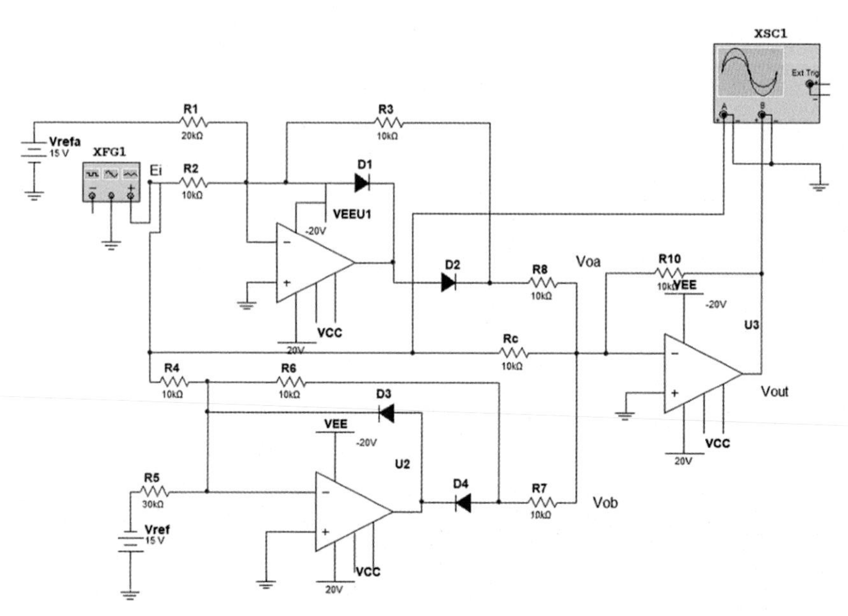

FIGURE 4.28 Electronic clipper circuits.

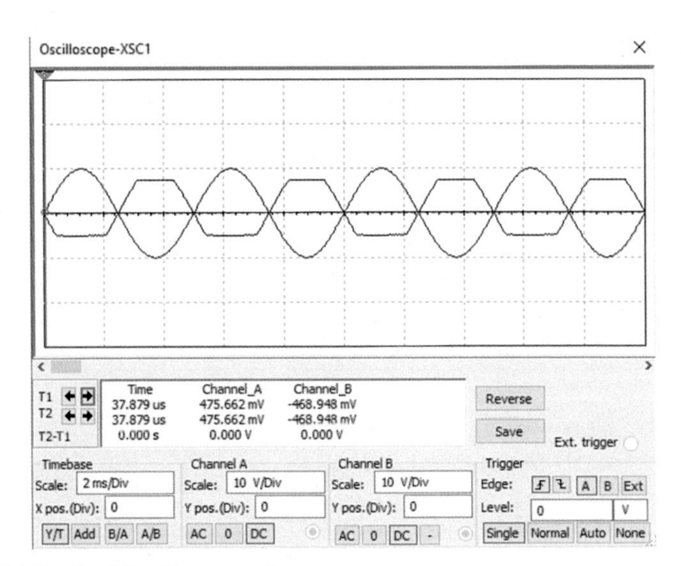

FIGURE 4.29 Precision clipper waveform.

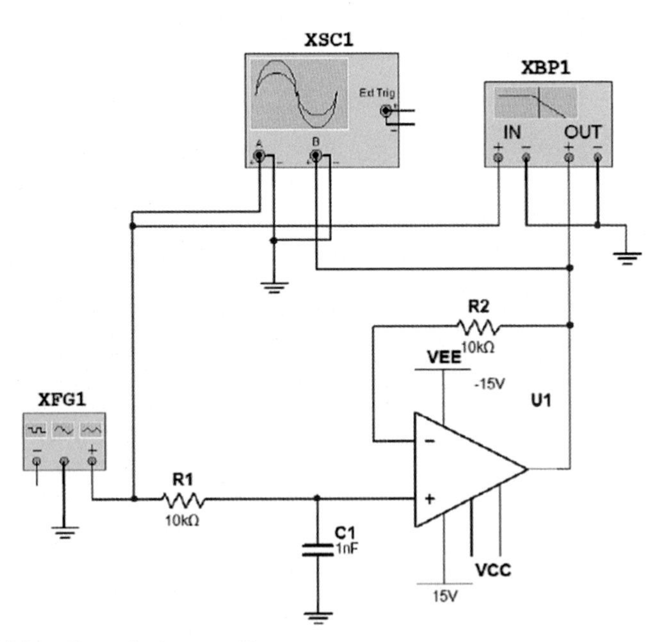

FIGURE 4.30 First-order low-pass filter.

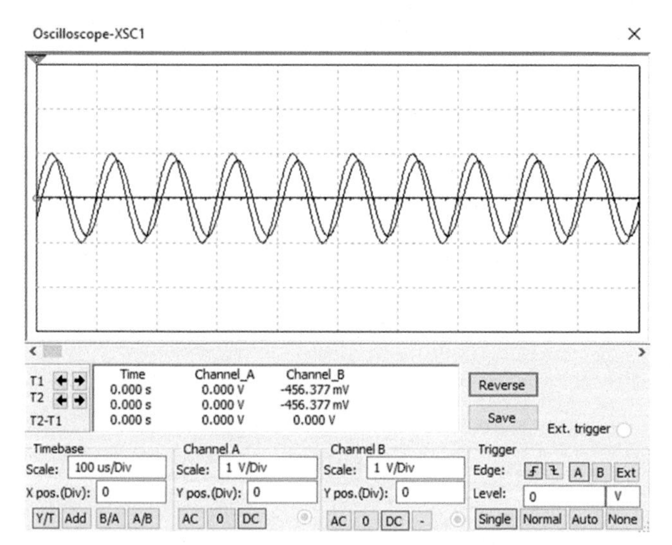

FIGURE 4.31 Primary low-pass strainer waveform.

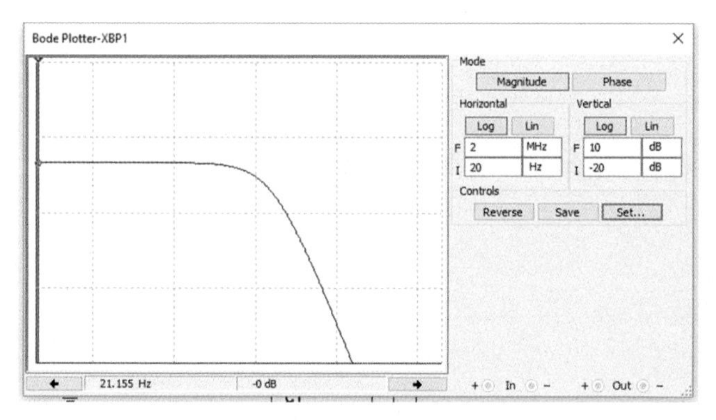

FIGURE 4.32 First-order low-pass filter frequency plot.

lot of higher current addition than every transistor taken independently (Fig. 4.43).

4.2.1.4 Two-tier common-emitter amp

This circuitry demonstrates a two-tier common-emitter amp. The output voltage of the primary transistor is coupled into the end of the second one. The final signal is amplified on the second transistor. The total voltage gain equals the product of the individual voltage gains (Figs. 4.44 and 4.45).

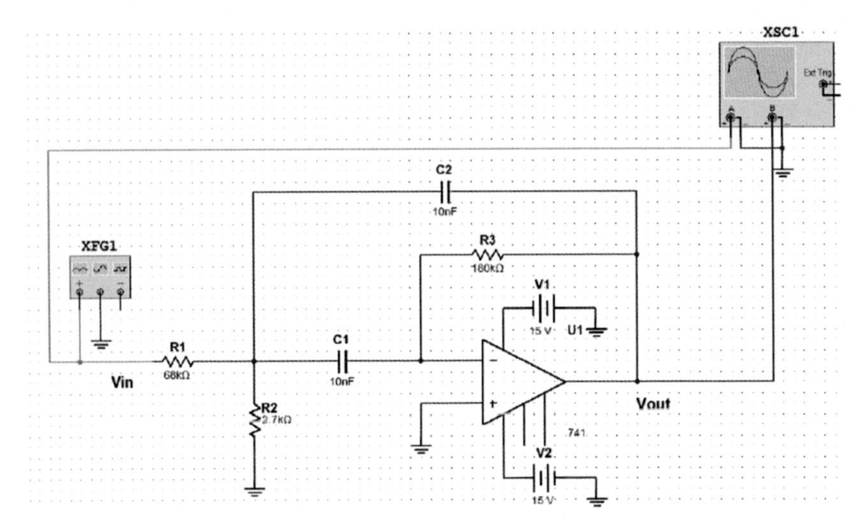

FIGURE 4.33 Active bandpass filter.

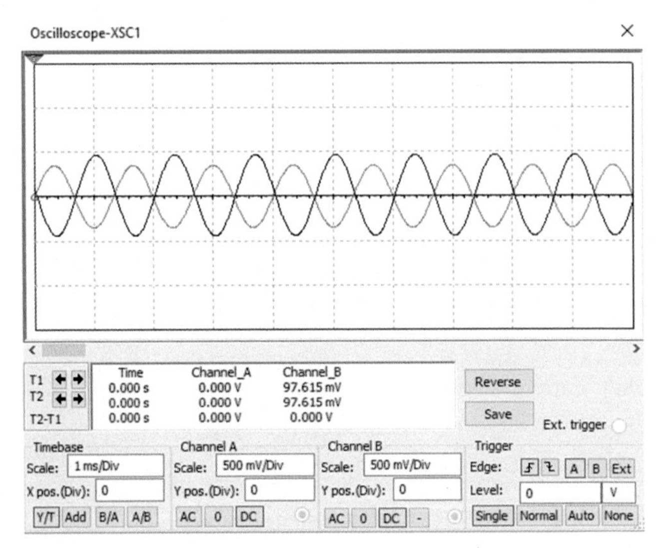

FIGURE 4.34 Active bandpass filter waveform.

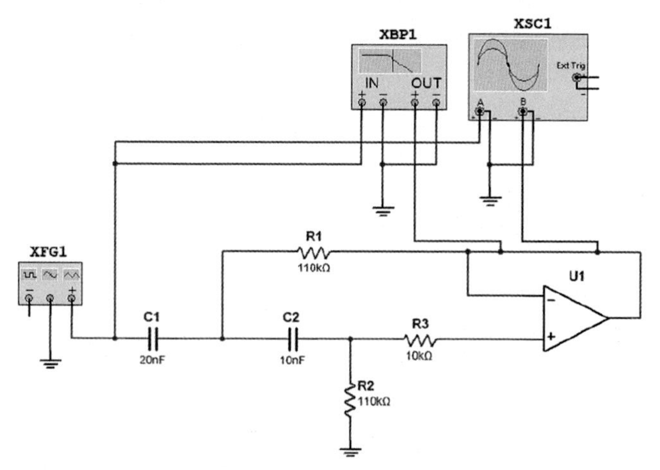

FIGURE 4.35 Long-wavelength active strainer.

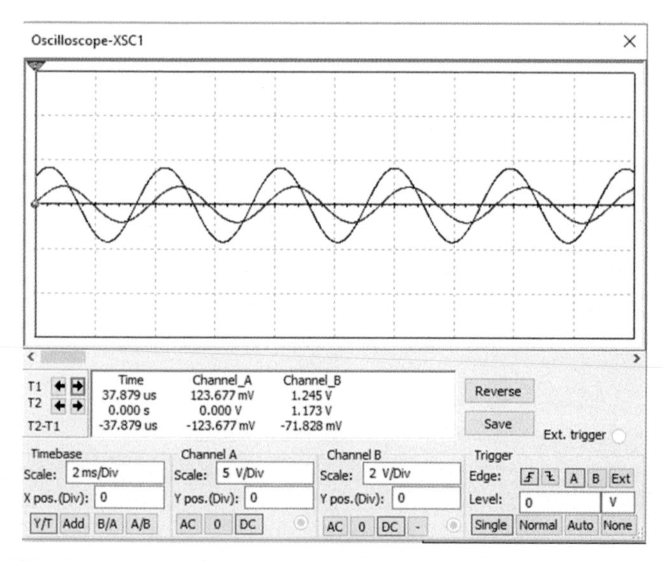

FIGURE 4.36 Long-wavelength active strainer waveform.

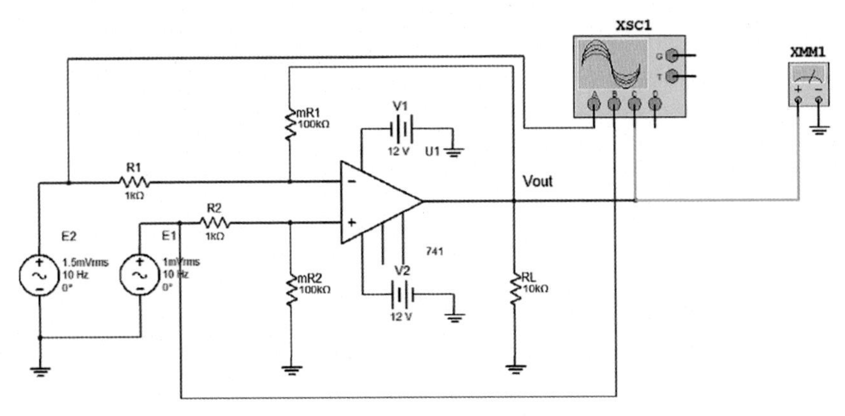

FIGURE 4.37 Basic differential amplifier.

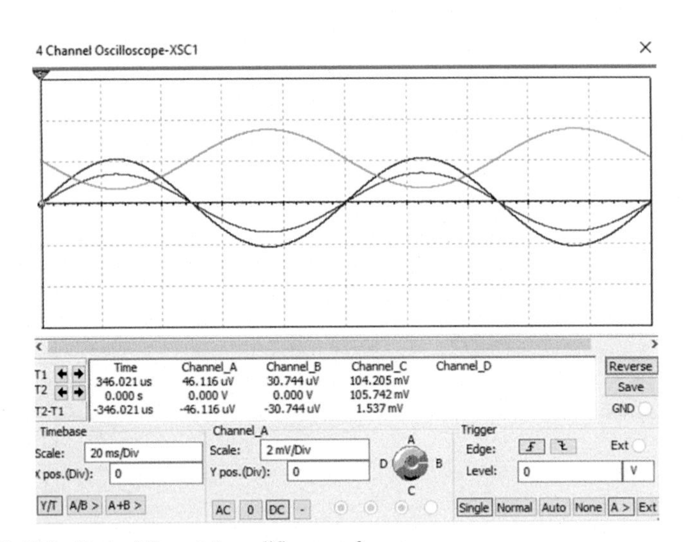

FIGURE 4.38 Basic differential amplifier waveform.

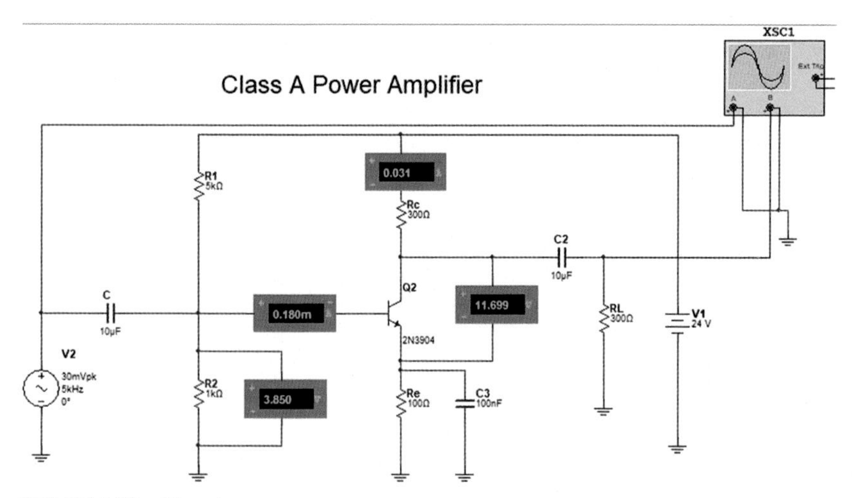

FIGURE 4.39 Class-A amp.

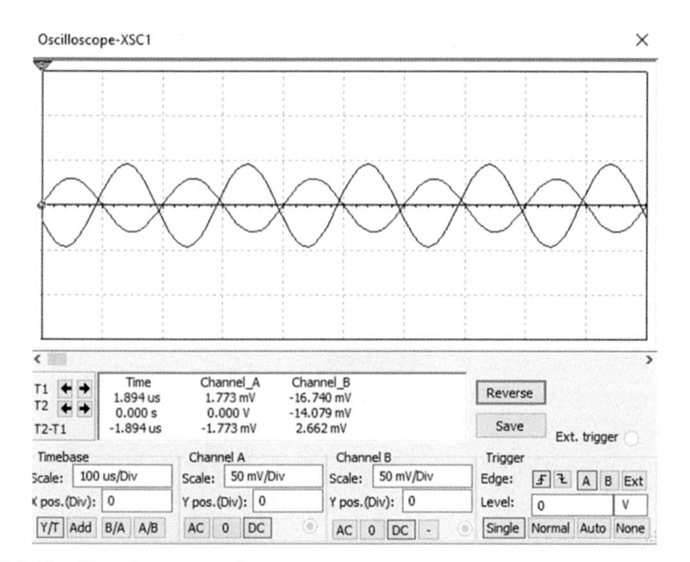

FIGURE 4.40 Class-A amp waveform.

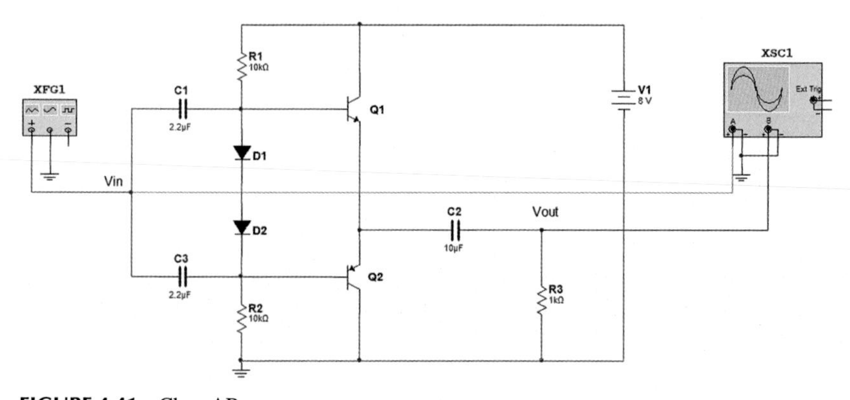

FIGURE 4.41 Class-AB amp.

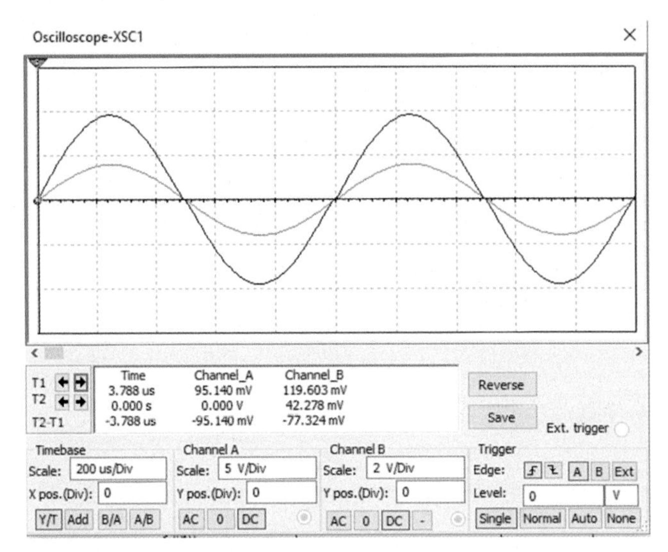

FIGURE 4.42 Class-AB amplifier waveforms.

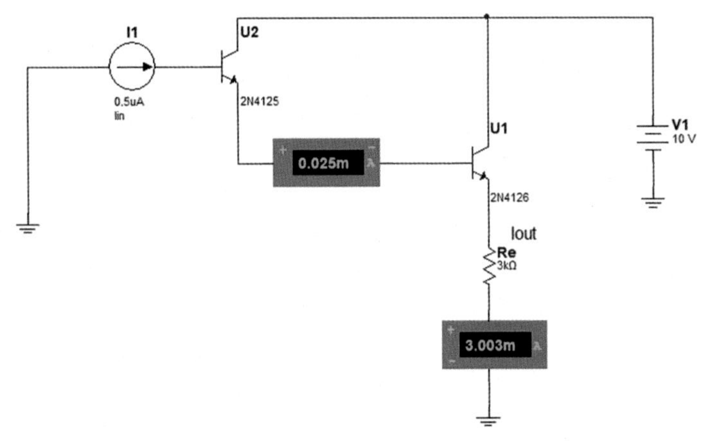

FIGURE 4.43 Darlington resistors.

4.3 Summary

The outcome of this chapter will possibly bring out a clear view of the tools accessible in Multisim and the job of Multisim in the simulating process field. Therefore the users become able to comprehend how to construct the circuitry relying upon the requirement. The comprehensive structure process would efficiently make the users construct and reproduce their own circuitries that could show the pathway of creating huge applications in engineering field even from the lower to higher level.

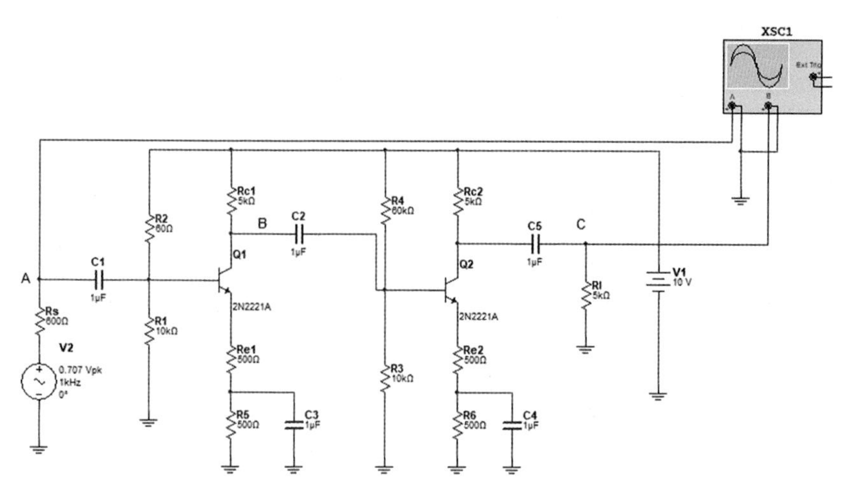

FIGURE 4.44 Two-tier common-emitter amp.

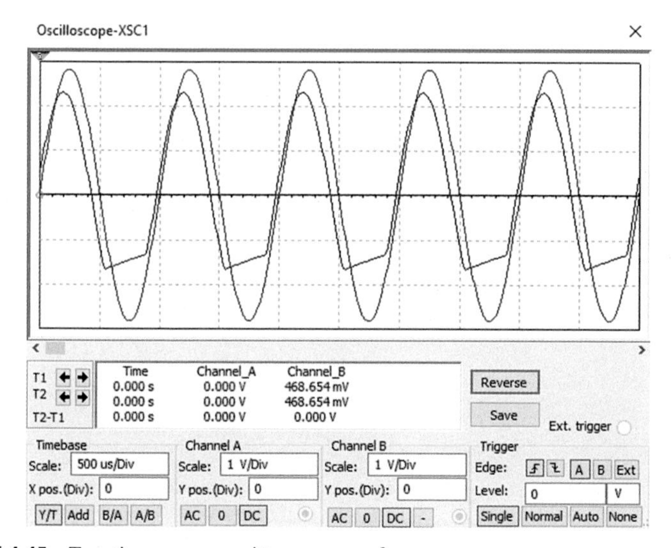

FIGURE 4.45 Two-tier common-emitter amp waveform.

4.4 Review questions

List out all the types of instruments available in Multisim software used for measurement of electrical quantities.

1. What are the types of circuit analysis available in Multisim software?
2. How to check electrical rules in any designed circuit using electrical rule checking option?

3. Explain the process of running simulation in the Multisim software.
4. Write the importance of grapher, plotter, and cursor in any simulated waveform analysis.
5. Design the boost converter circuit using Multisim and draw the waveforms?
6. Write the step by step procedure to design buck converter from file opening to obtaining output waveforms.
7. Do AC analysis on the flyback converter in Multisim environment.
8. Obtain the simulated output waveforms of the following circuits: (1) Clipper, (2) Clamper, (3) Low-pass filter, and (4) High-pass filter.
9. Design the two-tier common-emitter amp circuit utilizing Multisim and then analyze the outputs.

Chapter 5

Printed Circuit Board Design Tool—DesignSpark

Chapter Outline

5.1 Introduction to printed circuit board design software

5.1.1 Overview of printed circuit board design software

Printed circuit board (PCB) design software is a tool used for designing PCBs. PCB is used to support the circuit connection between the components with pads and traces through layers and also provides mechanical support. There are various paid and free versions available in the market. These tools can support PCB designs ranging from low-end to high-end complexity. This chapter focuses on the PCB design using the open source PCB design software named DesignSpark.

Software Tools for the Simulation of Electrical Systems. DOI: https://doi.org/10.1016/B978-0-12-819416-4.00005-3

5.1.2 Parts of the printed circuit board

The main parts of the PCB are

- Pads

  Pads are the copper areas from where the traces are connected. The basic shape of the pads can be round, rectangular, or square. There will also be custom designed shapes, which can be based on the design/component requirements.
- Traces/Tracks

  A copper line which makes electrical connections between components. It has a specific width and there is a need to maintain a certain distance between the traces.
- Vias

  Vias are pads which help the connectivity among the layers.
- Components

  Components are the electronic parts used in the design connected through traces and soldering. These include discrete components, input output connectors, and power supply. These components are of either leaded or surface mount type.
- Layers

The circuit is accommodated in terms of copper sheets/layers such as Top, Internal (Signal, Power, and Ground), and Bottom. These are called electrical layers. There are also nonelectrical layers which support protection (Top mask and Bottom mask), soldering (Paste layer), identification (Top silk and Bottom silk), and information about drill (Drill layer). Depending upon the complexity of the circuit, the number of layers will vary.

5.1.3 Printed circuit board design flow

PCBs are classified into three types. There are single-sided PCBs, double-sided PCBs, and multilayer PCBs. Single-sided board design is used for simple circuits and has the copper pattern on the single layer only. Double-sided board is designed for medium complex circuit. In this PCB the copper pattern is laid out on both sides of the PCB. Multilayered PCBs are designed for more complex design.

Fig. 5.1 shows the process flow involved in the PCB design.

5.1.4 Design guidelines

Here are a few of the most important basic PCB design guidelines to be followed while designing. These guidelines are normally driven by the requirements of PCB fabrication and may vary, depending on the fabrication house used.

- Selection of number of layers must be an even number.
- Large layout area can be obtained by using surface mount technology (SMT) components, as these are much smaller than the size of leaded ones and can handle high frequencies as well.
- Identify the highest-frequency components, place and route it first and then work toward the lower frequency components.
- Routing is the process of making connection between components using tracks on PCB. Routing is also known as "wiring." An electrical connection between two or more pads is known as a "net." Net length should be as short as possible. The longer the total track length, the greater its resistance, capacitance, and inductance.
- Tracks should be drawn at 45°. Trace corners have to be rounded for smooth signal flow. Route the track to the center of the pad using proper

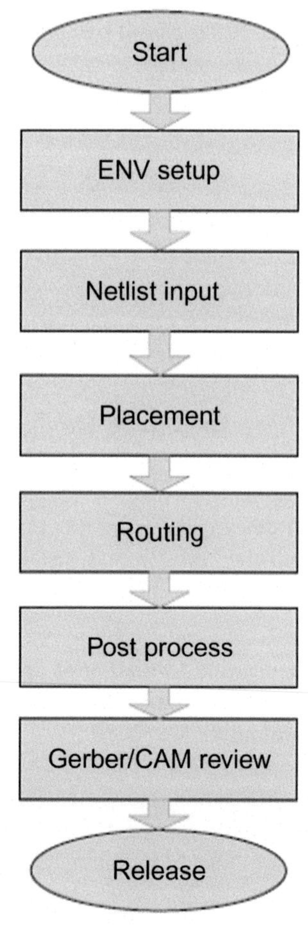

FIGURE 5.1 PCB design flowchart. *PCB*, Printed circuit board.

use of object snap. "Neck down" the track width between pads where possible. Route the critical tracks first and later other signals. Unconnected copper islands, also called "dead copper," has to be cleared or grounded.

- Minimum air gap should be no less than the minimum trace width.
- No copper entities should be placed within 0.050 in of a cutout area or nonplated through hole edge.
- Proper power and ground planes have to be laid out as the current always returns to its source through the path of least impedance which minimizes signal and power integrity issues and cross talk. Disconnection in the return plane should be avoided. The return path should be continuous under the trace. Local decoupling capacitors should be mounted to minimize the power/return loop area if the board has no plane layers. The power plane can be split if there is more than one power source. For example, ± 5, 1.8, and 3.3 V digital, and 3.3 V analog are all on the same plane. Similarly the ground plane can also be split between the analog and the digital grounds if the circuit has both.
- Try to avoid an electrically "floating" metal area as it can increase capacitive coupling between circuits or can radiate across the metal fill area to return the plane and connect the shield plane return planes with multiple vias.
- There are mainly three types of solder mask available: liquid photoimageable, wet screenable, and dry film. A bare board is generally best, but if solder mask must be used, do not cover sensitive signal traces.
- Depending upon the operating temperature different board material has to be chosen. For the operating temperature less than 100°C and from 100°C to 185°C, FR4 and polyimide can be used, respectively.

5.2 Printed circuit board design in DesignSpark

5.2.1 Overview of DesignSpark

DesignSpark is a free PCB design tool—there is no charge for the software. Users can download it from the website and install it with easy steps. The following major modules are available in this tool.

1. Schematic editor

   Schematic editor enables the formation of circuit diagrams by placing component blocks and making connections. The circuit diagram can be split up and drawn into multiple sheets.
2. PCB layout editor

   The connection/net information can be carried to the PCB layout editor where we need to place the footprint with PCB recommended dimensions or Association Connecting Electronics Industries (IPC) standard.
3. Autorouter

DesignSpark autorouter facilities the automatic routing by placing the traces according to the net connectivity present in the schematic circuit diagram.

The software can be downloaded and installed from the link https://www.rs-online.com/designspark/pcb-software.

5.2.2 User interface and management of DesignSpark work environment

The main window appears when double-clicking the DesignSpark icon. Fig. 5.2 shows the layout screen with schematic editor.

DesignSpark supports both English and metric units. Units and the desired precision can be set initially. In addition to this, default entities such as text, shape, pad track, vias, and nets can be set, as shown in Fig. 5.3.

The DesignSpark schematic has a set of useful tools to place components and text, to edit, find, and add properties, to generate necessary reports, library, etc. Other supporting features are setting units, grid, copy, paste, select, zoom, assigning color, shapes, measure, undo, redo, etc. Shortcut keys will help the designers to accomplish the process more easily. Each entity has been assigned with a property which can be set or edited by selecting and right-clicking it. Fig. 5.4 shows the property dialog.

All design rules can be set in a single window called "Design Technology" which is obtained by selecting "Settings/Design Technology" in the menu bar. Fig. 5.5 shows the setting for schematic entities and Fig. 5.6 shows the setting table for the PCB layout.

5.2.3 Schematic capture

Schematic capture is a process of laying the circuit design in the schematic editor, where we can generate the net or connection details which are required for PCB design.

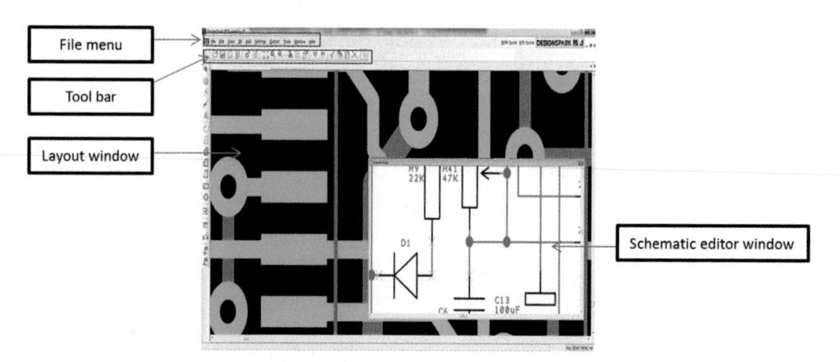

FIGURE 5.2 DesignSpark window.

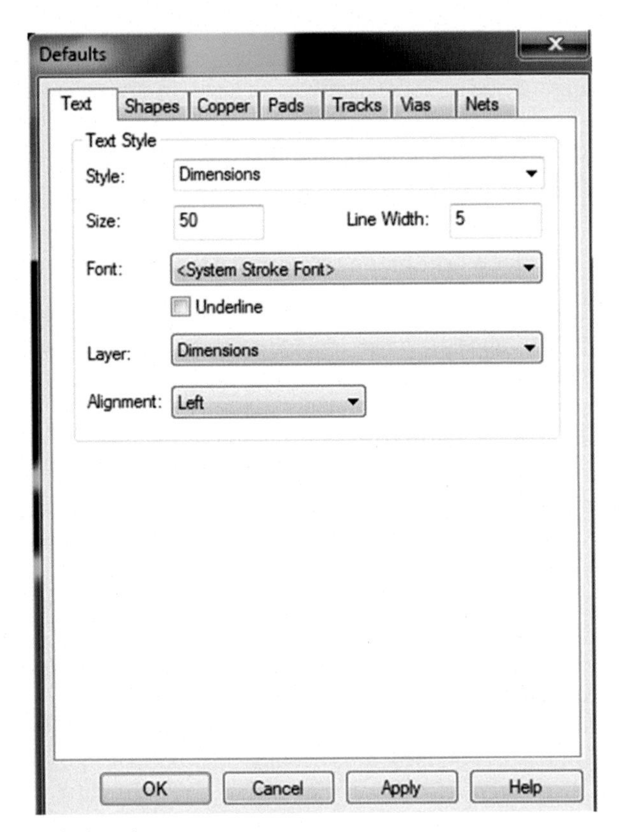

FIGURE 5.3 Default setting.

To create a new schematic design, go to File—New. New design window will appear as in Fig. 5.7.

Click on Schematic design and press ok. The new design name can be provided at this time or later while saving the design. The blank schematic editor screen will appear now.

Library manager helps to create or edit the schematic symbol, PCB footprint, and components.

Go to the menu File—Library. The Library Manager window appears as in Fig. 5.8.

The main tabs Schematic Symbols, PCB Symbols, Components are used to create new items, add items, and find them in the new library or the existing library.

Here we can create a new library by providing a library name. If you want to create a new schematic symbol, click on "New Item," and the editor opens. You can add terminals which will be called pads in footprint in the layout editor. Use the menu "Add/Pad" for placing the terminal. In the case

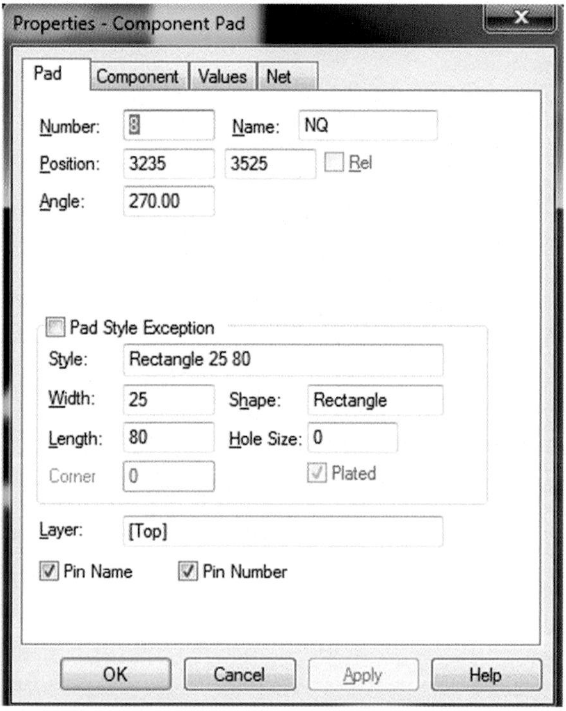

FIGURE 5.4 Property dialog.

of the resister symbol, the number of terminals will be two. Each terminal has a pin name, pin number, net name, and net sheets as its properties. The position and angle of the terminal can be controlled in the property window, as in Fig. 5.9.

The purpose of the terminal is to make connectivity in the circuit diagram. Similarly the PCB footprint can be created in the library manager in the tab "PCB Symbols" by clicking "New Item." The New Component editor will appear where you can create footprint by placing "Add/Pad," you can edit the pad information in the Property dialog.

Both symbol and footprint can be created via a wizard as well by clicking the Wizard button and following the simple instructions given. Figs. 5.10 and 5.11 show the dialog for creating a symbol and footprint using a wizard. After being created, it can be saved to the desired respective library.

Symbols can be added in the schematic editor after creating the symbols in the library. This can be done by the menu "Add/Component." After placing all the symbols, the connection among the symbols can be done by selecting "Add/Connection." Connectivity forms a net in the PCB layout. Similarly power and ground terminals can be placed and connected as per the circuit diagram.

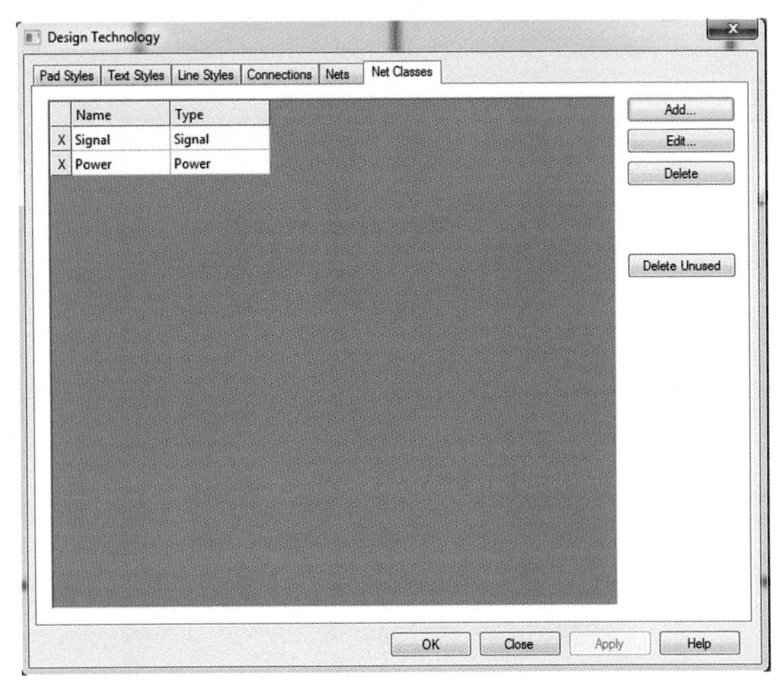

FIGURE 5.5 Design technology—schematic editor.

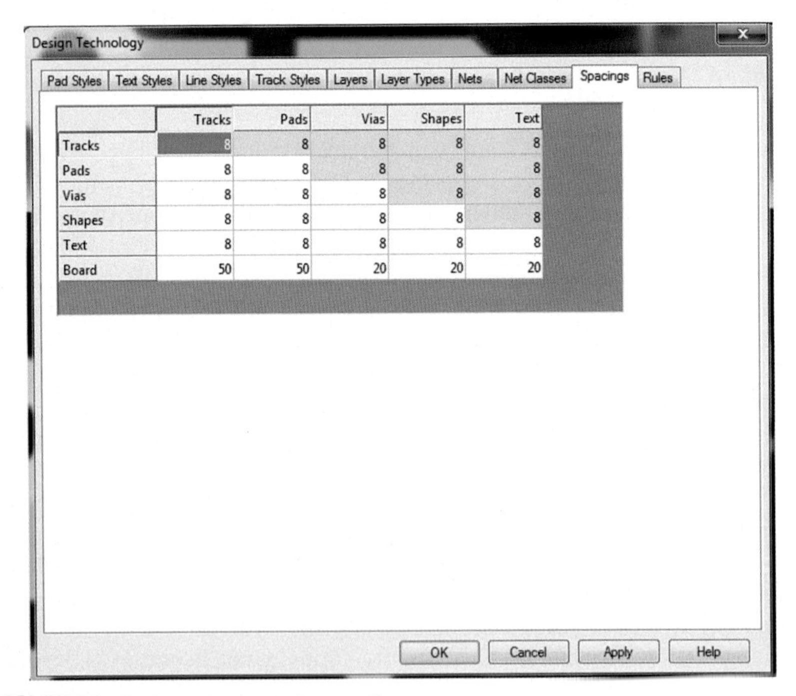

FIGURE 5.6 Design technology—layout editor.

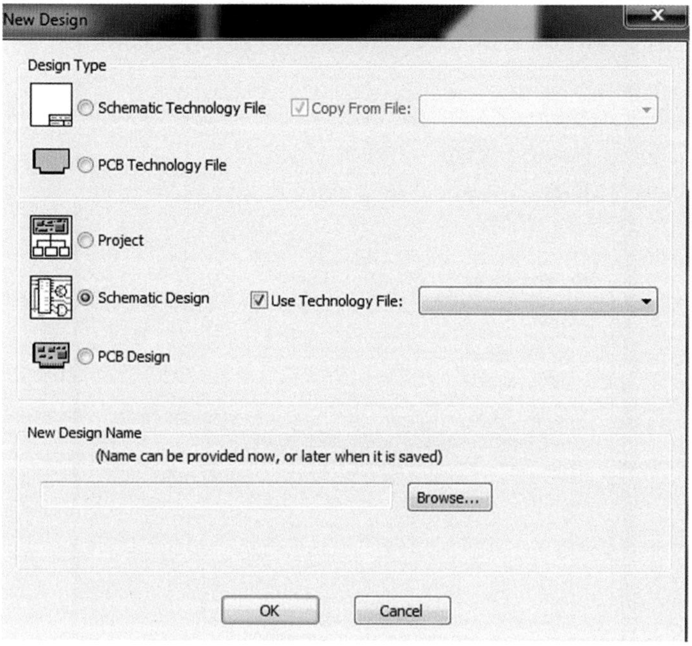

FIGURE 5.7 New design.

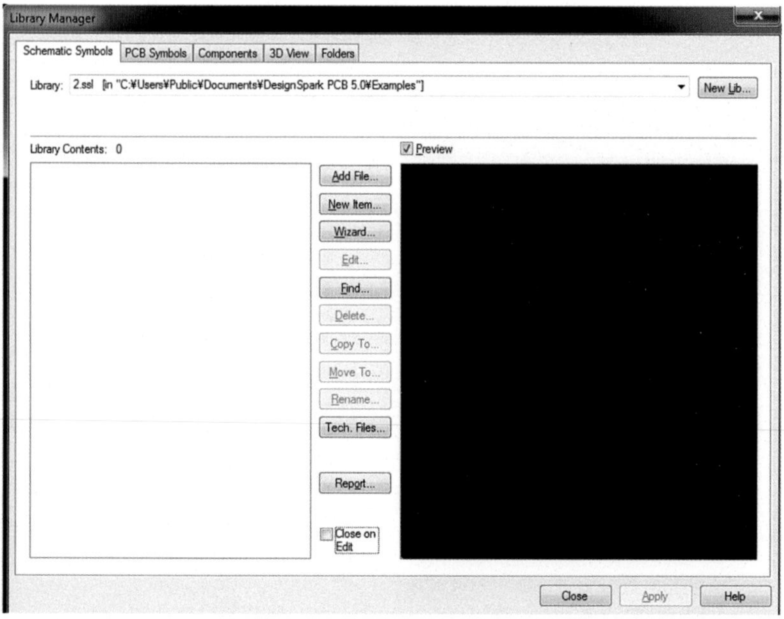

FIGURE 5.8 Library manager.

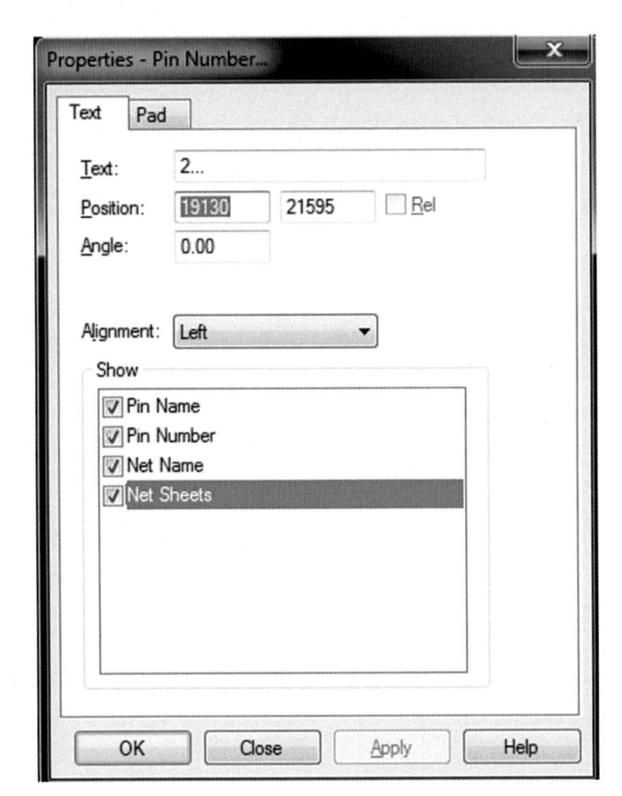

FIGURE 5.9 Properties—pin number.

FIGURE 5.10 Symbol creation wizard.

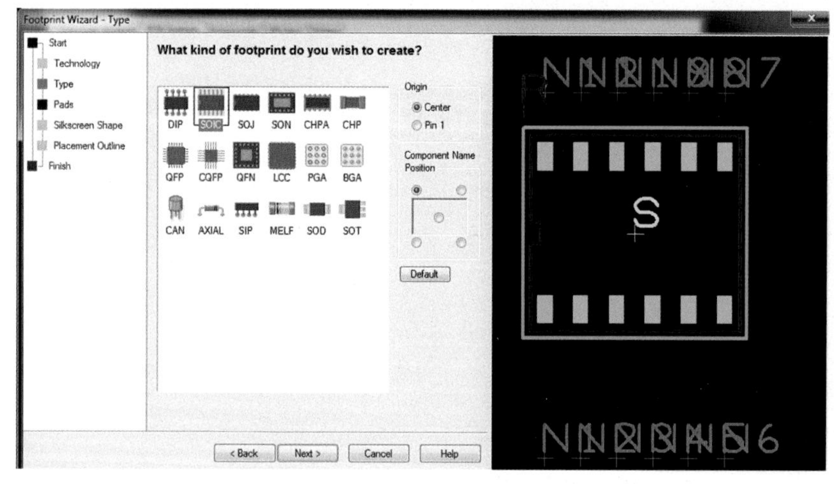

FIGURE 5.11 Footprint creation wizard.

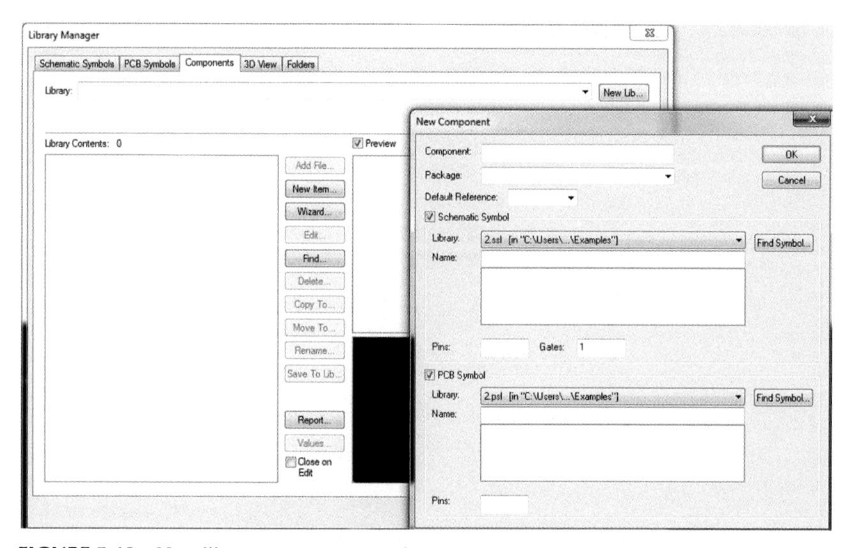

FIGURE 5.12 New library component creation.

5.2.4 Component creation

In the component tab of library manager, click "New Item" to create a new component. By entering the schematic symbol and PCB footprint information, a component can be easily created. Assigning both a schematic symbol and a footprint makes a component (Fig. 5.12).

It can also be created by using a wizard by clicking the Wizard tab, as in Fig. 5.13.

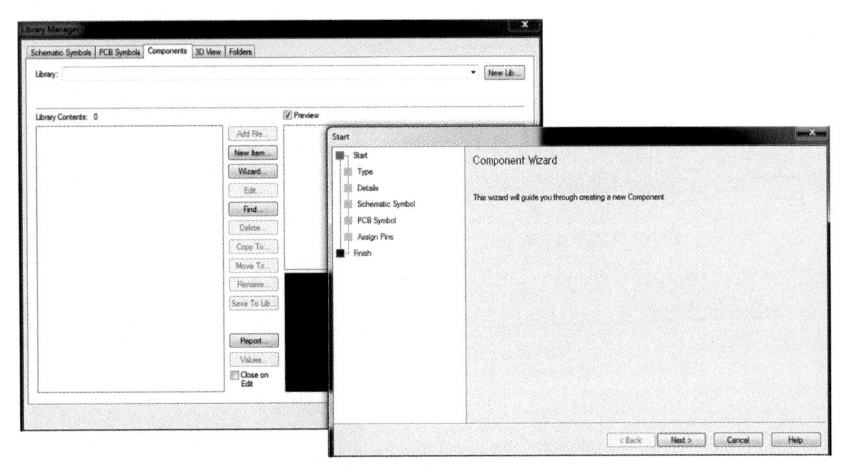

FIGURE 5.13 Library component creation using wizard.

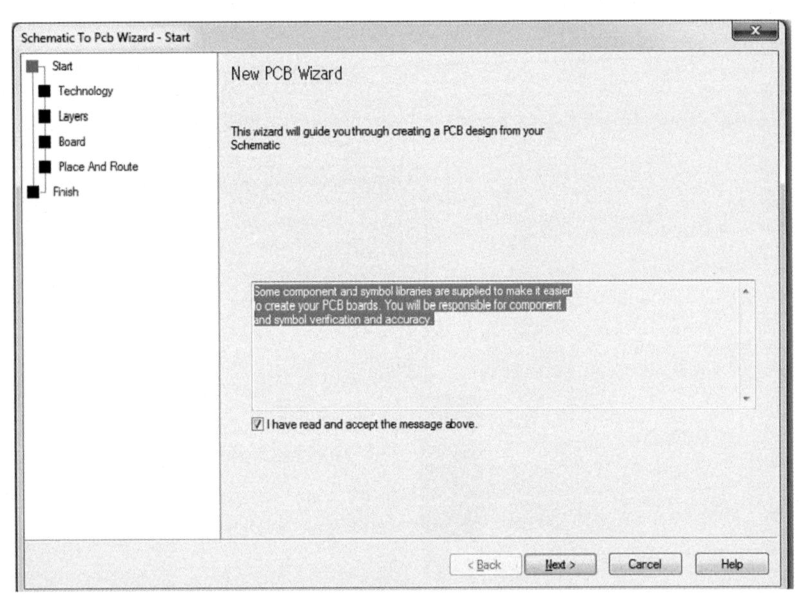

FIGURE 5.14 Schematic to PCB wizard. *PCB*, Printed circuit board.

5.2.5 Netlisting

For the schematic design, the net information or netlist can be translated into the PCB layout editor. This can be done using "Tools/Translate PCB." This will generate the new PCB dialog.

This comes with the settings such as technology file, layers, board dimension, place and route settings, and finally the name and location of the PCB file to be stored, as shown in Fig. 5.14.

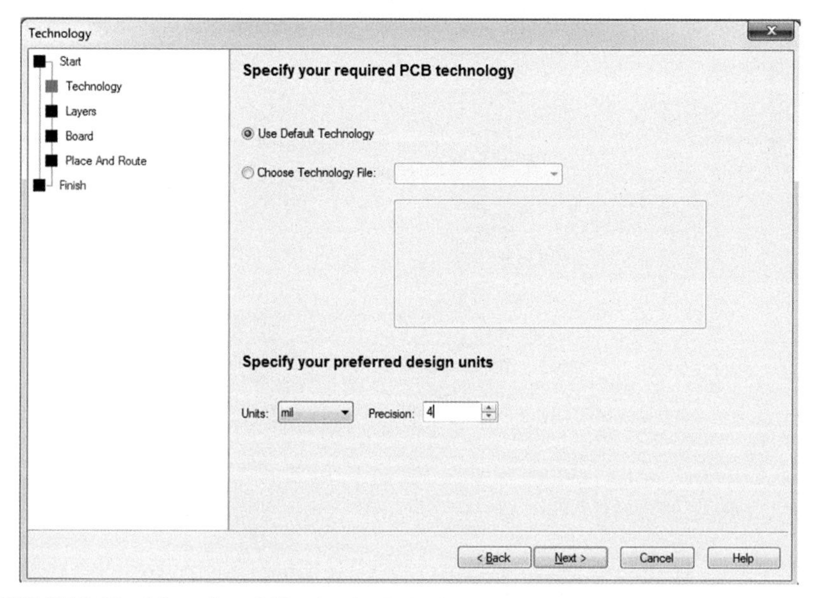

FIGURE 5.15 Schematic to PCB wizard—Technology. *PCB*, Printed circuit board.

As in Fig. 5.15, the designer can choose the default/new technology file and set the preferred design units with precision.

As in Fig. 5.16, the designer can define the number of layers required and other necessary nonelectrical layers.

As in Fig. 5.17, the designer can define the board outline.

As in Fig. 5.18, the "Place and Route" dialog allows us to place components automatically.

All the components present in the schematic are arranged outside the board outline area with the ratsnest connectivity (connections as per the schematic design) upon clicking the Finish button, as in Fig. 5.19. After that the components can be moved inside the board by dragging or considering the various placement criteria.

5.2.6 Component placement

Components can be dragged and placed in a single click. Fig. 5.20 shows the component placement with rastnest. Clearly isolate the analog, digital, and power supply components.

High-frequency, radio frequency (RF), high-voltage, heavy or high-heat components have to be taken care of while placing. Try to place most components on the top side only and if it is not possible, small-sized and low-heat dissipation component (like surface mount device (SMD)) can be on the bottom side. Decoupling capacitors should be placed in proximity to the

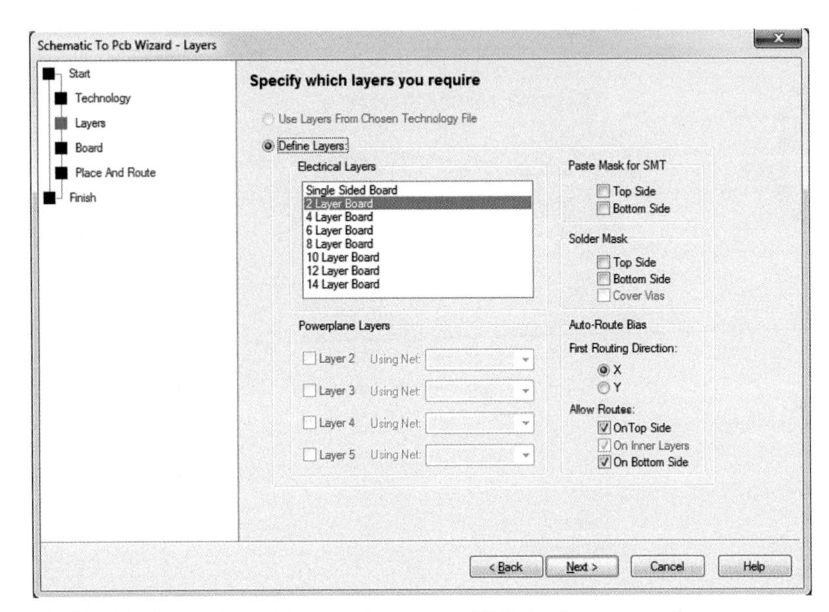

FIGURE 5.16 Schematic to PCB wizard—Layers. *PCB*, Printed circuit board.

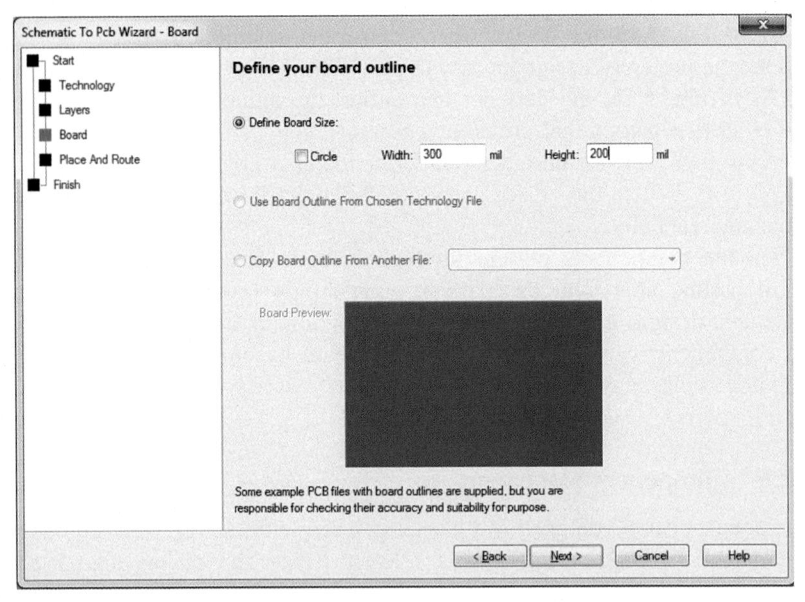

FIGURE 5.17 Schematic to PCB wizard—Board. *PCB*, Printed circuit board.

power supply (VCC) pin. For easy access, connectors, light emitting diode (LED), switches, and test points can be placed in proximity to the board edge.

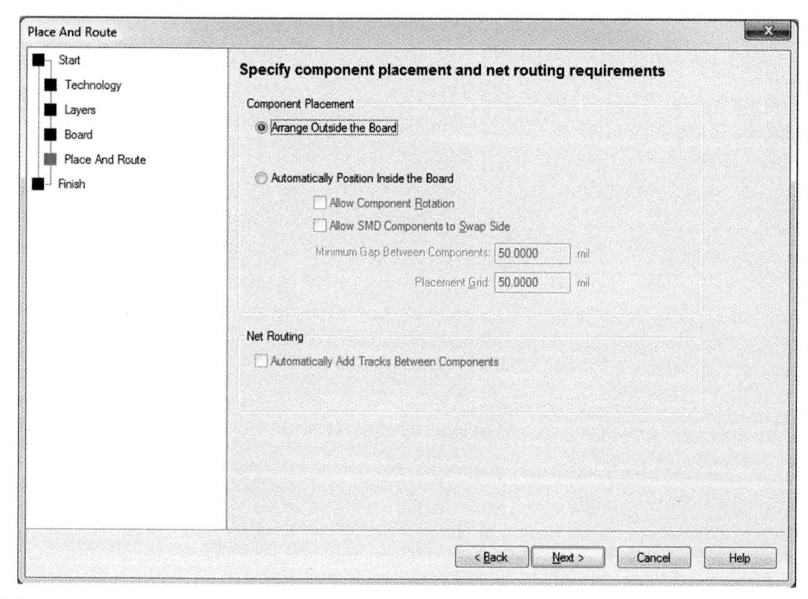

FIGURE 5.18 Schematic to PCB wizard—Place and Route. *PCB*, Printed circuit board.

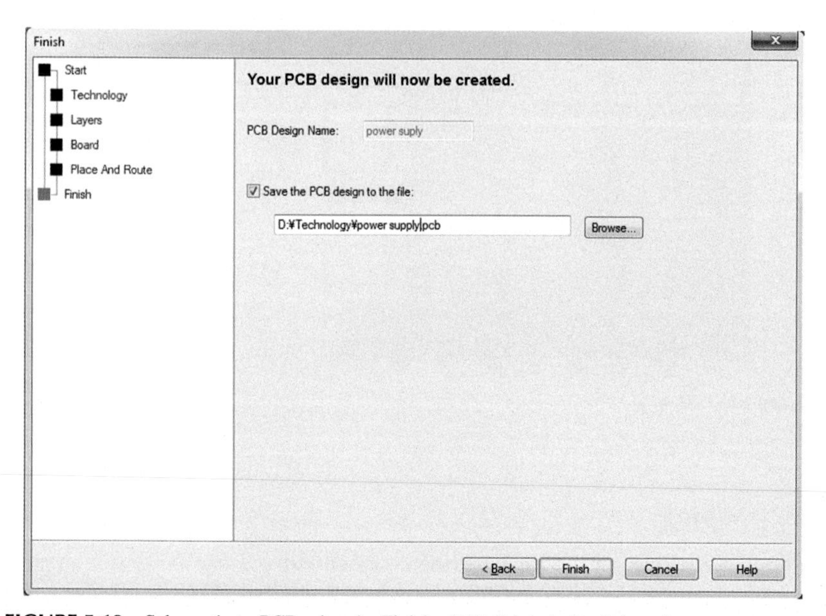

FIGURE 5.19 Schematic to PCB wizard—Finish. *PCB*, Printed circuit board.

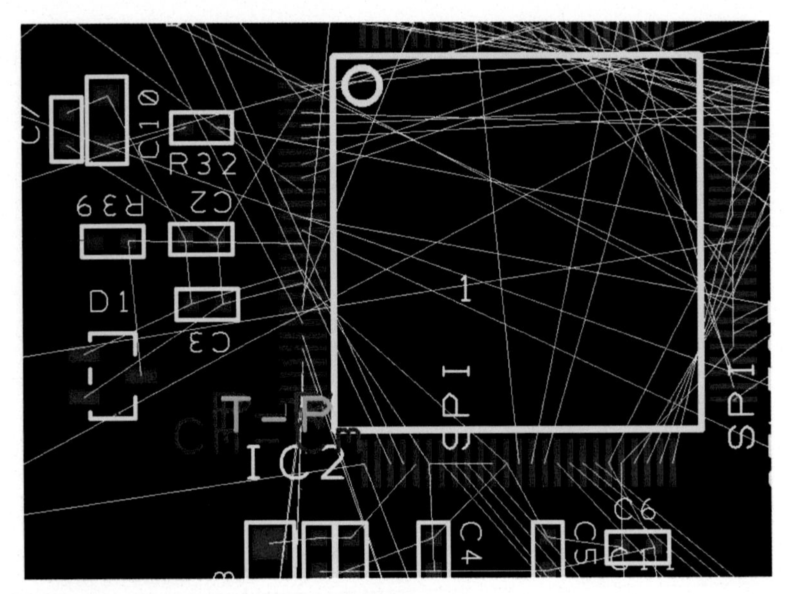

FIGURE 5.20 Component placement with ratsnest.

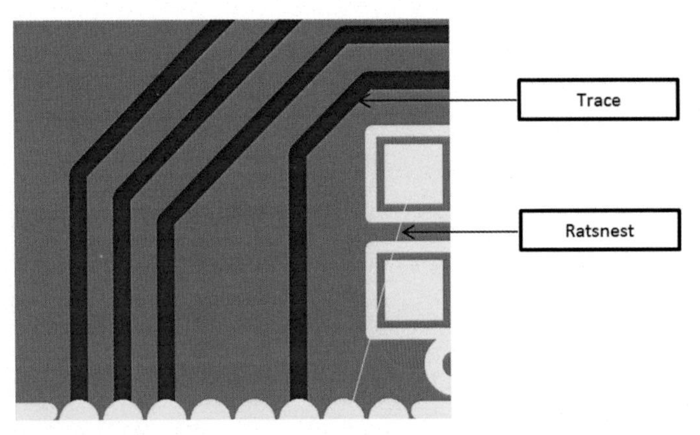

FIGURE 5.21 Routing.

5.2.7 Wiring

By double-clicking on the ratsnest, the connection/trace can be drawn manually. This is the easiest method for manual routing (Fig. 5.21).

The entire trace segment should be at 45°. If you want to remove the trace, first select the trace and right-click to select "Net/Unroutc ncts" (Fig. 5.22).

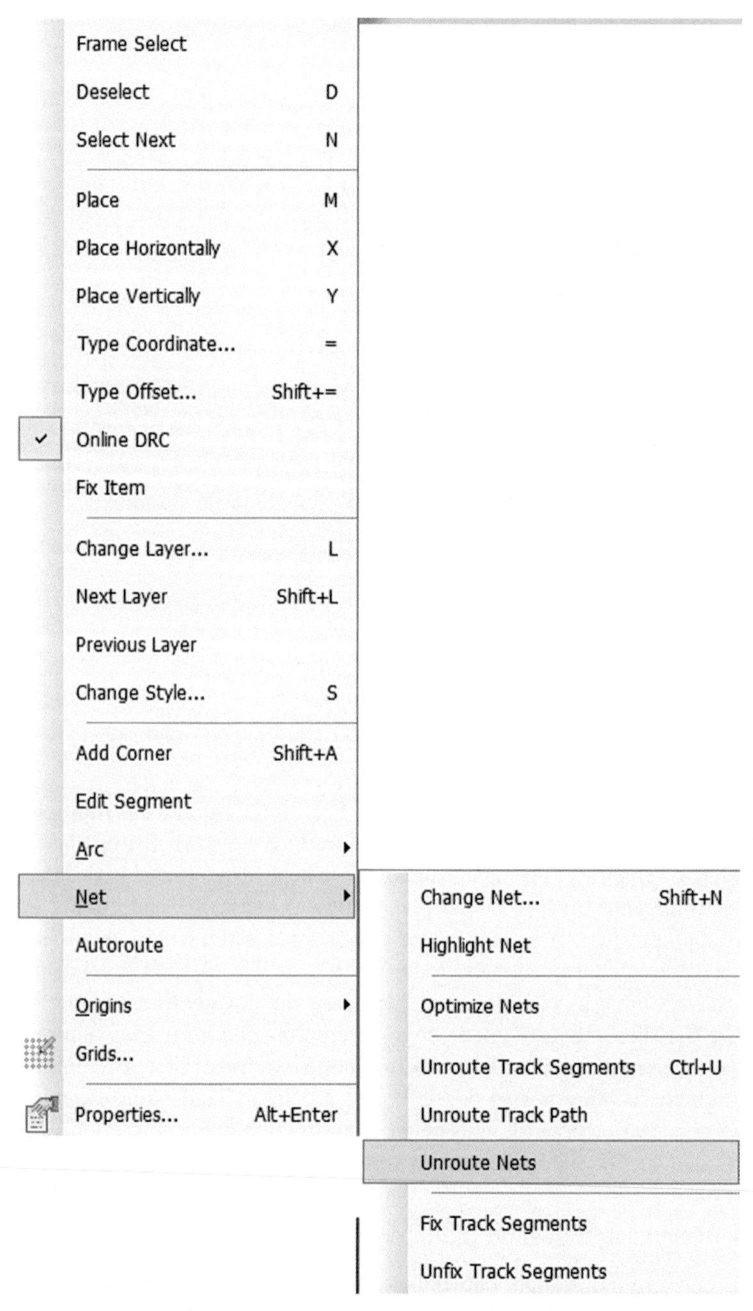

FIGURE 5.22 Unrouting the nets.

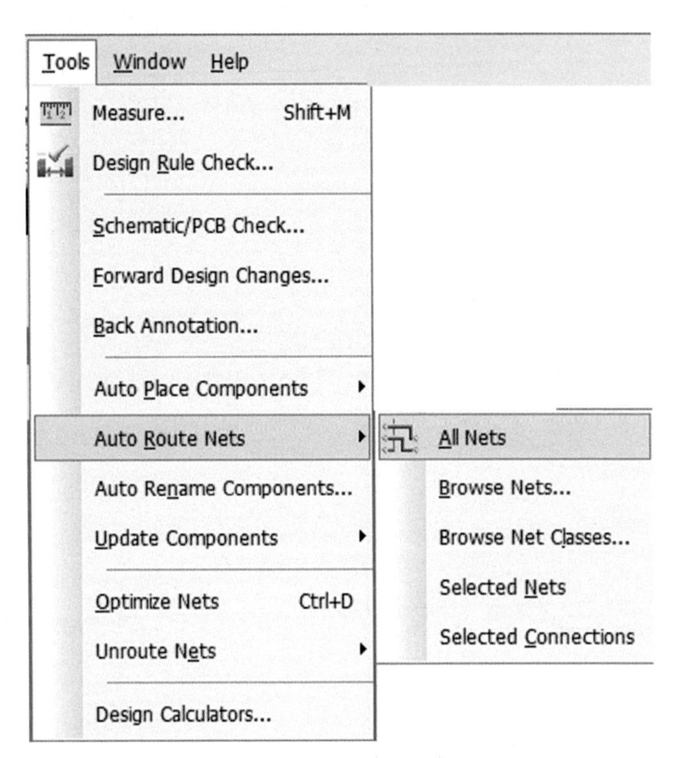

FIGURE 5.23 Autorouting the nets.

Keep trace length as short as possible. For high currents, use multiple vias when switching between layers. This will reduce your track impedance and improve the reliability. This is a general rule whenever you need to decrease the impedance of your track or power plane. If there are power and ground tracks, then route them first. Also make your power trace width as big as possible. In case of differential pair signals, route the two traces of a differential pair as close to each other as possible after they leave the device. Maintain a constant distance between the two traces of a differential pair over their entire path. Maintain the same electrical length between the two traces of a differential pair.

Automatic routing is also possible. By selecting "Tools/Auto Route nets," we are able to perform the automatic routing. A better completion rate can be optioned for the medium and low complex nets design (Fig. 5.23).

5.2.8 Power and ground plane creation

Power and ground signals can either be routed in a thicker width or connected using planes. Planes are copper pour where the necessary signal is connected. There are two types of planes available. They are positive and negative planes.

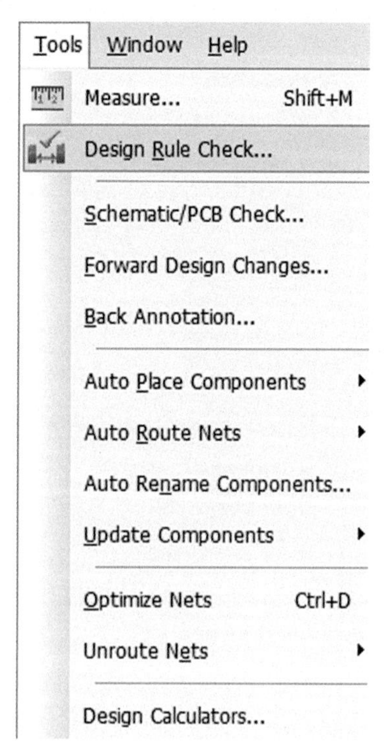

FIGURE 5.24 Design rule check.

- Positive plane

  Positive plane, a positive image of the plane, what you see is what you get. With a positive image, you see the actual plane.
- Negative plane

Negative plane, a negative image of the plane, what you see is the clearances between the plane and other features on that layer. A negative image is a much smaller size than a positive image.

Just copper pour is enough to create the planes. Before pouring the copper, we need to create the boundary by selecting "Add/Copper pour area." Right-click on the pour area and select Pour copper. This dialog will allow us to select the net to be assigned for the copper, minimum copper area, isolated islands, and thermal specification. Unconnected copper or dead copper should be grounded or removed.

5.2.9 Checking the design

Finally we need to check the design against the specification. This can be done by selecting "Tools/Design Rule Check" as in Figs. 5.24 and 5.25.

This dialog will check the spacing, net, and manufacturing. After running the check, the report will be generated.

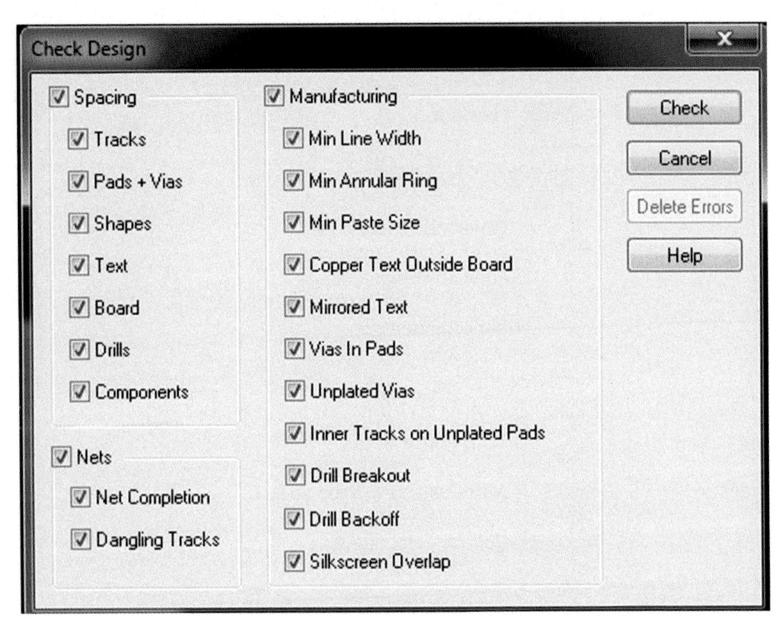

FIGURE 5.25 Design rule check—checking areas.

This can also be viewed in the layout editor of the "Goto" dialog by pressing F9 and selecting "Error" in the drop-down list as shown in Fig. 5.26. By selecting the particular error type and right-clicking will help us to see it in the design layout window.

The other detailed reports are available under "Output/Report." In this dialog, shown in Fig. 5.27, you can generate the existing standard report and user-defined report as well.

The three-dimensional (3D) view option provides the designer to view the PCB in a 3D view as shown in Fig. 5.28, by clicking "3D/3D View."

5.2.10 Gerber data output for manufacturing

DesignSpark allows us to output the design in the standard formats such as Gerber and Excellon drill under the menu "Output/Manufacturing plots." We can choose the layers to be plotted under the Plot tab. In the Output tab, we are able to choose the output type as shown in Fig. 5.29. By clicking on the "Run" button, output files can be generated.

5.3 Sample printed circuit board design—Schmitt Trigger

In this section we are going to design a simple PCB for a "Schmitt Trigger" circuit as shown in Fig. 5.30. The components required for the circuit are 555 timer IC, 100K resistor 2 nos, 0.01 μF capacitors 2 nos, power supply, and ground.

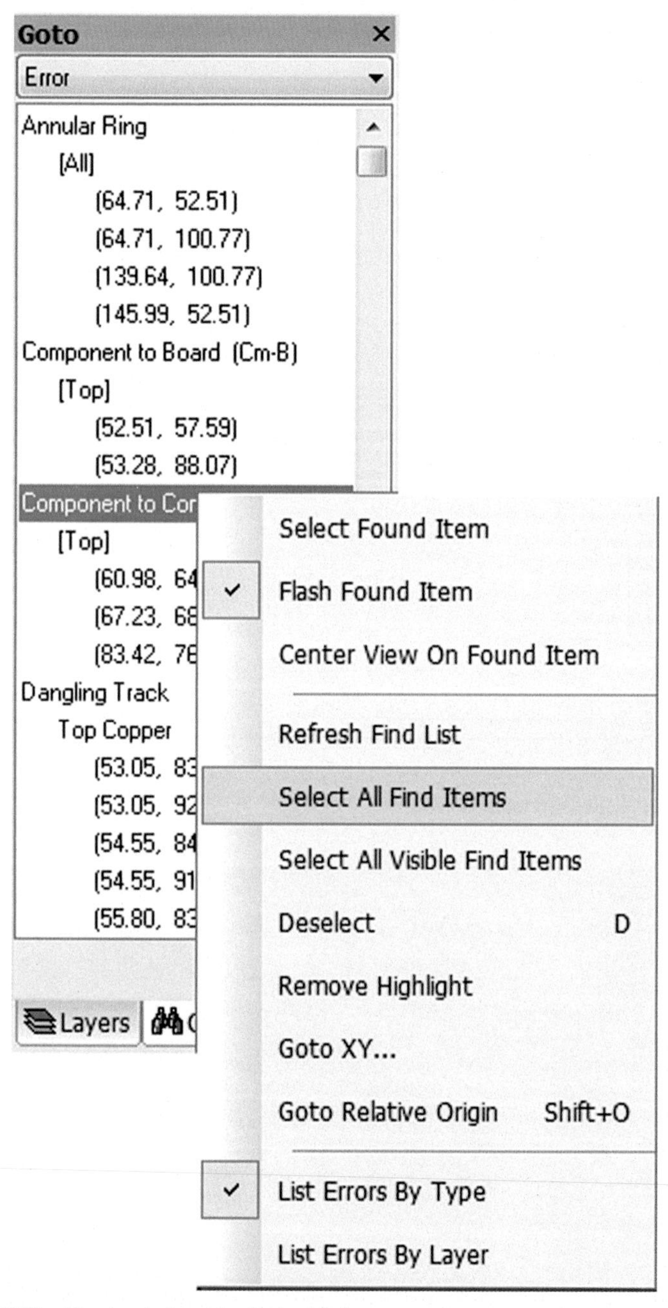

FIGURE 5.26 Viewing design rule check violations.

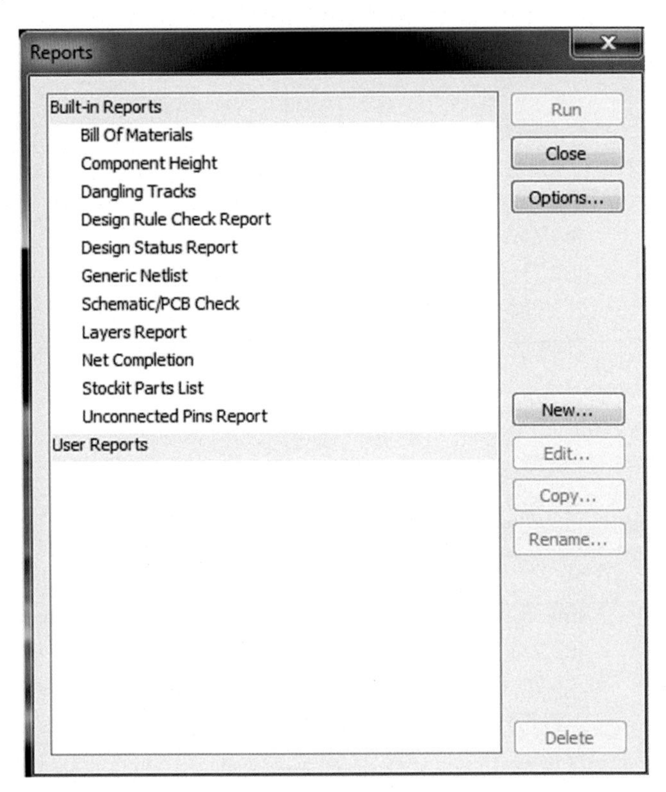

FIGURE 5.27 Design reports.

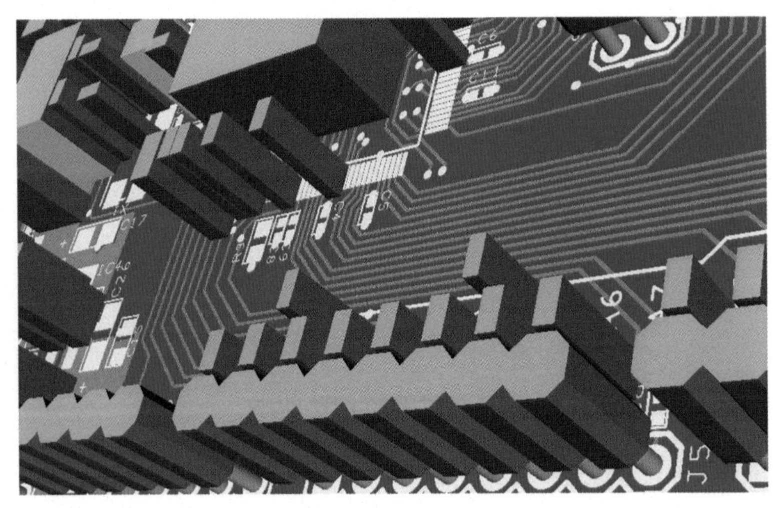

FIGURE 5.28 3D view of sample PCB. *3D*, Three-dimensional; *PCB*, printed circuit board.

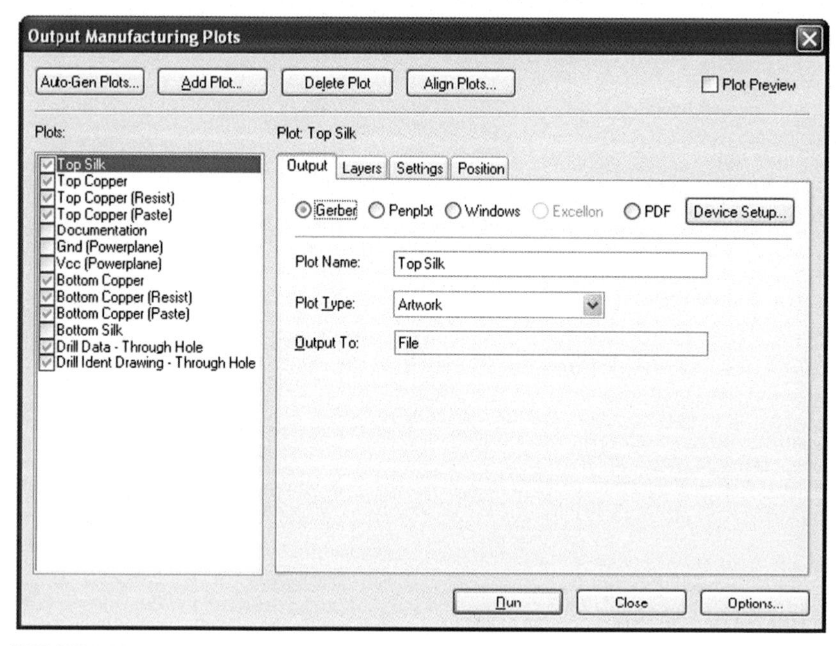

FIGURE 5.29 Design file output for manufacturing.

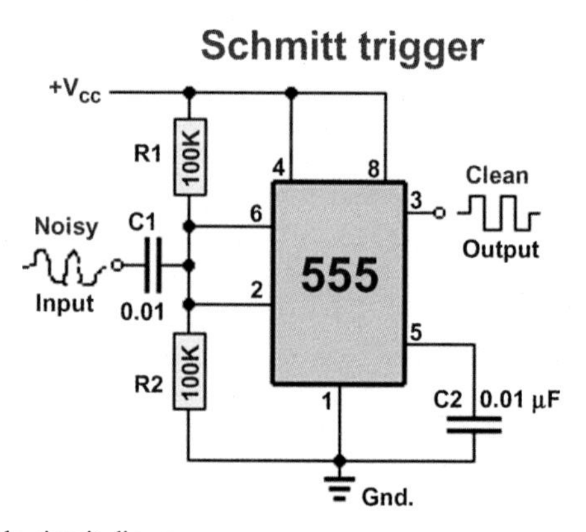

FIGURE 5.30 Sample circuit diagram.

5.3.1 Project creation

For creating a new project, double-click on the DesignSpark icon and go to File/New menu. The dialog box appears as shown in Fig. 5.31. Give a suitable project name under "New Project Name."

Upon clicking the ok button, the new project is created in the specified folder and the new project screen appears as in Fig. 5.32.

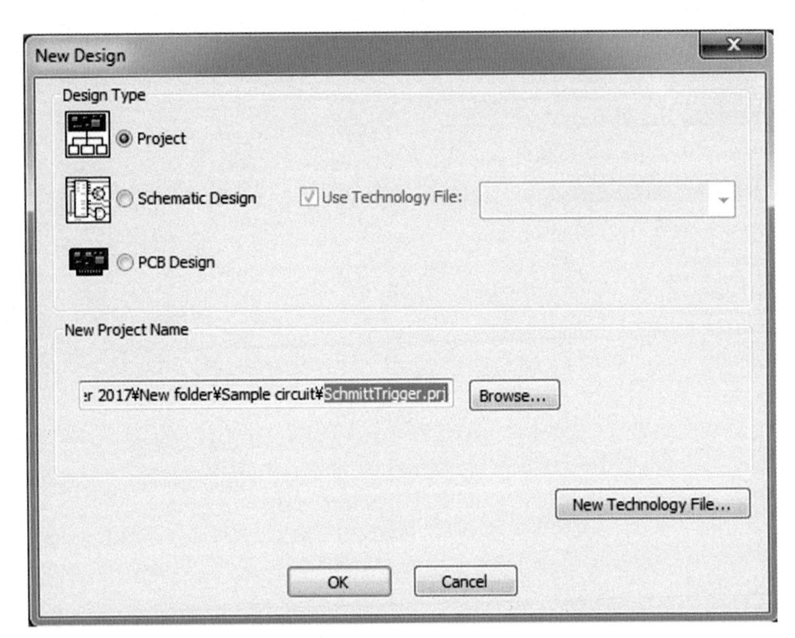

FIGURE 5.31 New project creation dialog.

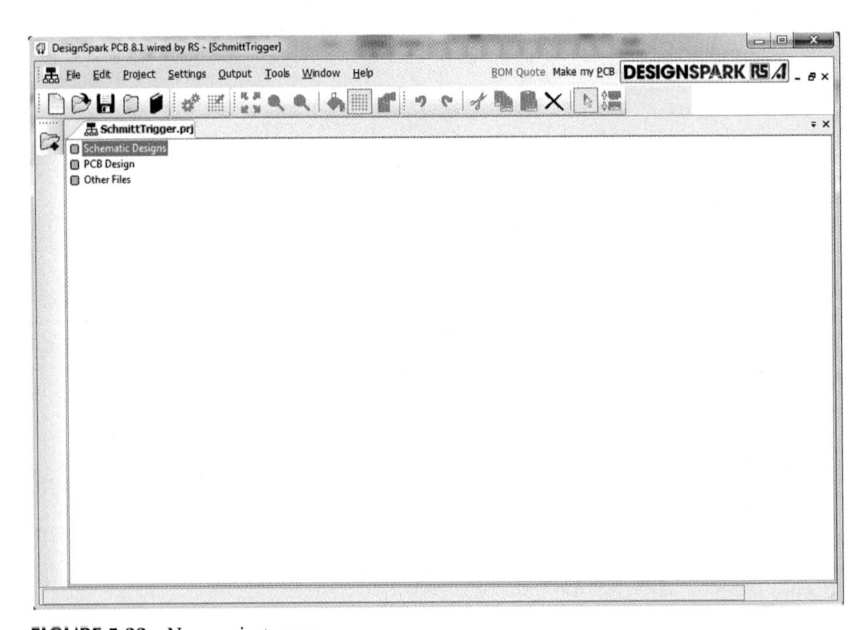

FIGURE 5.32 New project page.

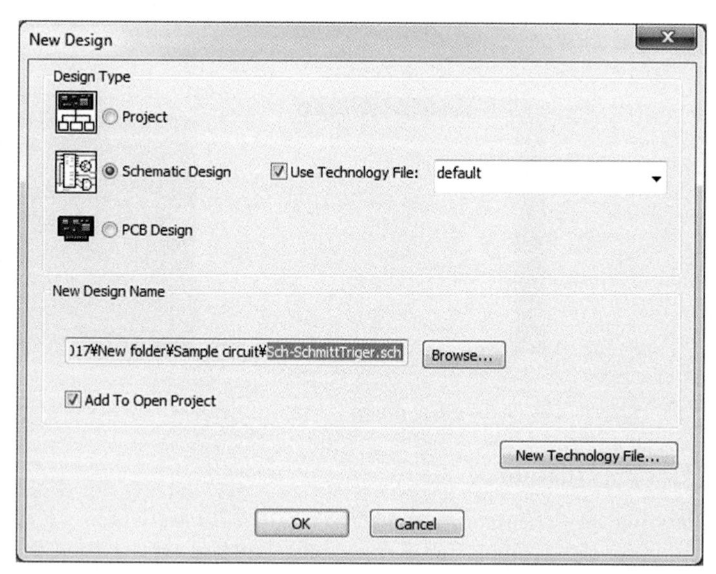

FIGURE 5.33 New schematic creation.

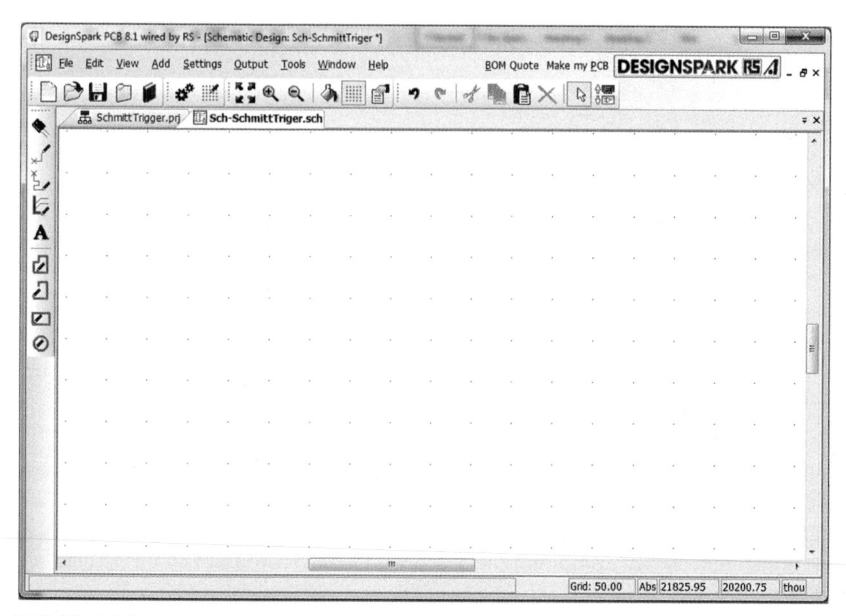

FIGURE 5.34 New schematic sheet.

Now, we need to create a new schematic design. To create the schematic file, click the New button and Fig. 5.33 will appear. Upon clicking the Ok button, a new schematic editor sheet will appear, as shown in Fig. 5.34, where you can draw the circuit.

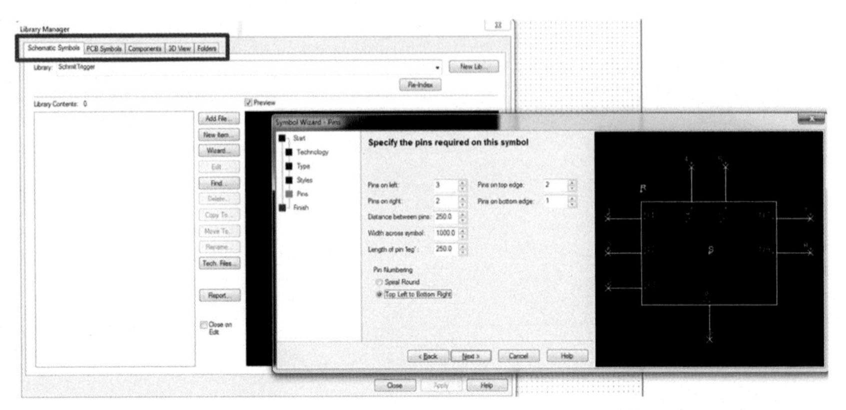

FIGURE 5.35 Schematic symbol creation dialog.

5.3.2 Library creation

Before drawing the schematic, we need to create a library where we need to create and store the symbols for components. Later it can be used in schematic editor and layout editor. The components required for the circuit have to be created in the library using Library manager. We can create a new library exclusively for this design by selecting the "New Lib" tab by giving a suitable library name. Here, the library name "SchmitTrigger" is created. In this library, the required schematic symbol, PCB symbol, and components can be created with the respective tabs as shown in Fig. 5.35.

By selecting the button "Wizard" under the tab "Schematic symbol," a new symbol can be created by following the simple instructions provided.

After finishing the symbol, the pin number can be edited if necessary by inserting and double-clicking on it.

To create the PCB symbol, we need to get the footprint details from the manufacturing part number. Based on the document, shown in Fig. 5.36, the dimensions of the lands need to be entered in the wizard under the tab "PCB symbol." Here a new PCB symbol can be created and saved in the same project library, as shown in Fig. 5.37.

Now the schematic symbol and PCB symbol are created for the 555 timer. Now both can be associated to define a component. Under the Library manager \Component tab, select "Wizard" and create a component by assigning the schematic symbol and PCB symbol in the dialog with the pin assignment and click the Finish button. By completing this process, a new component can be created and listed under the new library "SchmitTrigger," as shown in Fig. 5.38.

There are also inbuilt libraries, which come with the installation of the software by default. The inbuilt library has the set of schematic and PCB symbols. This can be accessed by selecting the library in the "Library manager" of all the tabs. We do not need to create symbols of our own every time. Use the "Copy to" button to copy the symbol from the inbuilt library

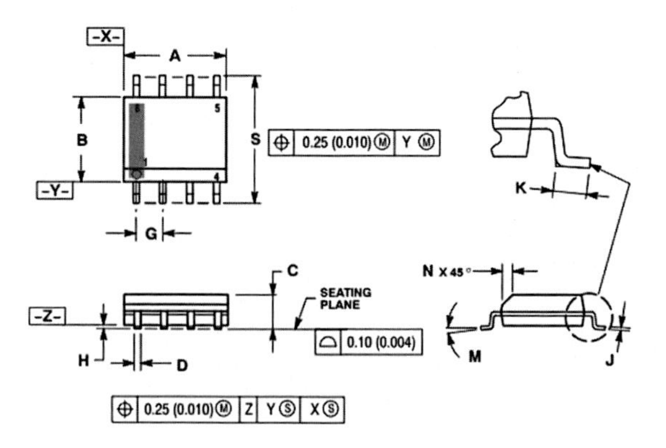

NOTES:
1. DIMENSIONING AND TOLERANCING PER ANSI Y14.5M, 1982.
2. CONTROLLING DIMENSION: MILLIMETER.
3. DIMENSION A AND B DO NOT INCLUDE MOLD PROTRUSION.
4. MAXIMUM MOLD PROTRUSION 0.15 (0.006) PER SIDE.
5. DIMENSION D DOES NOT INCLUDE DAMBAR PROTRUSION. ALLOWABLE DAMBAR PROTRUSION SHALL BE 0.127 (0.005) TOTAL IN EXCESS OF THE D DIMENSION AT MAXIMUM MATERIAL CONDITION.
6. 751-01 THRU 751-06 ARE OBSOLETE. NEW STANDARD IS 751-07.

	MILLIMETERS		INCHES	
DIM	MIN	MAX	MIN	MAX
A	4.80	5.00	0.189	0.197
B	3.80	4.00	0.150	0.157
C	1.35	1.75	0.053	0.069
D	0.33	0.51	0.013	0.020
G	1.27 BSC		0.050 BSC	
H	0.10	0.25	0.004	0.010
J	0.19	0.25	0.007	0.010
K	0.40	1.27	0.016	0.050
M	0 °	8 °	0 °	8 °
N	0.25	0.50	0.010	0.020
S	5.80	6.20	0.228	0.244

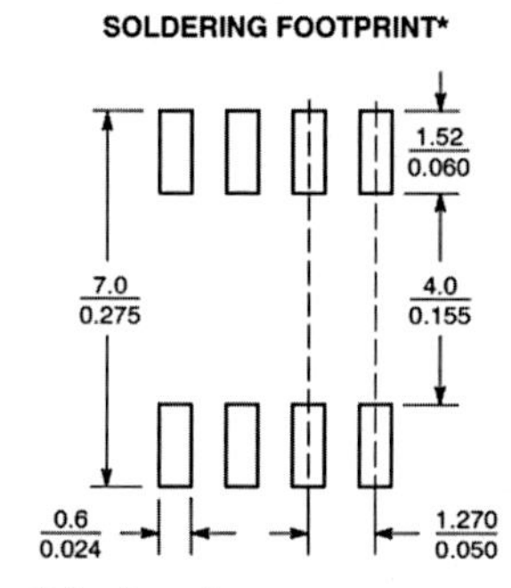

FIGURE 5.36 Component detail for footprint.

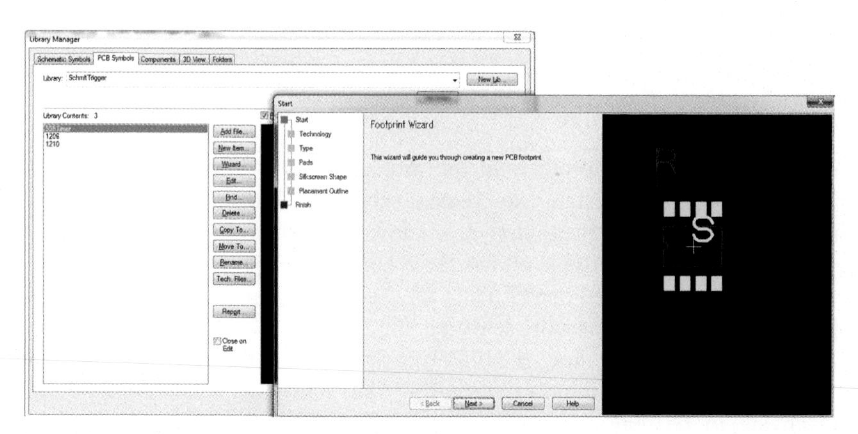

FIGURE 5.37 PCB symbol creation dialog. *PCB*, Printed circuit board.

to the design library "SchmitTrigger" and create the component using "Components\Wizard" as explained above. By this process, all the required components can be created as per the circuit diagram.

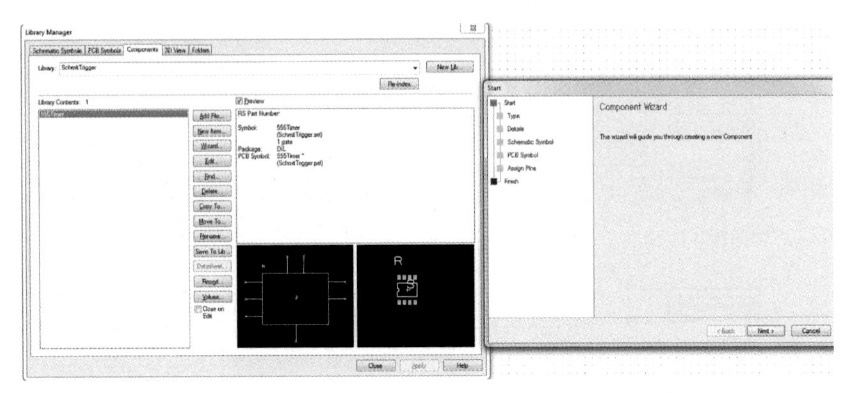

FIGURE 5.38 Component creation wizard.

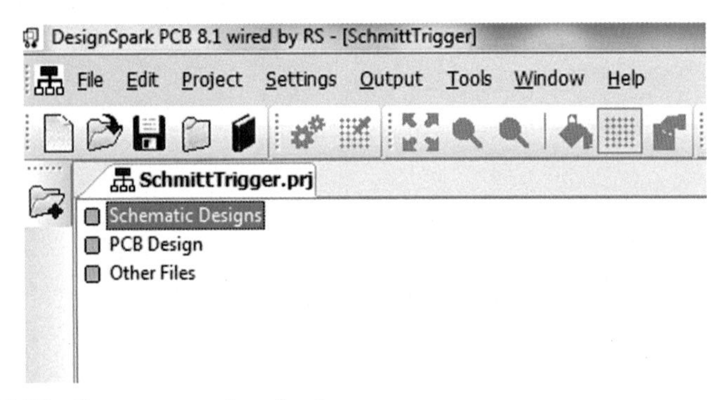

FIGURE 5.39 Component creation wizard.

5.3.3 Schematic design

Now open the created project. In the project window "SchmitTrigger.prj," click on "Schematic Designs" to create a schematic diagram which is shown in Fig. 5.39. This will open another window which is a schematic page created already (Fig. 5.34). This allows the schematic design to be drawn.

Now add the required components for the circuit which are present in the library. This can be done by selecting the menu "Add\Component" or by pressing the functional key "F3." Power supply and Ground symbols can also be added from the inbuilt library. Fig. 5.40 shows the added components which are to be connected according to the circuit Fig. 5.30. Use "Add \Connect" for making the connections.

Connection lines can be drawn according to the connectivity of the circuit.

Fig. 5.41 shows the completed schematic diagram. Make sure to check the connections are correct as per the circuit design.

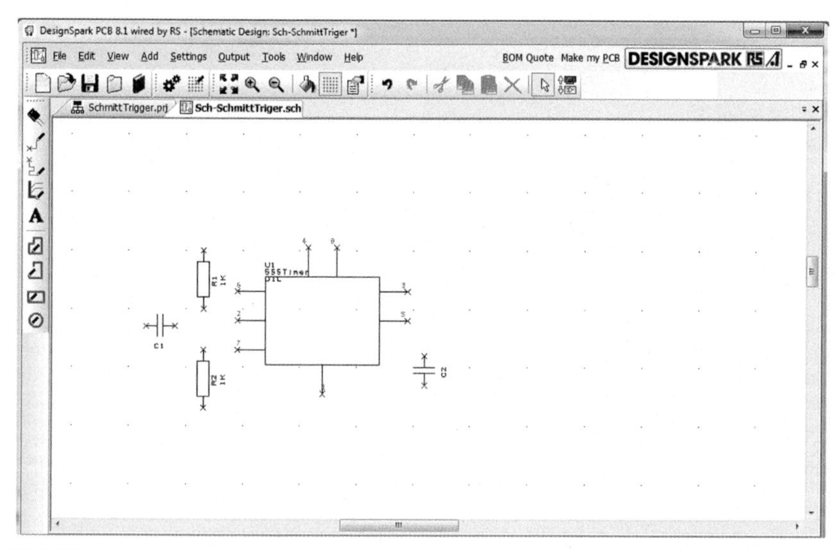

FIGURE 5.40 Schematic symbol placement.

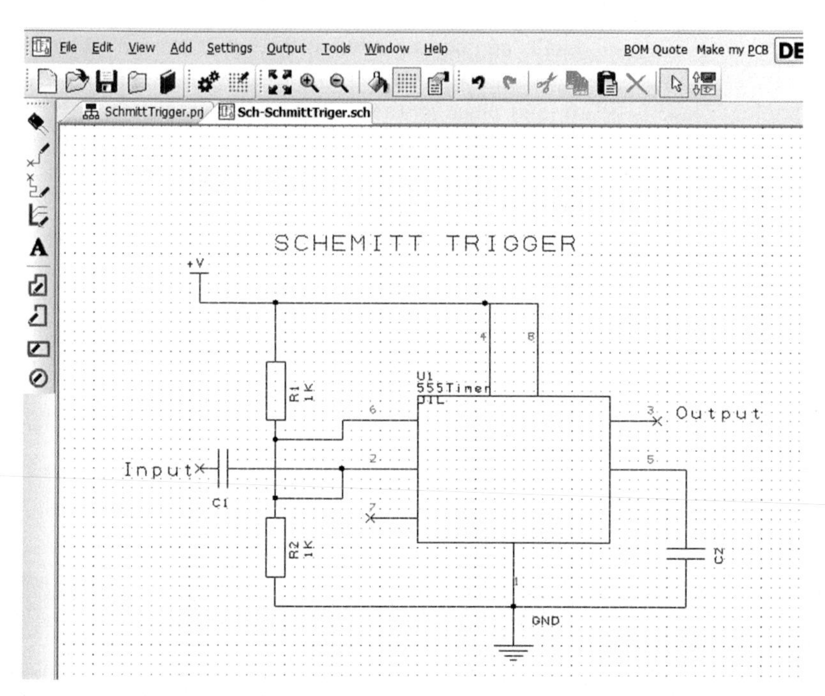

FIGURE 5.41 Schematic design.

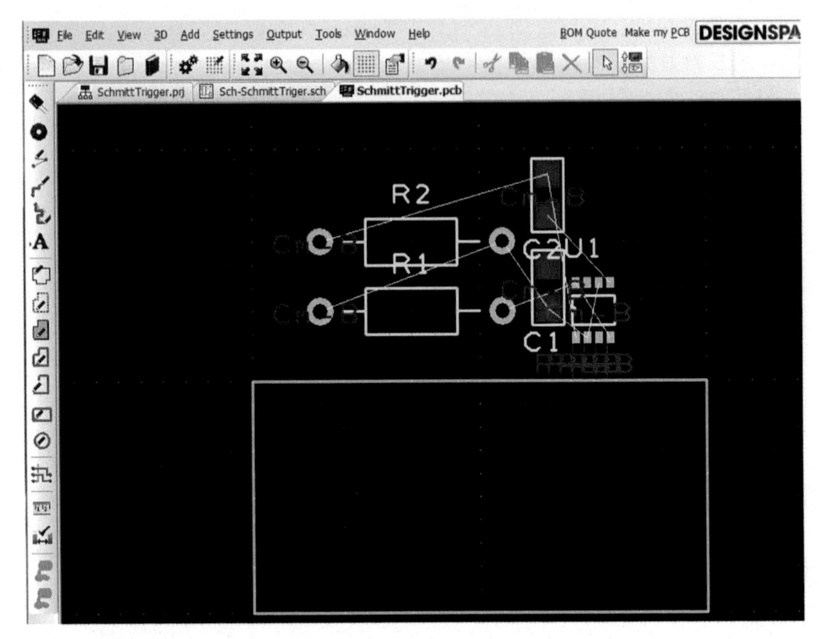

FIGURE 5.42 PCB design window. *PCB*, Printed circuit board.

5.3.4 Printed circuit board layout

The circuit connection details can be translated to the "PCB Design" window using the menu "Tools\Translate to PCB" by following a few dialogs like technology, layers, board, and place and route. Here we can choose units as thou (English unit) with precision, single layer, board size, component placement requirements, and finally PCB name. Now the window appears as in Fig. 5.42, with the components placed outside the board outline.

The components present outside the board can be moved inside the board by picking and dragging.

Fig. 5.43 shows the component placement inside the board area. The nets are shown as lines which have to be routed with the specified layers and trace widths.

The nets can be routed using the command "Add\Track." The routing can also be done using the AutoRoute mode. The completion rate may vary depending upon the complexity of the net and the criteria we set for the routing. Fig. 5.44 shows the PCB with routing.

5.3.5 Manufacturing file output

Once the routing is completed, we can check the design and whether we have done the board correctly. It can be verified using the menu "Output \Reports." After the analysis of the reports, the design with drill information

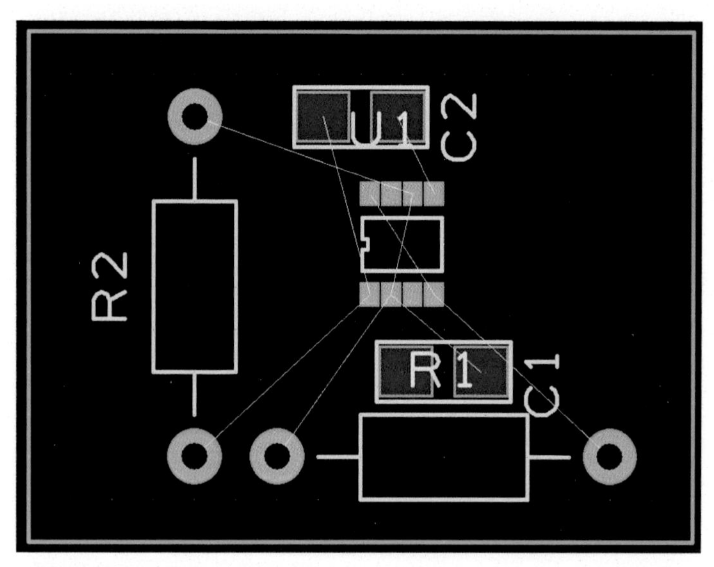

FIGURE 5.43 Unrouted PCB. *PCB*, Printed circuit board.

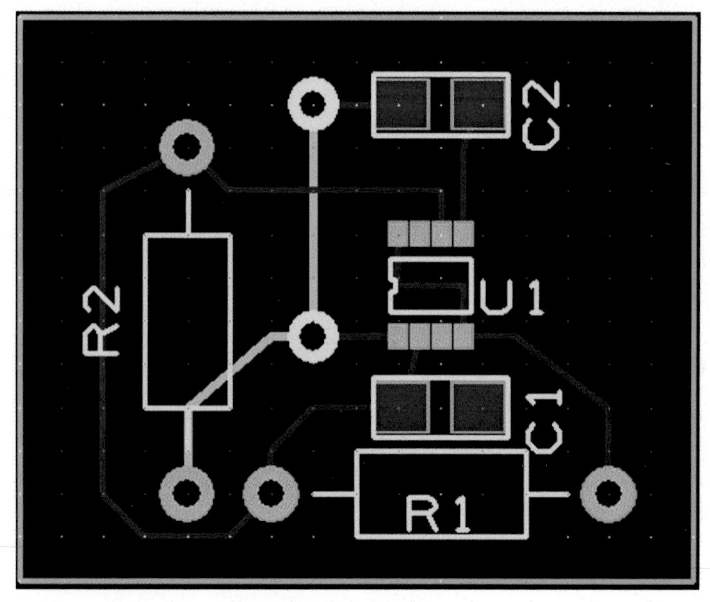

FIGURE 5.44 Routed PCB. *PCB*, Printed circuit board.

can be sent out to the manufacturing companies. The file format generally to be followed is Gerber 274X. The settings like file output, layers, and position are done in "Output\Manufacturing plots" as shown in Fig. 5.45. The output files will be stored in the folder where the design file is stored.

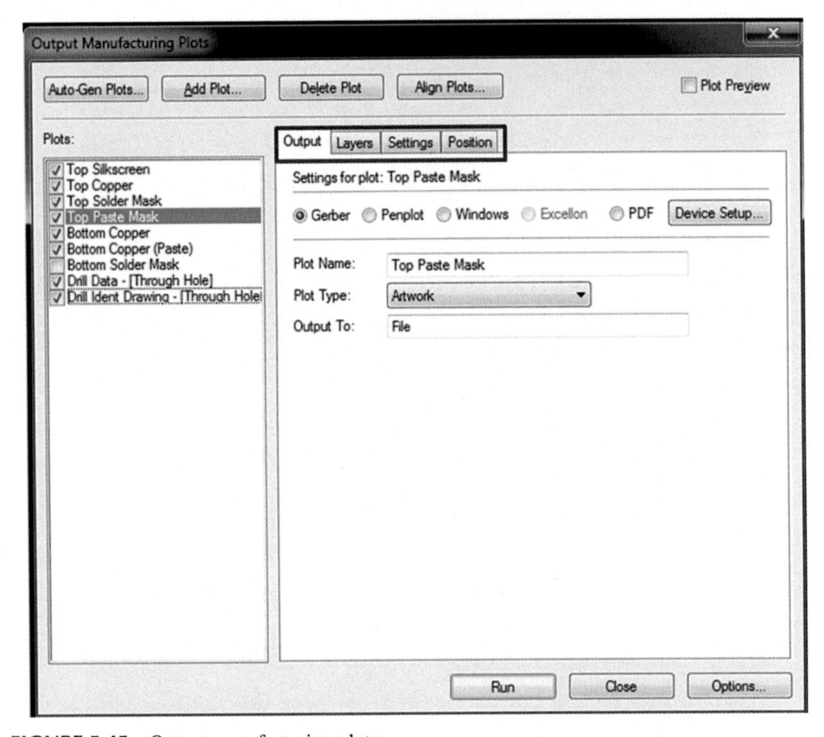

FIGURE 5.45 Output manufacturing plots.

5.4 Summary

The outcome of this chapter could enable a sound knowledge of PCB design. The users will get a brief idea about how to undertake schematics and how to undertake PCB design for the given electronic circuits.

5.5 Review questions

1. Give an overview of the PCB design software.
2. List the design guidelines followed in PCB design software.
3. Explain the user interface and management of the DesignSpark work environment.
4. What is meant by netlisting?
5. How do you place components to make one particular electronic circuit?
6. Write short notes on Gerber data output for manufacturing.
7. Explain library creation for Schmitt Trigger in detail.
8. How do you make a PCB layout for Schmitt Trigger?
9. Explain wiring, power, and plane creation step by step.
10. What do you mean by PCB design flow?

Chapter 6

Simulation of Hydraulic and Pneumatic Valves: Programmable Logic Controller

Chapter Outline

Software Tools for the Simulation of Electrical Systems. DOI: https://doi.org/10.1016/B978-0-12-819416-4.00006-5

6.1 Introduction to programmable logic controller

6.1.1 Programmable logic controller and its basic structure

Programmable logic controllers (PLCs) contains hardware and software elements, which is used to perform control functions and is often defined as miniature industrial computers. Precisely, PLC is used for control applications like machinery on factory assembly lines, amusement rides, or food processing in automation industries for elecromechanical processes. These PLCs were designed for multiple arrangements of digital and analog inputs and outputs with prolonged temperature ranges, immunity to electrical noise and resistance to vibration and impact. The PLC contains two major units: the central processing unit (CPU) and input/output (I/O) interface system (Fig. 6.1).

The system activity will be controlled over its processor and memory unit of the CPU. The CPU that contains microprocessor, memory chip, and other integrated circuits will do the control, monitoring, and communications. CPU contains various operating modes. The downloaded logic from a PC will be changed and the CPU in programming mode can accept that. The program will be executed and the CPU will operate the process during run mode. From the connected field devices, the input date is processed, and the control program that is stored in its memory system was executed or performed by the CPU. Repeatedly this program is processed by PLC. Scan time is defined as the time taken for one cycle through the program and it occurs very quickly. The program is stored in the memory of CPU and holds the status of the I/O.

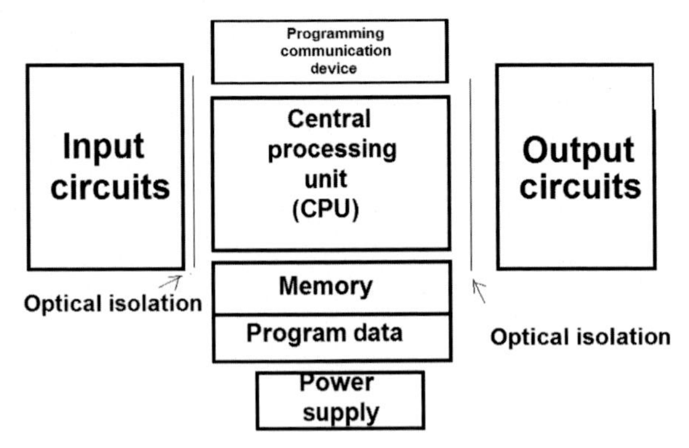

FIGURE 6.1 Basic block diagram of PLC. *PLC*, Programmable logic controller.

The interface between the CPU and its information providers and controllable devices are provided by the I/O system that is physically connected to the field devices. Once the CPU processes the input data, the required changes in the output will be done after the execution of the user program. All PLCs contains four basic steps of operation: input scan, program scan, output scan, and housekeeping. These steps continually take place in a repeating loop.

To accommodate all kinds of sensors and output devices a wide range of I/O modules are obtained by the PLCs. For example, object presence or events with devices such as proximity or photolectric sensors, push buttons and limit switches are detected by the use of discrete input modules. Similarly, loads such as motors, lights, and solenoid valves are controlled (ON/OFF) by discrete output modules. Process instrumentation such as flow, pressure, temperature, and level transmitters gave signals to the analog input modules. These modules can interpret the signal and preset value within a range determined by the devices' electrical specifications. Loads having varying control signal like panel meters, variable frequency drives or analog flow valves will be commanded by analog output. Specialized modules like high speed I/O or motion control, and serial or Ethernet communications are offered by more PLCs.

The best advantage of systemizing with a PLC is the capacity to reprise or change the operation procedure while gathering essential data. Cost, power, speed, and communication are the couple of numerous contemplations while picking the right PLC for our required activities.

6.1.2 History of the programmable logic controller

The PLC has modernized the automation industry. Nowadays, PLCs are found everywhere from factory equipment to vending machines, earlier 1968's (New Year Day) the programmable controller did not exist. Instead of PLC, a solution was existed for every unique set of challenges. Before understanding the history of PLC, we must know the problems that are existed by the programmable controllers.

Relays are the only way to control machinery before the invention of PLCs. Relay coil should be energized and produces magnetic force to pull the switch to the ON or OFF position. The switch releases and returns the device to its standard position (ON/OFF) once the relay gets deenergized. For example, if you want to control a motor, you could connect a relay between the source and the motor. After that, you can able to control when the motor is getting power by either energizing or deenergizing the relay. Without power, the motor should not run and hence everyone has the relay to control the motor. Hence, these types of relays are called power relay. The relay is said to be control relay (CR) because they control the relays that control the switch that turns the motor ON and OFF. By considering modern factories, we need more number of motors and ON/OFF power switches to control a single machine. It has been more complicated by installing and maintaining these types of large relay control systems.

6.1.3 Birth of the programmable logic controller solution

According to Dick Morley, the undisputed father of the PLC, "The programmable controller was detailed on New Year's Day, 1968." The popular forum PLCDEV.com outlines a list of requirements that GM engineers put out for a "standard machine controller." It is this request that Dick Morley and his company, Bedford and Associates, were responding to when the first PLC was proposed. Further interchanging the relay system, the requirements listed by GM for this controller includes:

1. A solid-state machine that was versatile like a computer but fairly priced with a relay logic machine of the same nature.
2. Easily maintained and configured according to the currently agreed way to do things relay ladder logic.
3. In an urban area it had to operate consisting of water, humidity, electromagnetism and vibration.
4. It had to be flexible in shape to allow for quick part exchange and upgradability.

The PLC's programming look needed repair electricians and plant technicians to quickly understand and use it. As relay-based control systems developed and has become more complex, the use of physical device position wiring diagrams also developed into the relay logic being represented in a ladder fashion. The control power hot wire will be the rail left, with the control power neutral as the rail right. The different relay contacts, push buttons (PBs), toggle switches, limit switches, relay coils, engine starter coils, solenoid valves, etc., shown in their logical order will shape the rungs of the ladder. The PLC had been asked to be configured in this ladder logic format (Fig. 6.2).

6.1.4 Programmable logic controller applications, disadvantages, and advantages

6.1.4.1 Applications

- PLC controller act as a central part of a process control system in automated system.
- To run processes that are more complex is possible to connect more PLC controllers to a central computer.

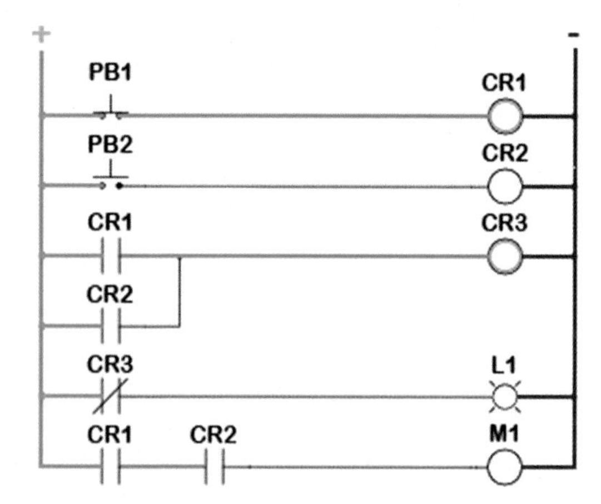

FIGURE 6.2 Ladder logic fashion.

6.1.4.2 *Disadvantages*

1. The linking of wires needs so much effort.
2. Changes or replacements challenges.
3. Difficulty finding mistakes; skillful work force needed.
4. Hold-up time is infinite, usually long when a question arises.

6.1.4.3 *Advantages*

1. Rugged and engineered to withstand turbulence, humidity, temperature, and noise.
2. Have interfaces already within the controller for inputs and outputs.
3. Quickly programmed, and provide a programming language easily recognized.

6.1.5 Major types of industrial control systems

In various industries, industrial control system or ICS include various control systems that are currently in operation. These control systems include PLC, supervisory control and data acquisition (SCADA), and distributed control systems (DCS) and various others:

PLC: PLCs are based on Boolean logic operations however some models use timers and some have continuous control. These devices are computer based and are used to control various process and equipment's within a facility. The primary components in smaller control configurations of DCS and SCADA are controlled by PLCs.

DCS: DCS contains decentralized elements and all the processes are controlled by these elements. Labor costs and injuries were minimized by human interactions.

Embedded control: In this control system, with the help of a network, small components are attached to the industrial computer and control is exercised.
SCADA: SCADA refers to a centralized system, and this system is composed of various subsystems like remote telemetry units, human machine interface, programmable logic controller, or PLC and communications.

6.1.6 Hardware components of a programmable logic controller system

Processor unit (CPU), memory, I/O, power supply unit, programming device, and other devices.

CPU: A microprocessor-based CPU allows arithmetic operations, logic operators, block memory moves, computer interface, local area network, functions, etc. The CPU will discover the eventual errors early due to the routine number of check-ups of the PLC controller. To execute the control program, store logic data, and communicate to external devices the CPU contains a microprocessor, memory storage and other integrated circuits. In today's industrial applications the serial ports are very important, since many of the existing networks use these standards, but Ethernet becomes the most suitable method of all. To program or monitor its logic, a recent addition is a universal serial bus, which is very useful when connecting to a CPU. Just as important, as the ports available are the protocols the CPU can support Ethernet/IP, Modbus TCP, etc. Some PLCs use registered communication and others use open standards. Both the way, make sure the CPU you select has the communication capabilities you need.

Memory size is one of the important aspect of CPU. To handle the required amount of tasks you are going to assign, for that, our CPU has enough storage space and it does not hurt to have a little extra for future needs. The user can avail more memory, and nowadays can have 50 MB and more. The large memory capacity in these CPUs allows for almost limitless programming, ample space for program documentation and impressively quick scan times. Another popular device is a removable memory card support. Without a PC, these memory cards allow extended data logging and easy program downloads by adding an additional 1–32 GB of memory storage.

6.1.6.1 Memory

The operating system and the fixed data, which is used by the system, is permanently stored by the system (ROM), and RAM for data. The devices like I/O, timers, counters, and other internal devices store the information on the status. EPROM for ROM's can be programmed and then the program made permanent.

I/O sections: The field devices, such as switches and sensors are monitored by the inputs. Likewise, other device like motors, pumps, solenoid valves and lights is controlled by the outputs. The PLC and the I/O modules affected the controlled current state, and their respective end devices will allow this. The input and output modules contains different types but this can be classified as analog, discrete or specialty I/O.

With these classifications, discrete I/O is the simplest and delivers the PLC with ON/OFF control. This I/O device provide the CPU with a yes/no, true/false indication, and permit simple full ON or full OFF response with both AC and DC voltage ranges. Devices such as photo eyes provide the input signals, proximity switches, E-stop PBs, float switches, etc. Our command choices are either ON or OFF for discrete outputs which is commonly used for stack lights, alarms, relays, solenoids, etc. with nothing in between them.

In discrete I/O a thing should be considered whether we need a sink, source, or relay configuration. While completing the circuit the PLC provides the reference voltage (typically 0 V) with sinking I/Os. Sourcing I/Os PLC provide source voltages of 12 VDC, 24 VDC, 240 VAC, etc., which is opposite to sinking. Relay types do not provide either. They function just as a relay contact, which is connected to a load once activated by an external source.

Analog I/O deals with the gray area between full ON and full OFF that discrete I/O ignores. It provides the PLC with the data it needs for precision control of a process. One important factor to remember with analog modules is the resolution they provide.

Power supply: Most PLC controllers work at either 24 VDC or 220 VAC. Some PLC controllers have electrical supply as a separate module, while small and medium series already contain the supply module. The power supply unit of the PLC is included with the base or it may have a separate unit. It supply a restricted amount of amperage and contains multiple voltage ranges including 12–24 VDC and 110/220 VAC. The sufficient power supply is supplied to the CPU and I/O modules from the chosen power supply.

Programming device: The required program is entered into the memory of the processor by the use of programming device. Once the program is programmed in the device, then it is stored in the memory unit of the PLC.

6.2 Ladder logic

PLCs have become useful in the controls marketplace and are utilized all through the globe. After some time, they progressed to turn out to be more effective and more affordable. Various kinds of programming dialects have likewise been produced for PLCs, yet the most often utilized is the ladder diagram.

6.2.1 The origins of ladder logic: relay logic

The electromagnetic switch boards comprise of various electromechanical electromagnetic switches that are joint together to play out a specific capacity within the plant. The straightforward opening and shutting of electromagnetic switch contacts on the board provides the system with the ON or OFF controls it requires in the assembling procedure.

Utilizing this combination of switches, electromagnetic switches, loops, and contacts is mentioned to the relay logic control. electromagnetic switch logics is a reliable controls technique still presently in constrained use. Be that as it may, the expense related with it as far as tedious logics changes, mechanical disappointments after some time, and broad wiring along with the space necessities has constrained numerous enterpriscs to reexamine their control requirements. What they found was none other than the PLC.

6.2.2 The structure of ladder logic

The composition behind ladder diagram depends on the electrical ladder logics that were utilized with relay controls. These diagrams recorded how associations between devices were based on relay panels; they are designated "ladder" logics since they are built in a manner that looks like the ladder diagram with two vertical rails and rungs among them. Therefore, the positive source rail (on the left-side) streams to those negative source rails (on the right-side) via the physical devices associated with the rung part. The model beneath displays a ladder logic with PB, CRs, and motor (M) along with the lights (L) (Fig. 6.3).

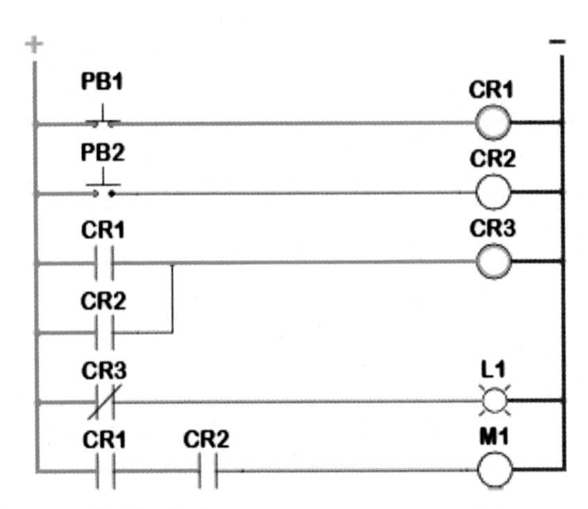

FIGURE 6.3 Structure of ladder logic.

6.2.3 Similarities with ladder diagrams

Ladder logic was designed to have the same look and feel as electrical ladder diagrams, but with ladder logic, the physical contacts and coils are replaced with memory bits (Fig. 6.4).

In this, the relay logic control's ladder logic is copied with ladder diagram; not any more hard-wired logics, yet memory areas. A portion of these memory areas are utilized inside and others are utilized with external inputs and outputs. To monitor and control real world devices, they should be wired to I/O modules.

For this specific PLC, these data sources and outputs are allocated to X and Y memory tends to be similar the X001 displayed with PB1. So this ordinarily open the state of the contact is used from the inputs on the I/O modules in which the physical PBs are associated. Then again, every Y bit has an output device connected to it as observed with the light constrained by Y001. The entirety of different areas is relegated to internal bits that we may utilize as required.

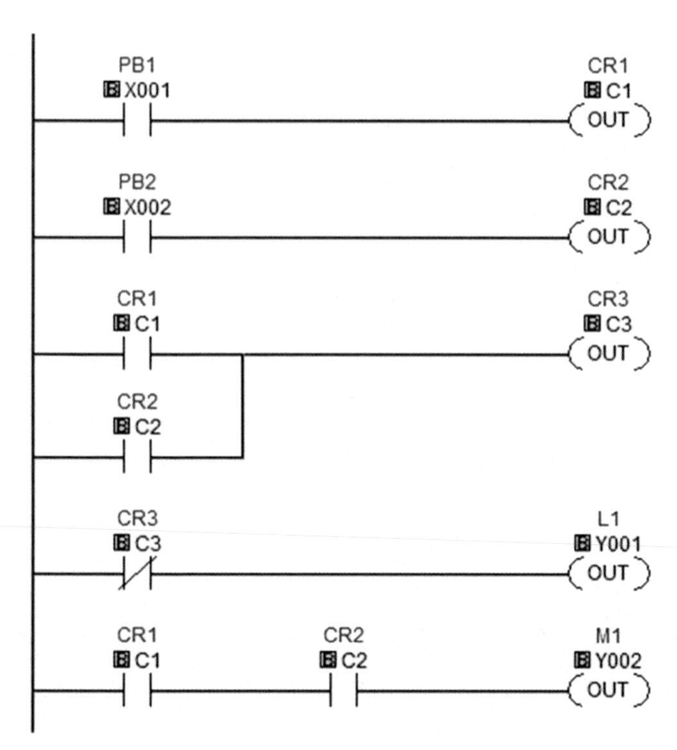

FIGURE 6.4 Ladder logic with memory bits.

6.2.4 Execution of ladder logic

Normally before beginning to code the logics, the processing unit peruses the physical data sources attached to the I/O modules to refresh their status within the processing unit's memory tables. At that point, beginning at the upper left-side of the program, the processing unit works the way down the trail executing every rung or subrungs from left-side to right-side. So when PB1 is pressed, the processing unit will switch ON CR1. Because CR1 has modified the states, within rung 3 the processing unit will initiate CR3. Now, CR3's normally closed (NC) state is utilized in rung 4, hence, the processing unit will at that point switch OFF L1.

Despite the fact that regardless still refer to coils and contact in the ladder diagram, recollect that they are, in actuality, memory representations, not actual devices. When the CPU arrives at the last rung that will refresh this present outputs, at that point loop back and operate everything once more. This procedure will proceed up until the processing unit is controlled and within the RUN module.

The time that consume the processing unit to execute the single pass and loop back to the start is called scanning time. This scanning time can be essential to apps in which the timing is basic. Subroutines and particular reason I/O modules may also be utilized to help diminish the scanning time if necessary.

6.2.4.1 The logic behind the ladder

With the increasing demand for functionality and ease of use, many of today's PLCs integrate function blocks with ladder logic. The organization of the program is still ladder with the more difficult instructions being function blocks.

6.2.5 Ladder logic instructions: the basics

Open contacts in series (AND gate) or parallel (OR gate) as seen below (Fig. 6.5A and B).

Table 6.1 contains all of the available ladder logic elements in the CLICK programming software with descriptions of their functions.

The compare contact can be programmed to look at two different numerical values; either or both values can be a variable or fixed value.

If the evaluation condition is met, then the result will be true, passing the logic path along to the next instruction or turning the rung output ON (Fig. 6.6).

The output of the temperature sensor generates an analog signal, which is wired into a PLC. The PLC is designed to take out the analog signal and move it to Fahrenheit degrees. If the temperature registers above 32°F, we

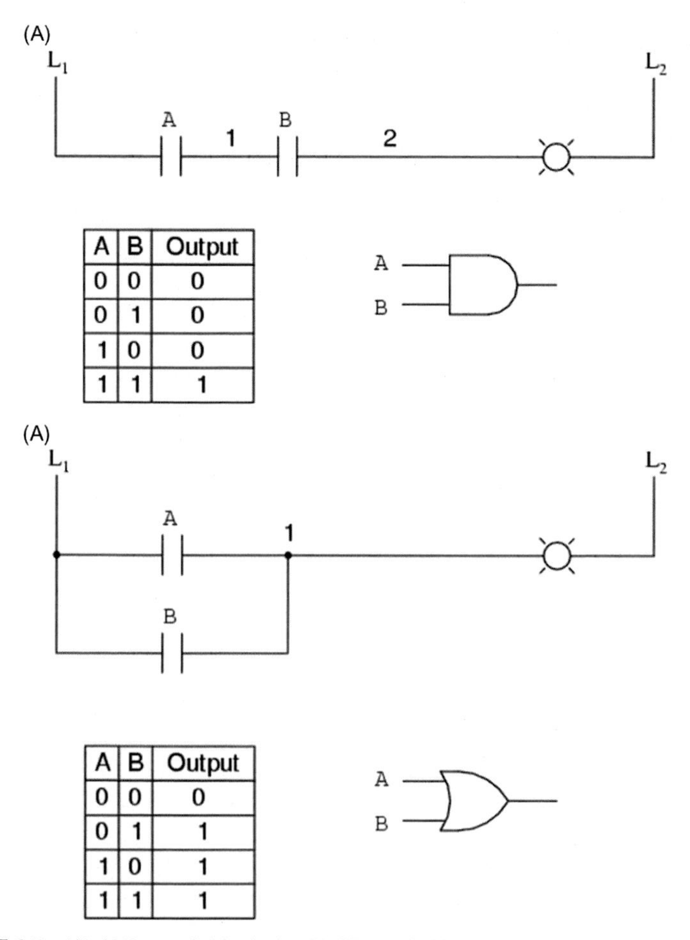

A	B	Output
0	0	0
0	1	0
1	0	0
1	1	1

A	B	Output
0	0	0
0	1	1
1	0	1
1	1	1

FIGURE 6.5 (A) AND gate ladder logic. (B) OR gate ladder logic.

can sound an alarm horn to warn us that the frozen fish will thaw and spoil soon. We may equate our real freezer temperature (stored in memory position DF16) to the constant 32, which reflects 32°F (Fig. 6.7), and use the communication instruction dialog box.

The communication dialog box compare helps us to pick one of six various methods of correlation. In this example, we have chosen better than that if our freezing temperature is ever higher than the number 32, then our comparison communication will become real.

Then, as shown below, an out coil is programmed from our comparison touch and provide the signal wired to the outside alarm bell, which alerts us to a problem with our freezer and possibly gives us time to escape catastrophe (Fig. 6.8).

TABLE 6.1 List of ladder symbols.

Ladder symbol	Title	Type
X001	Normally open contact	Bit instruction
X001	Normally closed contact	Bit instruction
X001 ↑ X001 ↓	Edge contact	Bit instruction
DS1 = 100 DF1 ≥ 2 XD1 < 200h TXT1 ≠ "ADC"	Compare contact	Word instruction
Y001 (OUT)	Out coil	Bit instruction
Y001 (SET)	Set coil	Bit instruction
Y001 (RST)	Reset coil	Bit instruction

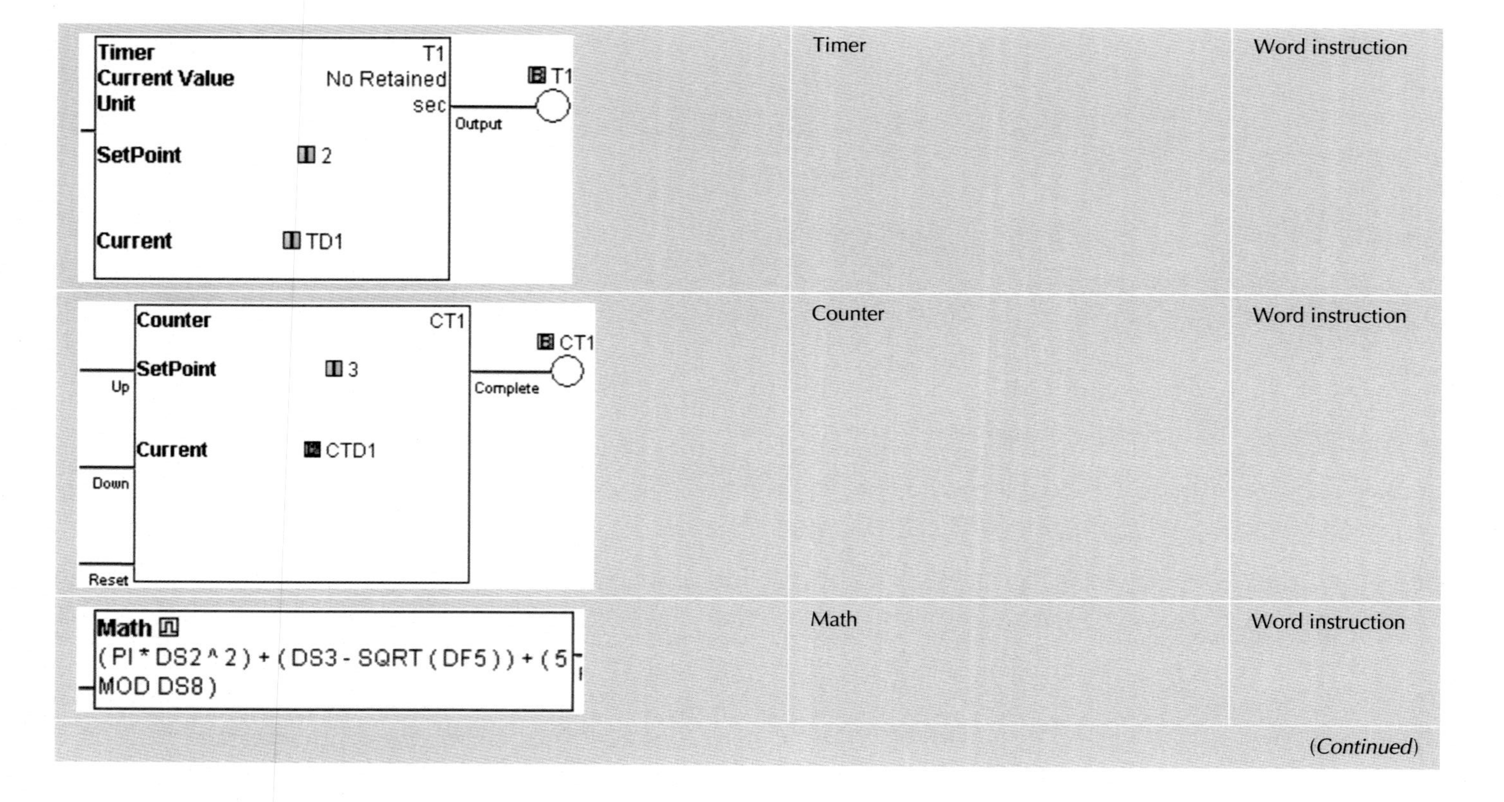

Timer T1 / Current Value No Retained / Unit sec / SetPoint 2 / Current TD1 / Output T1	Timer	Word instruction
Counter CT1 / Up / SetPoint 3 / Down / Current CTD1 / Reset / Complete CT1	Counter	Word instruction
Math (PI * DS2 ^ 2) + (DS3 - SQRT (DF5)) + (5 MOD DS8)	Math	Word instruction

(*Continued*)

TABLE 6.1 (Continued)

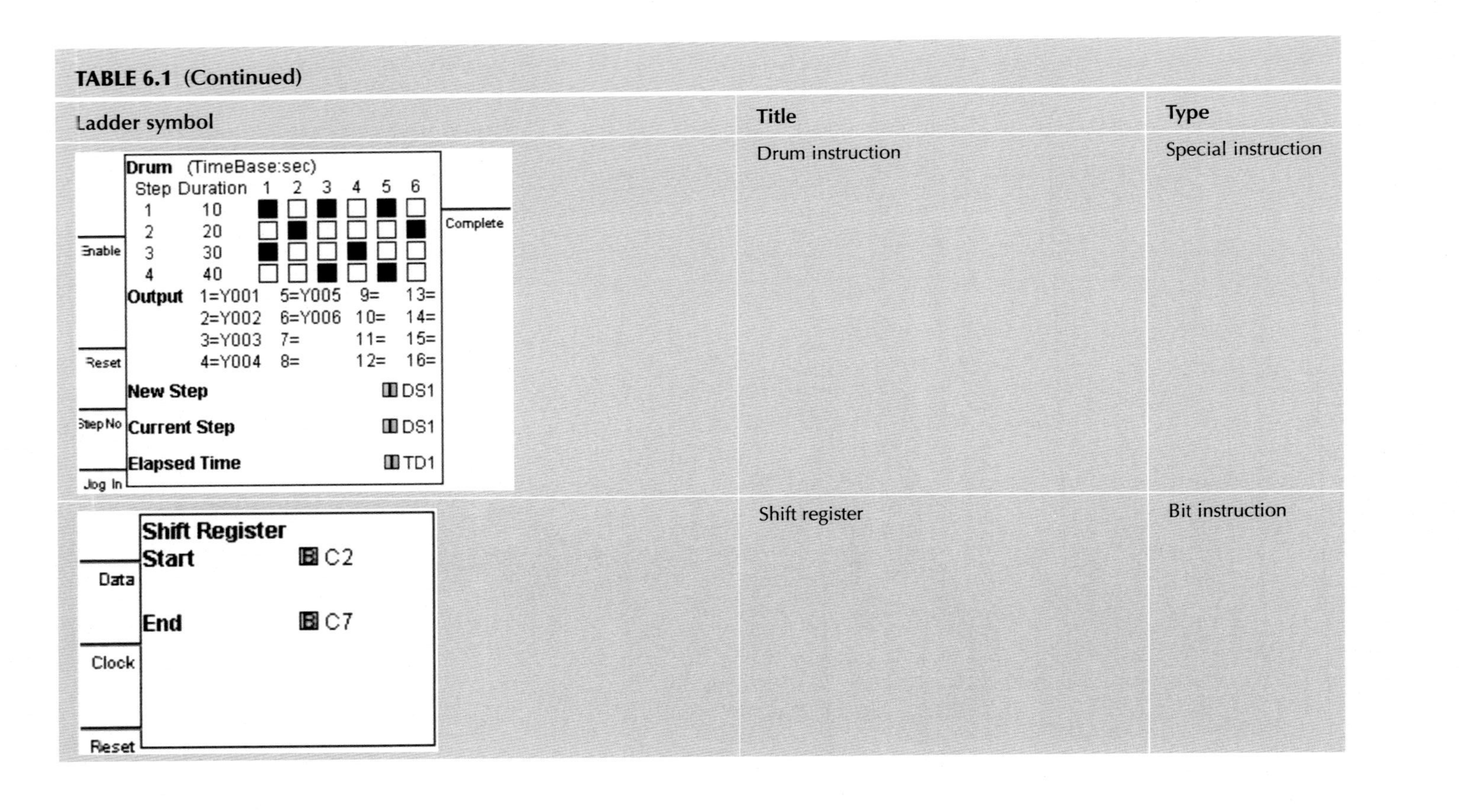

Ladder symbol	Title	Type
Drum (TimeBase:sec)	Drum instruction	Special instruction
Shift Register	Shift register	Bit instruction

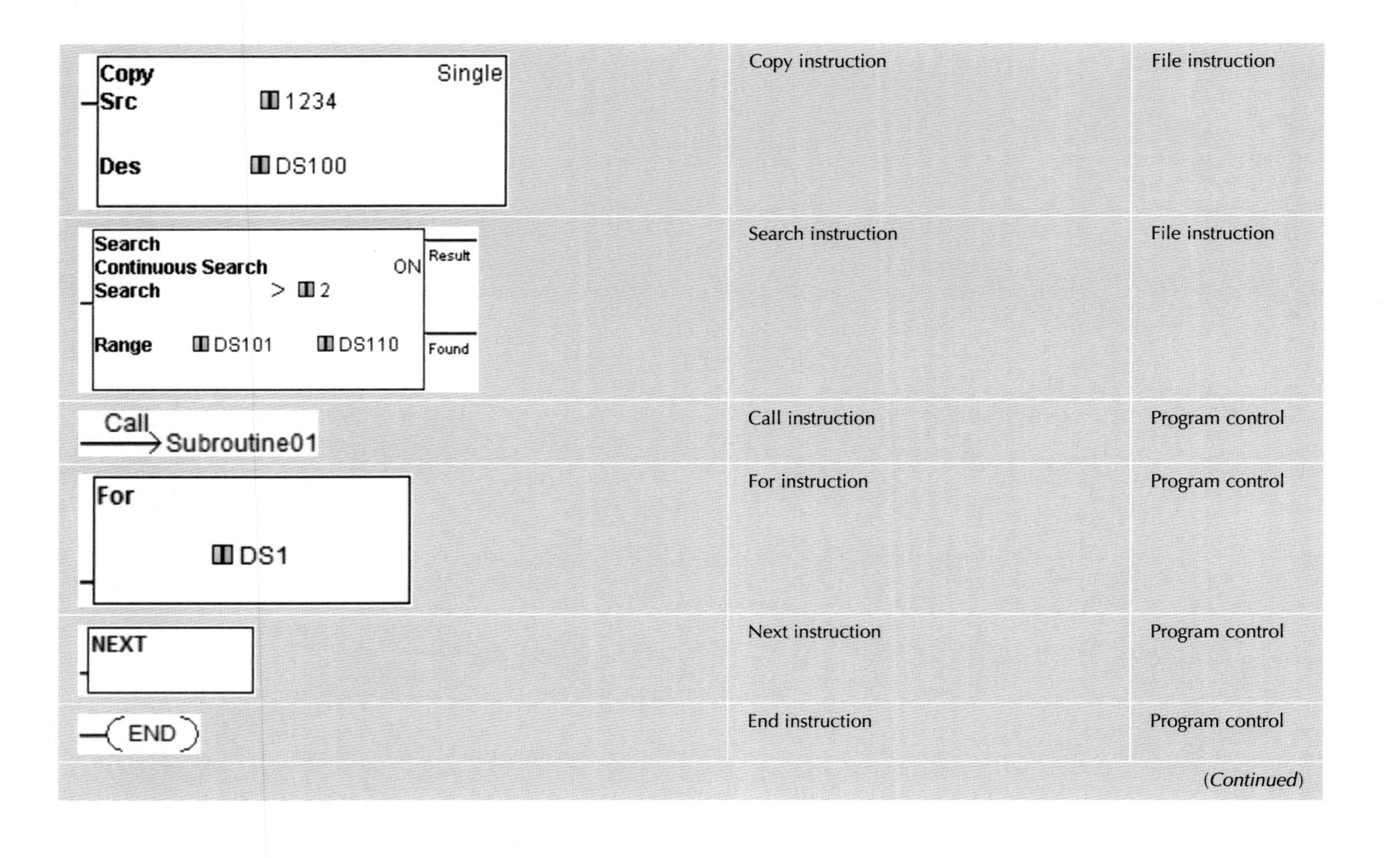

Copy Single Src 1234 Des DS100	Copy instruction	File instruction
Search Continuous Search ON Search > 2 Range DS101 DS110 Result Found	Search instruction	File instruction
Call Subroutine01	Call instruction	Program control
For DS1	For instruction	Program control
NEXT	Next instruction	Program control
END	End instruction	Program control

(*Continued*)

TABLE 6.1 (Continued)

Ladder symbol	Title	Type
Receive (Port:2) MODBUS Slave ID 1 Modbus Function Code 01 Slave Addr 9 NO. of Bits 32 Master C16	Receive instruction	Communication
Send (Port:2) MODBUS Slave ID 3 Modbus Function Code 05 Slave Addr 1 Master C54	Send instruction	Communication

6.2.6 Examples for ladder logic

Example 1: Two switches labeled A and B are wired in parallel controlling a lamp as shown in Fig. 6.9 Implement this function as PLC ladder logic where the two switches are separate inputs.

The action of a switch circuit is defined as "When switch A or switch B is closed (on) the lamp is on." All possible combinations of these two switches and the consequent lamp action is shown as a truth Table 6.2.

The PLC ladder logic notation assumes that the inputs are connected to discrete input channels (Fig. 6.10). Also, the actual output (lamp) is connected to a discrete output channel controlled by the coil. The control for the

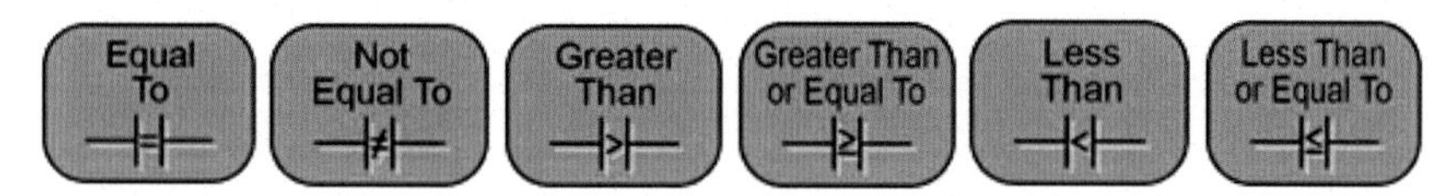

FIGURE 6.6 Compare contact.

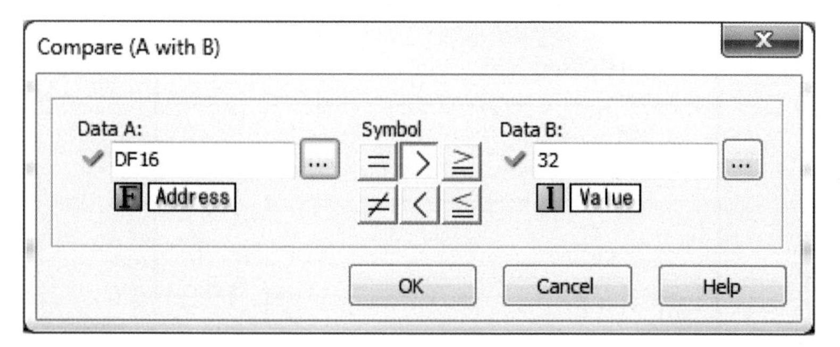

FIGURE 6.7 Fahrenheit conversion.

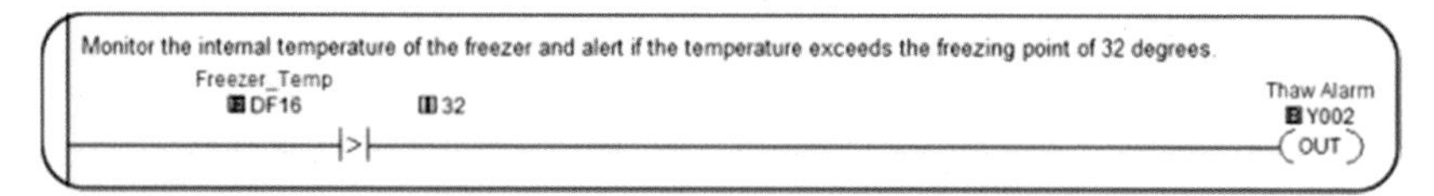

FIGURE 6.8 Fahrenheit conversion using compare contact.

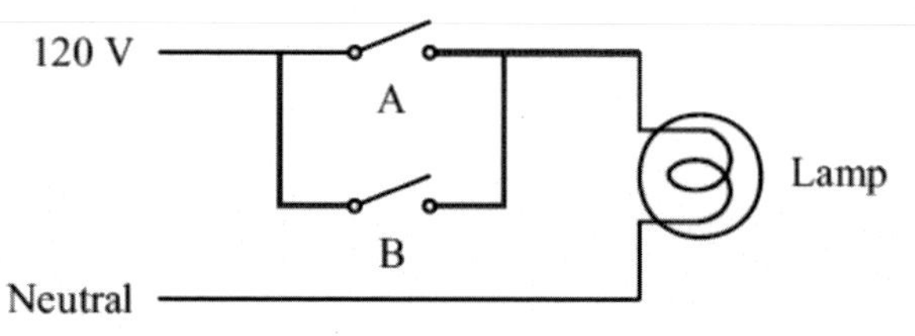

FIGURE 6.9 Parallel controlling a lamp.

coil driving the contact is shown as a label above the corresponding contact symbol. Also, the rung output arises on the extreme right-side with the power assumed to flow from left to right. The PLC ladder logic rung is inferred as: "When switch (input) A is on OR B is on then the lamp is on," which is the same as the statement describing the switch circuit in Fig. 6.11.

TABLE 6.2 Parallel controlling a lamp.

A	B	Lamp
Off	Off	Off
Off	On	On
On	Off	On
On	On	On

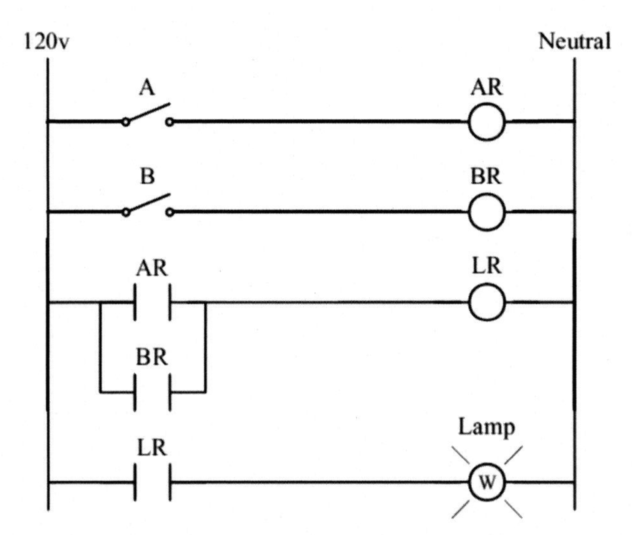

FIGURE 6.10 PLC ladder logic. *PLC*, Programmable logic controller.

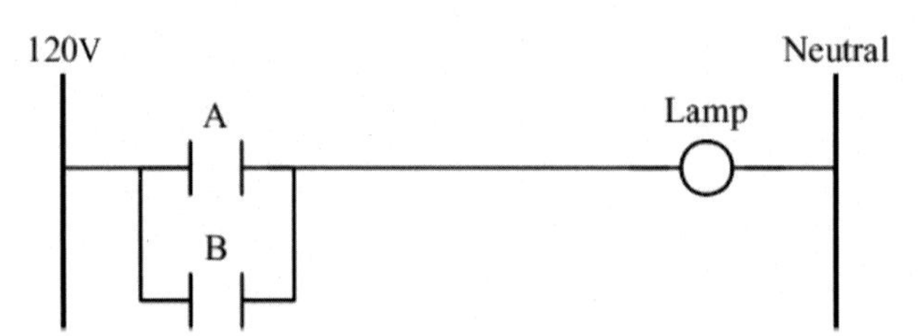

FIGURE 6.11 Parallel control ladder logic.

Example 2: Two switches controlling a lamp are labeled as A and B which are wired in series as shown in Fig. 6.12. Implement this function as PLC ladder logic with the two separate inputs from each of the switches.

The switch circuit action is defined as "The lamp is on when switch A is on and switch B is on." The truth Table 6.3 shows all possible combinations of the two switches and the consequent lamp operation. The only modification from previous example to wire the normally open contacts of CRs AR and BR in series to control the light is to enforce this feature using relays (Fig. 6.13). The cabling of both the A and B switches and the lamp does not shift (Fig. 6.14).

Example 3: Consider the implementations of a logical NOT function. Suppose a lamp needs to be turned off when switch A is off (open) and

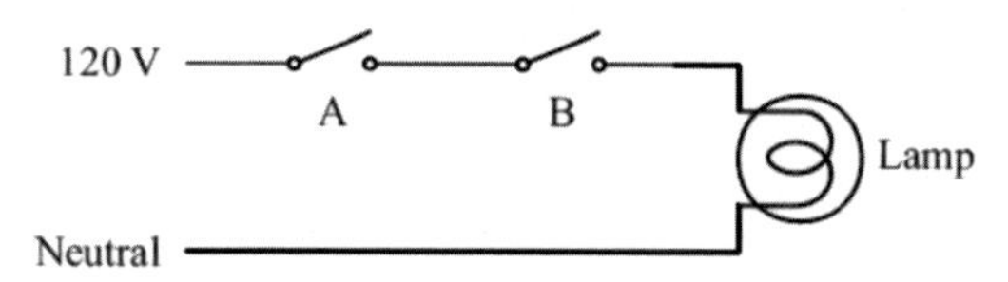

FIGURE 6.12 Series controlling a lamp.

TABLE 6.3 Series controlling a lamp.

A	B	Lamp
Off	Off	Off
Off	On	Off
On	Off	Off
On	On	On

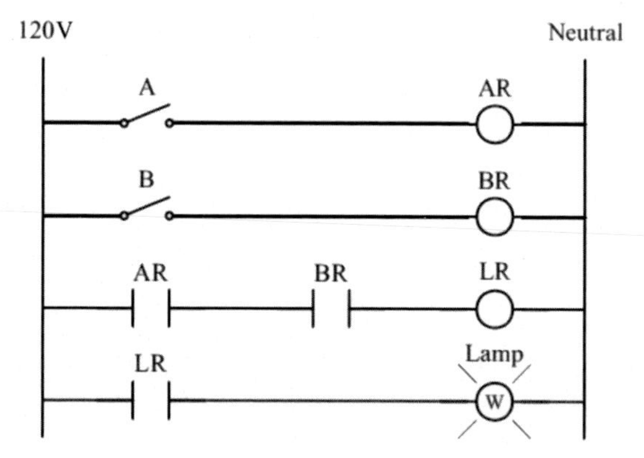

FIGURE 6.13 Series control ladder logic.

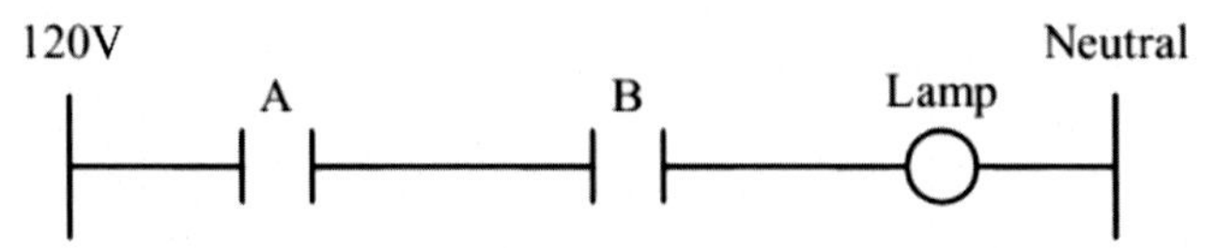

FIGURE 6.14 PLC ladder logic. *PLC*, Programmable logic controller.

TABLE 6.4 Logical NOT function.

A	B	Lamp
Off	Off	Off
Off	On	Off
On	Off	On
On	On	Off

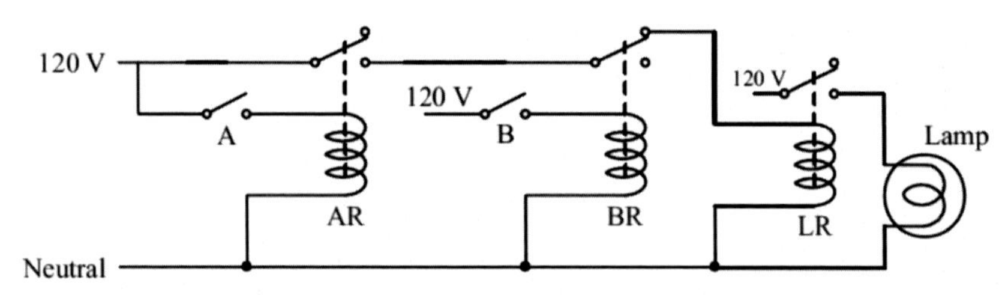

FIGURE 6.15 Logical *not* implementation.

switch B is on (closed). Implement this function as PLC ladder logic with the two separate inputs from each of the switches.

Table 6.4.shows the truth table, relay implementation and ladder logic for this example. The only difference between the relay implementation in AND and OR is the wiring of the relay BR contacts. The logical NOT for switch B is accomplished with the NC contact of relay BR (Fig. 6.15). Only the second contact symbol is different from above in the PLC ladder logic. The PLC ladder logic is inferred as "When switch (input) A is on (closed) and B is off (open) then the lamp is on." This specific example is impractical to implement with a combination of only two normally open (NO) switches and no relays (Fig. 6.16).

Example 4: Draw a ladder diagram that will cause the output, pilot light PL2, to be on when selector switch SS2 is closed, push-button PB4 is closed, and limit switch LS3 is open.

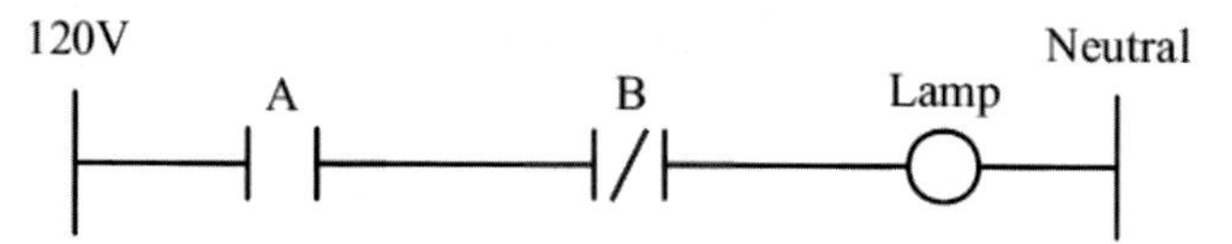

FIGURE 6.16 PLC ladder logic. *PLC*, Programmable logic controller.

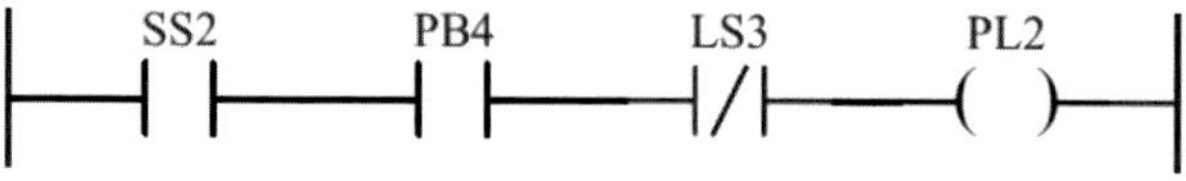

FIGURE 6.17 Ladder logic.

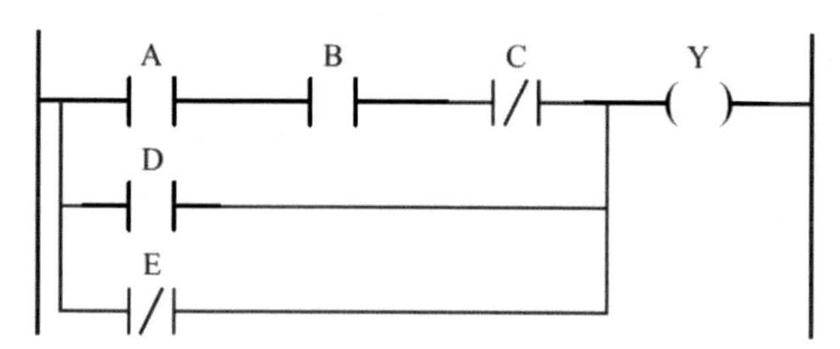

FIGURE 6.18 Ladder logic.

Since the output is PL2, the corresponding coil is put on the right-side of the rung. Secondly, the type of connection of contacts to be used is considered. To turn on the pilot light all the three switches must be connected in a certain position serially. Thirdly, the type of contact is determined by the switch position to turn on the pilot light (Fig. 6.17).

Example 5: Draw a ladder diagram which is the same as the following descriptions.

Y is on when (A is on and B is on and C is off) or D is on or E is off (Fig. 6.18).

6.2.7 Normally open contact of programmable logic controller

The PLC representation of NO contact is given in Fig. 6.19. This contact, at the specified bit address, scans for the signal state ON (1). If the scanned bit address has a signal state ON (1) then the power flows through NO contact. This contact is used for scanning the signal state of output or input devices or other internal program elements.

The PLC representation of NC contact is given in Fig. 6.20. This contact, at the specified bit address, scans for the signal state OFF (0). If the scanned bit address has a signal state OFF (0) then the power flows through NC contact. This contact is used for scanning the signal state of output or input

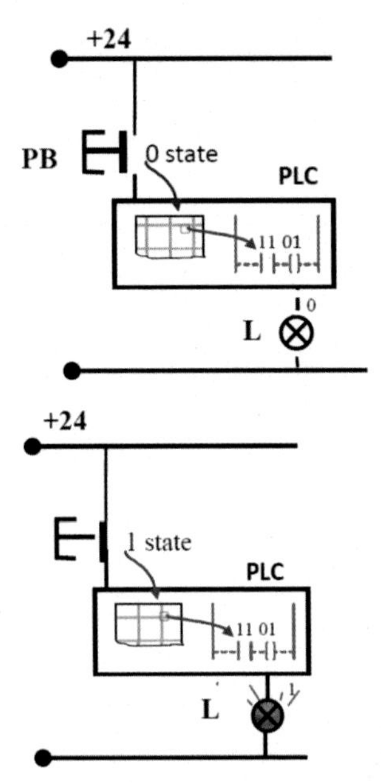

FIGURE 6.19 PLC with NO contact position using NO push-button: (A) open and (B) close. *NO*, Normally open; *PLC*, programmable logic controller.

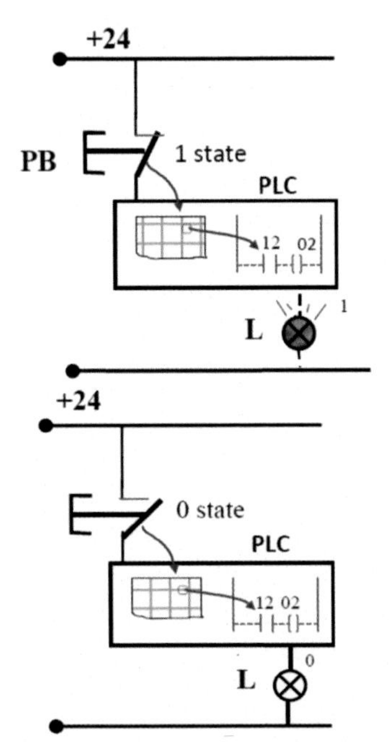

FIGURE 6.20 PLC with NO contact position using NC push-button: (A) close and (B) open. *NC*, Normally closed; *NO*, normally open; *PLC*, programmable logic controller.

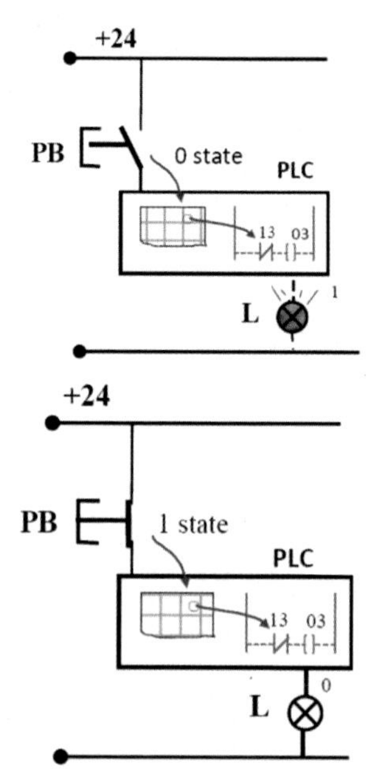

FIGURE 6.21 PLC with NC contact position using NO push-button: (A) released and (B) pressed. *NC*, Normally closed; *NO*, normally open; *PLC*, programmable logic controller.

devices or other internal program elements. Fig. 6.21 shows PLC circuit with NC contact position using NO push-button. Fig. 6.22 shows PLC circuit with NC contact position using NC push-button.

6.2.8 Programmable logic controller timers

Many control tasks require the programming of time. For example, if the cylinder 1 is retracted, then the cylinder 2 will be extended after a delay of few seconds. The timers of a PLC are reorganized in the form of software modules and are based on the generation of digital timing. To store the values of the delay time memory space is allocated in system memory. The representation of the timer address varies from manufacturer to manufacturer. For easier understanding we can represent the timer addresses as T1, T2. The typical numbers of timers available in commercial PLC are 64, 128, 256, 512, or even more. To explicitly reset timer, a result of logic operation (RLO) of 1 has to be applied at the reset port.

There are two types of PLC timer

PLC on-delay timer: The timer will be ON when it receives a start input signal and when the preset timing is achieved, the output signal status switches from 1 to 0. The signal state of the output shifts from 0 to 1 when the predetermined time is reached by changing the RLO from 0 to 1 (ON) at the starting input. Fig. 6.23 displays the operational diagram.

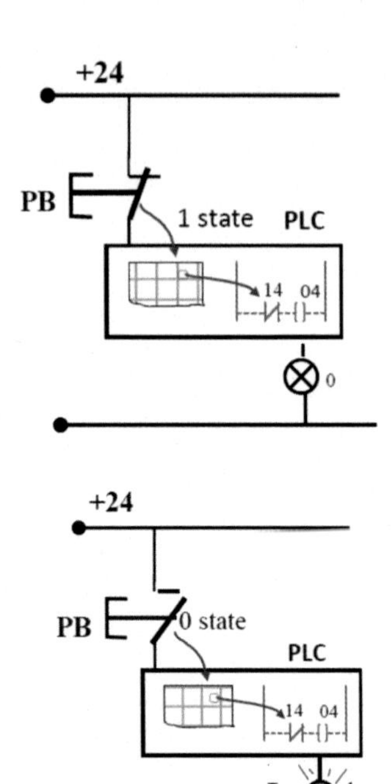

FIGURE 6.22 PLC with NC contact position using NC push-button: (A) released and (B) pressed. *NC*, Normally closed; *PLC*, programmable logic controller.

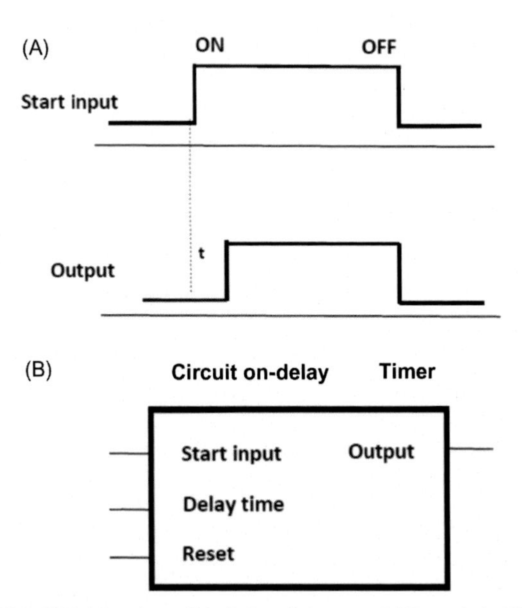

FIGURE 6.23 PLC OFF-delay timer (A) timing diagram and (B) symbol. *PLC*, Programmable logic controller.

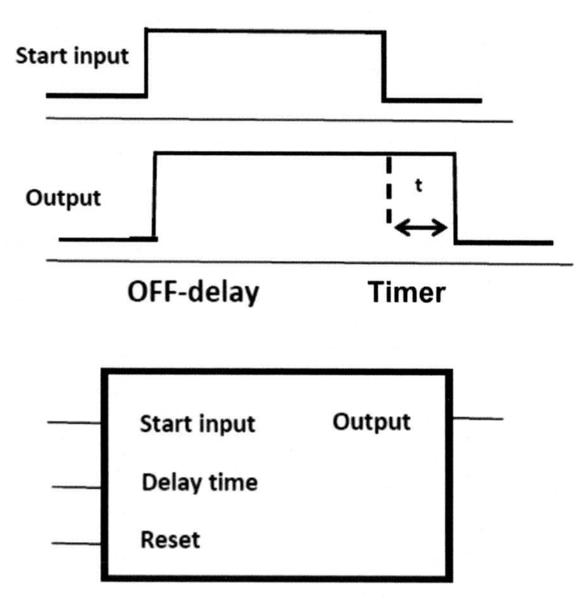

FIGURE 6.24 PLC OFF-delay timer (A) timing diagram and (B) symbol. *PLC*, Programmable logic controller.

PLC off-delay timer: The timer will be ON when it detects a start input signal and once the preset timing is achieved, the signal output state switches to 1 from 0. The output signal state shifts from 1 to 0 when the preset time is achieved with respect to RLO shift from 1 to 0 (OFF) at starting input. Fig. 6.24 displays the functional diagram.

The pieces of numbers and events are detected by use of counters. Controllers recurrently have to operate with counters in practice. For example: a counter in circuit is required if exactly 20 identical components are to be conveyed to a conveyor belt via a sorting device.

In general, there are two type of counters: (1) up counter and (2) down counter.

When the input to up counter goes true the accumulator value will be increased by 1 (despite of how long the input state is true). The counter bit will be set once the accumulator value reaches the preset value. A down counter will decrease the accumulator value until the preset vaule is reached. Symbols are shown in Fig. 6.25.

6.2.9 Programmable logic controller memory elements

They use memory elements to store intermediary values. Memory feature is accomplished using flags and device memory (bit memory locations). You

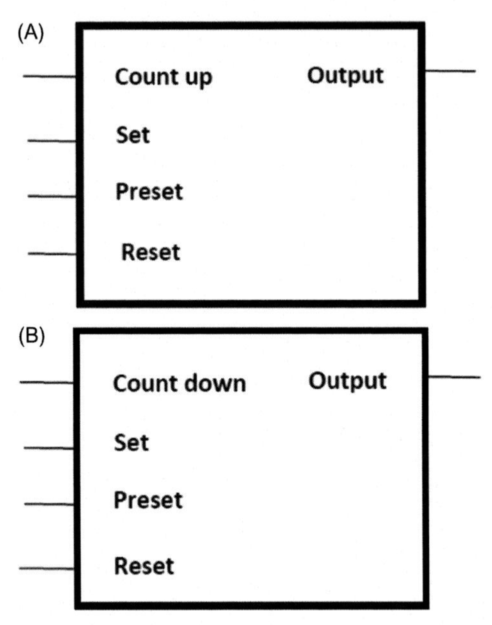

FIGURE 6.25 (A) Up counter and (B) down counter.

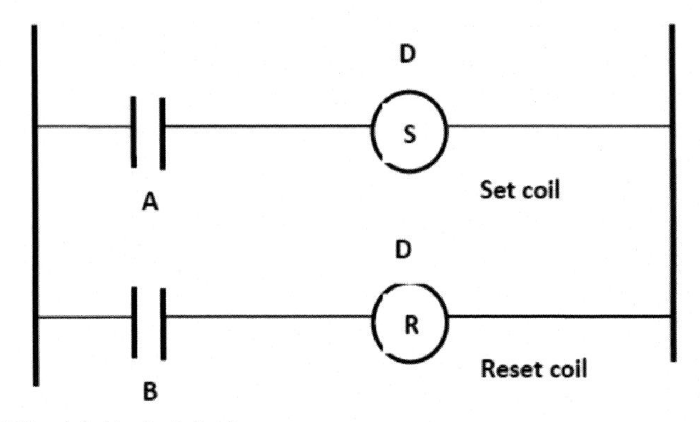

FIGURE 6.26 A ladder logic latch.

can customize or reset the required bit memory using a set coil. As seen in Fig. 6.26, a lock in ladder logic uses one lock instruction, and a second unlatch instruction.

Both set coil and reset coil can be combined in one box as shown in Fig. 6.27. The specified memory address is set to signal state 1 if the power to the set coil flows either momentarily or continuously. The memory address is reset to signal state 0 if power flows momentarily or continuously to the reset coil to the particular memory address. The memory address remains unaffected if there is no power in the set or reset input. Either NO or NC program element the output of the memory function can be accessed through.

Set and reset functions are combined in one memory box as shown in Fig. 6.27. They can be further classified as

1. Memory box with set priority
2. Memory box with reset priority

The memory box functions analogous to the memory coils. In the memory box with set priority, the associated memory address is set when signal state 1 appears simultaneously at both the set and reset inputs. At both the reset and set inputs of the memory box with reset priority, the related memory address is reset when signal state 1 appears concurrently.

6.2.10 Simple pneumatic examples

Example 1: The machine operations are performed using double-acting cylinder. By pressing two push buttons concurrently the pneumatic cylinder is advanced. If any one of the push-button is released, cylinder comes back to start position. Draw the pneumatic circuit, ladder diagram and PLC wiring diagram to implement the above task.

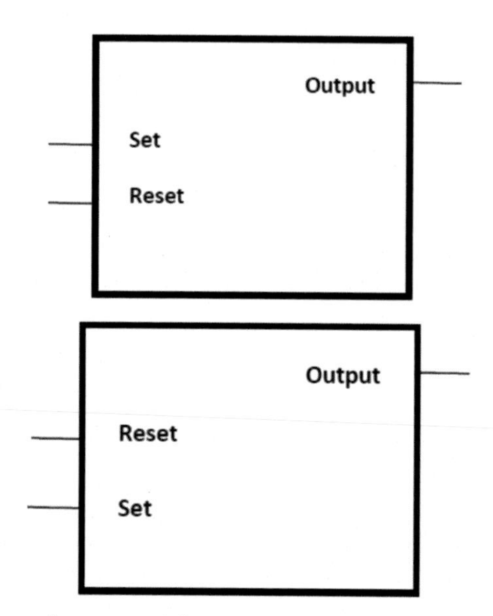

FIGURE 6.27 Memory box representation.

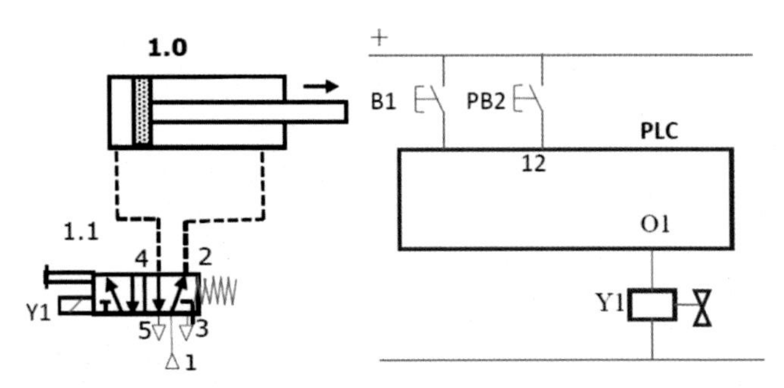

(A) Pneumatic diagram. (B) Wiring diagram.

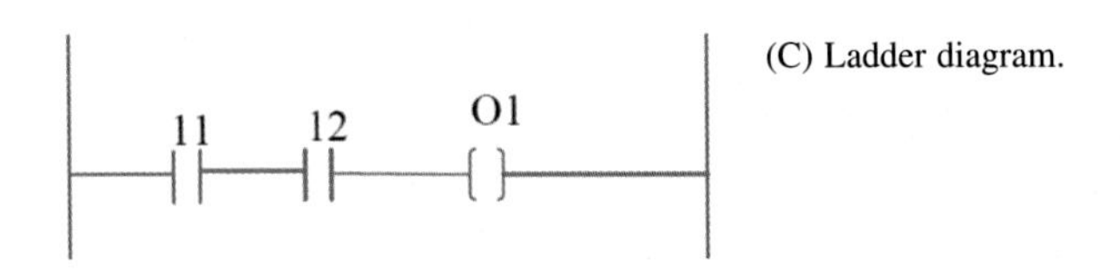

(C) Ladder diagram.

The PB1 and PB2 push buttons are attached at memory address I1 and I2 as shown in the PLC wiring diagram. In ladder diagram, I1 and I2 are connected in series to unlock this AND logic work. When the push buttons PB1 and PB2 are pressed concurrently, the I1 and I2 addresses switch to state 1 from state 0, as a result of which the power passes through the coil and the output will be at coil 01. Output at coil 01 worked the solenoid coil and cylinder goes forward to do the operation necessary.

Example 2: The forward and return motions are performed by using double-acting cylinder. By pressing push buttons PB1 the corresponding pneumatic cylinder is advanced. By pressing the push-button PB2 the cylinder is returned. Draw the pneumatic circuit, ladder diagram and PLC wiring diagram to implement the above task.

PLC wiring and ladder diagrams are shown in the figure below. The address I1 turns to 1 and thus there will be output 01 once the PB PB1 is in pressed state. The output of 01 operates the solenoid Y1 and cylinder moves forward, when the cylinder reaches the extreme forward position, and Push-button PB2 is operated, the state of address I2 turns to 1 and thus there will be output 02. The cylinder will return back to its initial position when the output of 02 operates the solenoid Y2.

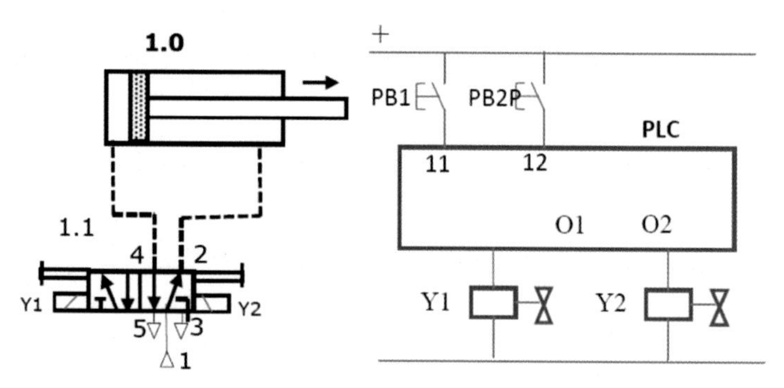

(A) Pneumatic diagram. (B) Wiring diagram.

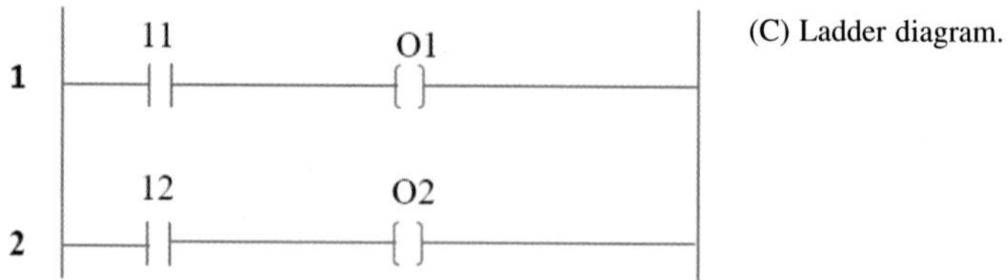

(C) Ladder diagram.

Example 3: The forward and return motions after reaching the extreme forward position is automated by use of a double-acting cylinder. By pressing the push buttons PB1, the pneumatic cylinder is advanced. Draw the pneumatic circuit, ladder diagram and PLC wiring diagram to implement the above task.

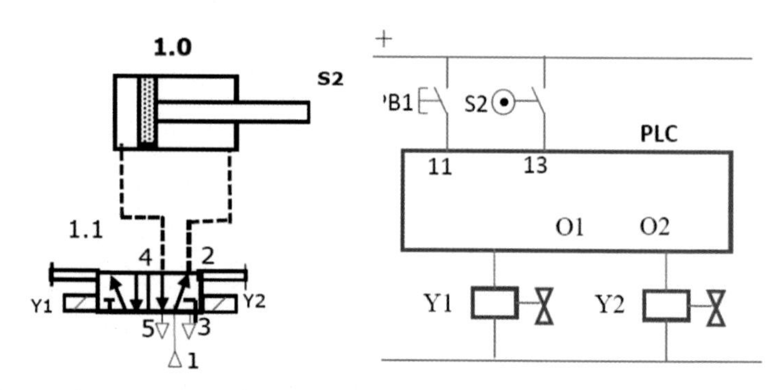

(A) Pneumatic diagram. (B) Wiring diagram.

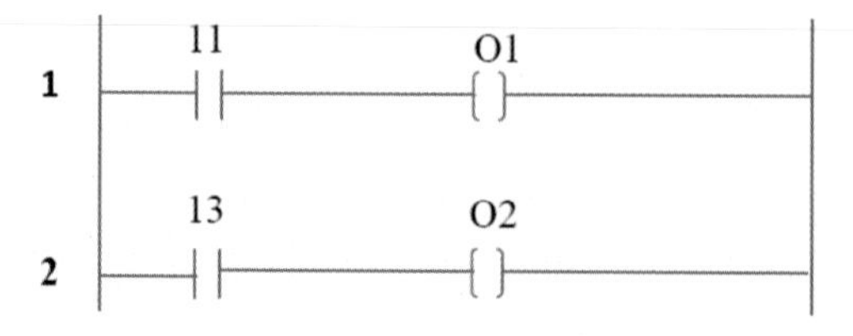

(C) Ladder diagram.

PLC wiring and ladder diagrams are shown in the figure below. There will be output 01 once the push-button PB1 is in pressed state of the address I1 turns to 1. The output of 01 operates the solenoid Y1 and cylinder moves forward, when the cylinder reaches the extreme forward position, and limit switch S2 is operated, the state of address I3 turns to 1 and thus there will be output 02. The cylinder return back to its initial position when the output of 02 operates the solenoid Y2.

Example 4: The continuous to-and-fro motion is performed by using a double-acting cylinder. When PB1 button is pressed, the cylinder has to move forward and once the to-and-fro reciprocation starts it should continue until the stop button PB2 is pressed. Limit switches are used for end-position sensing. Draw the pneumatic circuit, ladder diagram, and PLC wiring diagram to implement the earlier task.

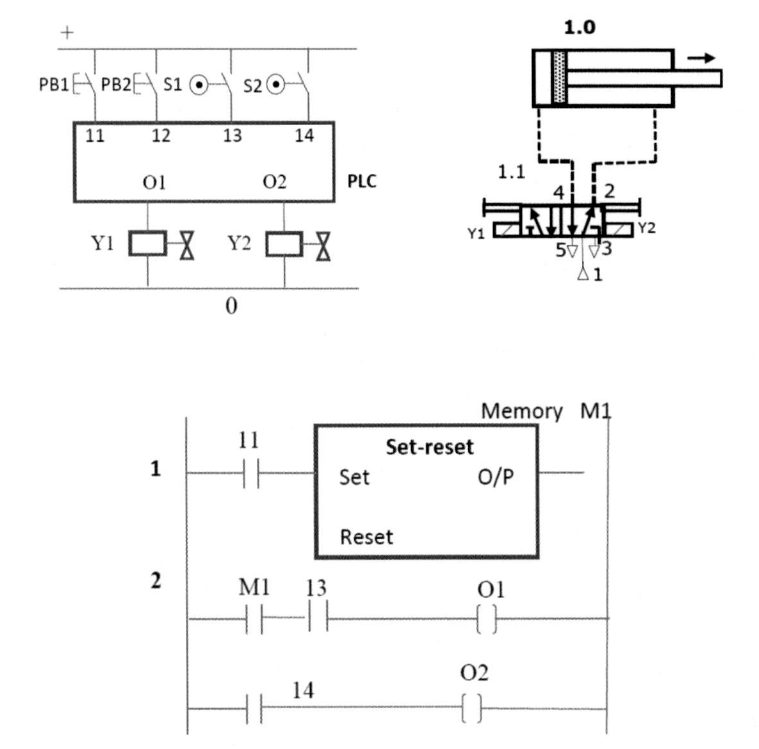

Using memory flag with address M1 that is set by PB1 and reset by PB2, the start and stop operations can be implemented. The state of the memory element M1 is scanned through an NO contact and to get start and stop controls it is further combined in series with the state of sensor S1.

Example 6: Double-acting cylinder is used to perform to-and-fro operation. The cylinder has to move forward when PB1 button is pressed and continue to-and-fro motion until 10 cycles of operations is performed. Draw the pneumatic circuit, PLC wiring diagram, and ladder diagram to implement this task.

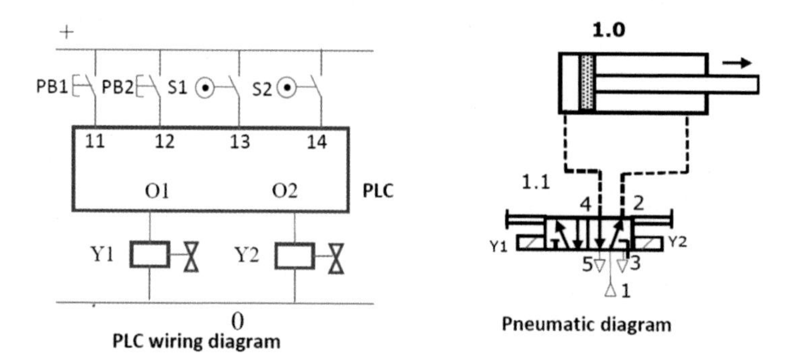

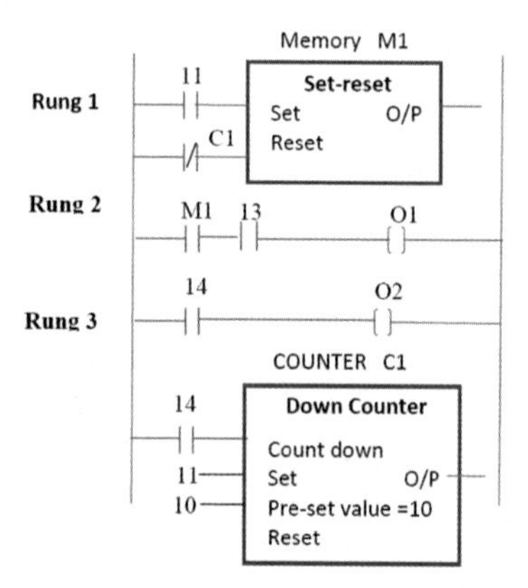

The complete auto process of cylinder could be got as former by means of limiting switches S1 and S2. start off and stopping action can be done by applying memory flag with tag of M1 that is set by PB1 at I1 and retune by NC contact of a down counter. The status of memory flag M1 which scan by an NO contact (rung 2) is pooled in succession with the state sensor S1 to obtain start and stop actions.

Case 7 illustrate the pneumatic circuit, PLC single line drawing and ladder chart to apply A + B + B − A − series.

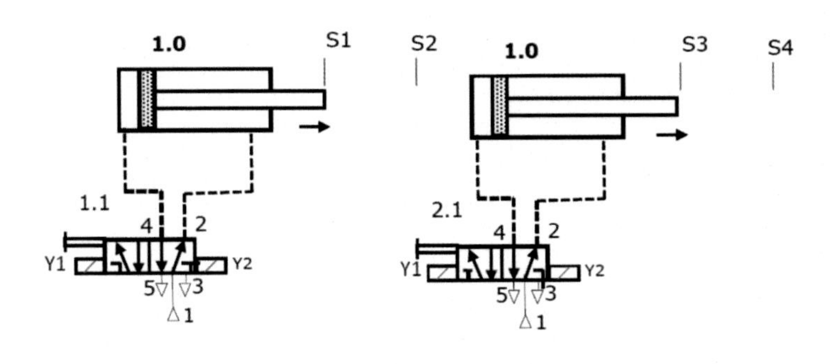

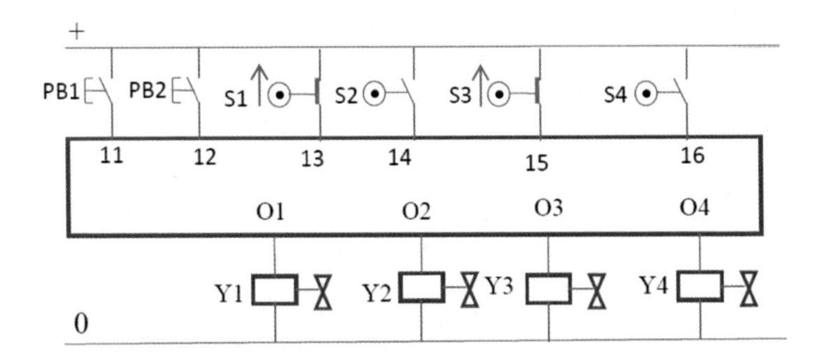

In this series circuit, PB2 is applied to start the program. Switching the PB2 makes the final memory condition M4 to position and every memory flags M1, M2, and M3 to reset. To begin, S1 and S3 are switched on to give resultants.

State 1: Switching PB1 set memory flag M1 and M4. Solenoid Y1 is gets energy. Cylinder A extend (A +). Sensor S1 is comes back to original condition once A moves and S2 is energizes when last part location is attained.
State 2: While S2 is in action, memory M2 acts and memory flag M1 is retuned. Solenoid Y3 gets power. Cylinder B extend (B +). Sensor S3 comes back to original condition once A moves and S2 is energized when last part location is attained.
State 3: While S4 is in action, memory M3 acts and memory flag M2 is retuned. Solenoid Y4 gets power. Cylinder B comes back (B +). Sensor S4 comes back to original condition once B moves and S3 is energized when last part location is attained.
State 4: While S3 is in action, memory M4 acts and memory flag M3 is retuned. Solenoid Y2 gets power. Cylinder A comes back (A −). Sensor

S2 comes back to original condition once B moves and S1 is energized when last part location is attained

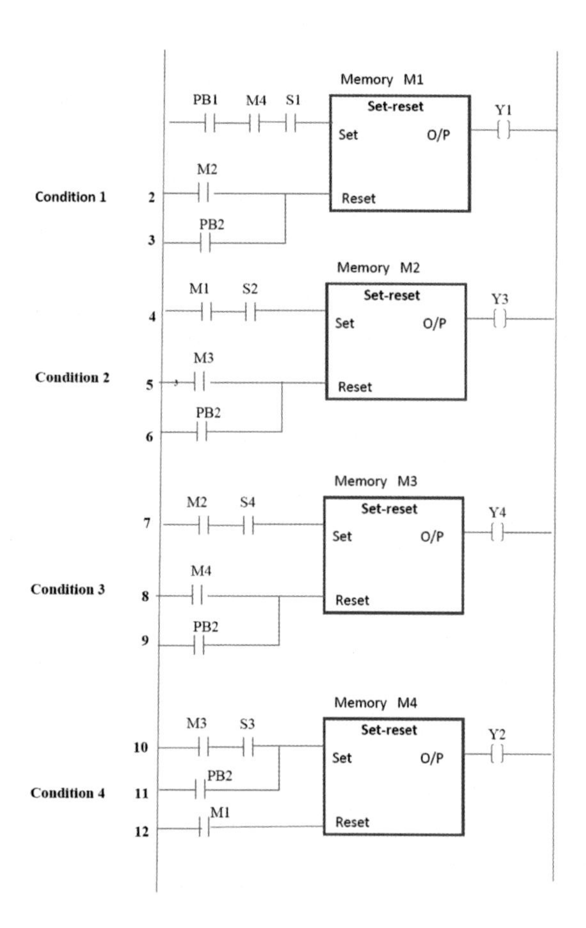

6.2.11 Areas of application of a programmable logic controller

Each scheme or appliance has a controller based on the kind of know-how used, controller is having the types like pneumatic, hydraulic, electrical, and electronic controllers. Often it is applied by the blend of diverse technologies. The demarcation is prepared involving a direct-wiring run program and program-run logic controller. The first variety is applied where any reprogramming by the consumer depends on the job size requirements and the improvement of a special controller. The characteristic purpose for these controllers can be established in automatic washing machines, video cameras, and automobiles.

Controllers used in the PLC is capable of modifying the program at any point of time, including an extra counter or set of timers can be easily implemented using this on-chip memory controllers unlike the conventional hard wiring method. The PLC represent a universal controller. It could be applied for diverse applications and through the program in its memory, which gives the user a way of altering, extending, and optimizing the control process.

PLC are extensively applied in industries owing to the following reasons.

- The price of PLC computerization is lower and PLC is very adaptable.
- They can be commissioned and applied with no trouble.
- Programming is fairly easy. Ladder encoding is supple.
- It is not hardware controlled. It could be programmed and reprogrammed to contain recurrent modifications in programs.
- Supervision of online work processes is simple, and therefore, snag rectification and repairs of PLC is simple.
- It can be subdivided as lower-priced automation of equipment.
- It could be applied in inconsiderate surroundings where humidity and temperature are soaring. Their operational ability is not influenced by vibration and shock.
- It could be applied to carry out multipart mathematical algorithms, servomotor control, stepping control, axis control, self-diagnosis, online monitoring, condition monitoring, system problem solving, and communication with other PLCs, data acquirement, networking, storage, and report production.
- It is most appropriate for cheaper automation, where recurrent modification to the control obligation would be required in their functioning life, such as in batch-type of manufacturing systems.

6.3 Electropneumatics using programmable logic controller

6.3.1 Introduction

Electropneumatics is now normally used in engineering industries and low-priced automation. They are applied all-encompassing in industries such as fabrication, health care, and chemical and promotional material systems. There are important alterations in controls systems. Relays have been arranged by the programmable logic controller for the increasing needs for much supple automation. Electropneumatic controls exist for electrical activity scheme operatives with air-powered systems which has solenoid valves as convergence of electrical and pneumatic systems. Equipments like limit switches and proximity sensors are utilized as feedback equipments.

Electropneumatic control consists of pneumatic and electrical components and its being used in most of the industrial applications. An electropneumatic control is an AC or DC electrical signal and the work is carried out by compressed air. Operational voltages is between 12 and 220 V are widely applied. The ultimate control valve is excited by solenoid propulsion. The reset of the valve is by spring (single solenoid) or applying another solenoid (double-solenoid valve). Normally the valve set/reset is activated through pilot-aided solenoid efforts to cut down the size and expenditure of the valve control condition of the electropneumatic scheme is done by applying relays and contactors or by PLC. A relay is frequently used to change signaling stimulus with detectors and switches through a number of output signals. Signal processing can be easily achieved using relay and contactor combining. A PLC can be handily utilized to get the results as per the needed logic, time hold, and consecutive activity. Eventually, the end signals are rendered to the solenoids actuating the control valves and controls motion of different cylinders. The positive side of electropneumatics is the consolidation of different kind of proximity sensors (electrical) and PLC for effectual control. The electrical signaling rate is higher, and the interval time can be decreased with the signal sent over extended lengths.

In electropneumatic controls, three essential courses include:

- Signal input equipment: The generation of items such as switches and contactors, different types of contact, and proximity sensors.
- Signal processing: Application of aggregation of contactors or applying PLCs.
- Signal outputs: Used for activation of solenoids, indicators, or alarms.

6.3.2 Seven basic electrical devices

Seven elementary electrical equipment normally used in the criterion of hydraulic power systems include (1) nonautomatic operated push buttons, (2) limit switches, (3) pressure switches, (4) solenoids, (5) relays, (6) timers, and (7) temperature switches.

Other electrohydraulic equipment include proximity sensors and electric counters.

6.3.3 Push-button switches

Push button is an input device and it acts as a switch to close or open a circuit. They are principally applied to start or hold in place a running machine. They are actuated by pushing a button into the housing. They are subdivided as (1) momentary push buttons, which come back to their normal place when they are free and (2) maintained contact or detent push buttons, which use latching mechanics to hold the button in a chosen point.

The contact of the push buttons, subdivided on its functions:

1. NO,
2. NC, and

3. changeover (CO).

The cut view of different kind of push buttons in the OFF and ON conditions and their symbolization are in Fig. 6.28. In the NO type, the contacts are open in the natural position, stopping the energy flow through them. But in the actuated position, the contacts are closed, permitting the signal flowing via the NC type, and the contacts are unopened in the regular position, allowing signal flow. Contacts are open in the ON condition, disallowing the signal flow. A changeover contact is a coalition of NO and NC contacts.

6.3.4 Limit switches

Any switch that is energized owing to the ON/OFF condition of a hydraulic control part (typically a piston rod or hydraulic drive). The energizing limit switch gives an electrical jolt that causes a suitable scheme reaction.

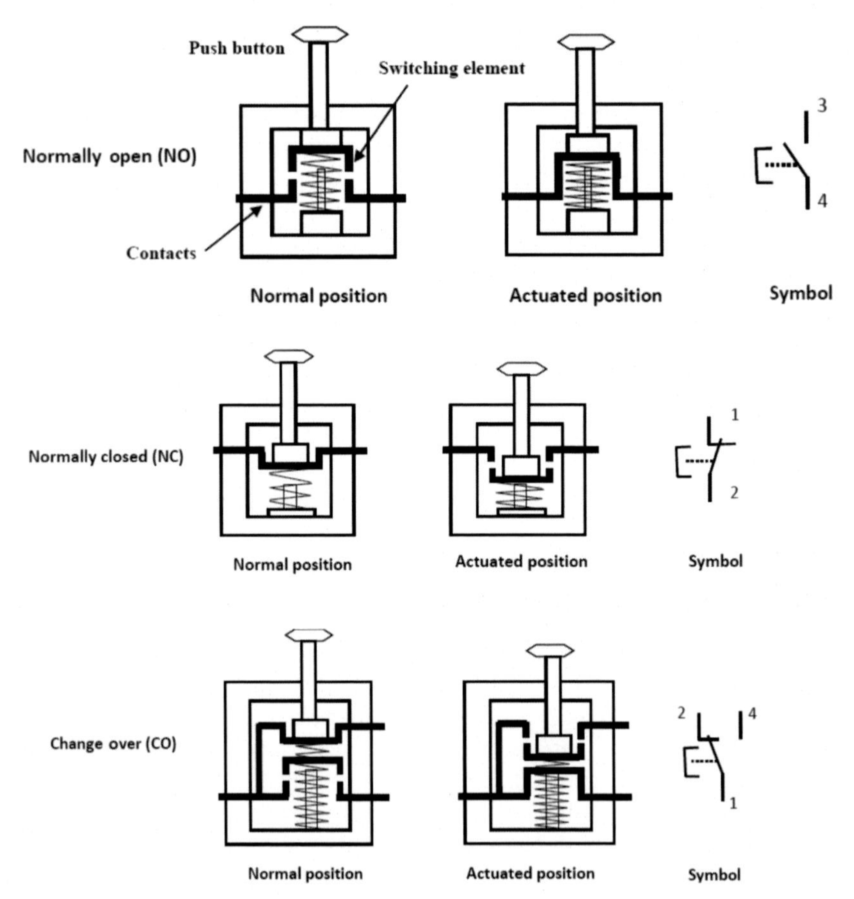

FIGURE 6.28 Push buttons and their symbols.

Limit switches carry out the similar purpose as push buttons. They are manually actuated, whereas limit switches are mechanically actuated.

There are two classifications:

1. lever-actuated contacts, and
2. spring-loaded contacts.

In lever-type limit switches, the contacts are actuated gradually. In spring-type limit switches, the contacts are energized quickly. Fig. 6.29 exhibits a simple cross-section analysis of a limit switch and its symbol.

6.3.5 Pressure switches

A pressure switch is an air-electric signal changer. Load switches are applied to interact with a variation in pressure, and it will open or close an electrical switch when a preset pressure is attained. A bellow or diaphragm is applied to interact with the alteration of pressure, and is utilized to enlarge or become small in reaction to amplify or reduce the pressure. Fig. 6.30 exhibits a diaphragm kind of pressure switch. When the pressure is functional at the inlet and when the preset pressure is reached, the diaphragm expands and pushes the spiral-loaded needle to make or break contact.

6.3.6 Solenoids

Electrical directional control valves outline the edge between the two spares of an electropneumatic control. The vital task of electrically operated directional control valve (DCVs) comprise of

1. switch supplying air on/off, and
2. extension/retraction of cylinder drives.

Electrically directional control valves are switched on with the help of solenoids and divided into

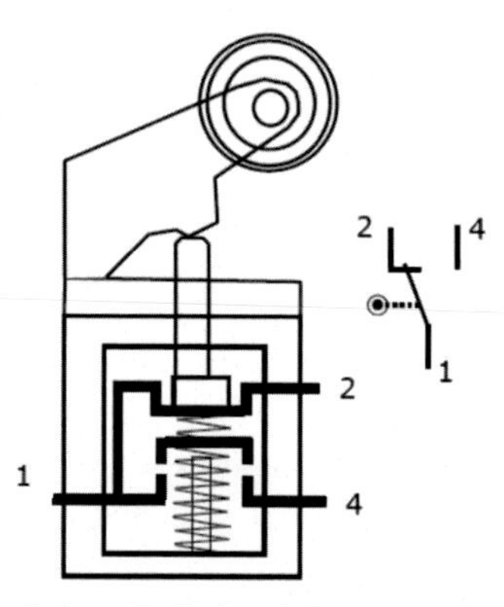

FIGURE 6.29 Cross-sectional view of a limit switch.

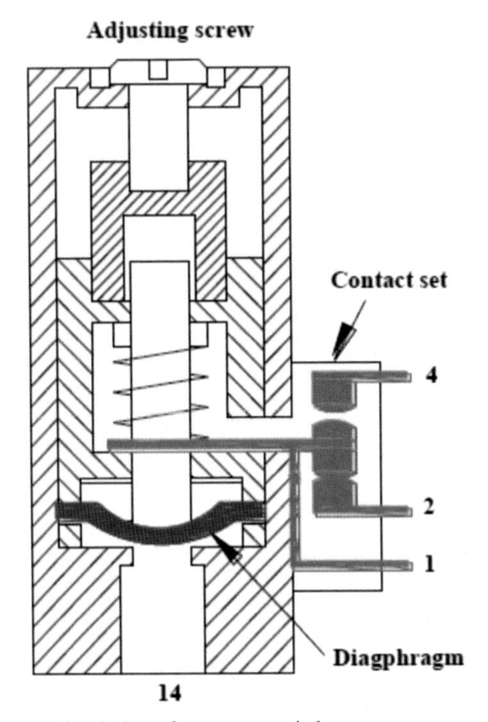

FIGURE 6.30 Cross-sectional view of a pressure switch.

1. spring comeback valves remain in the energized state until the current flow in the solenoid; and
2. double-solenoid valves keep the final switched state in spite of no current flows in the solenoid.

In the first condition, every solenoid of electric-powered DCVs has no power and the solenoids are immobile. A double valve has no position initially, as it does not have a return spring. The probable voltage for solenoids is 12 V DC, 12 V AC, 12 V 50/60 Hz, 24 V 50/60 Hz, 110/120 V 50/60 Hz, and 220/230 V 50/60 Hz.

6.3.6.1 3/2 Way single solenoid valve, spring return

The sectional analysis of 3/2 way single solenoid valve in the closed and energized position are depicted in Fig. 6.31(A). In the normal situation, path 1 is clogged up, and path 2 is coupled to 3 through the back slot (as in circle). When set voltage is given to the coil, armature is actuated en route for the midpoint of the coil and the armatures is lifted away of valve seat. The high pressure air now flow from port 1 to 2, and 3 is closed. During the voltage disconnected, the valve proceeds to the normal condition. Fig. 6.31 depicts a 2/2 solenoid operated valve.

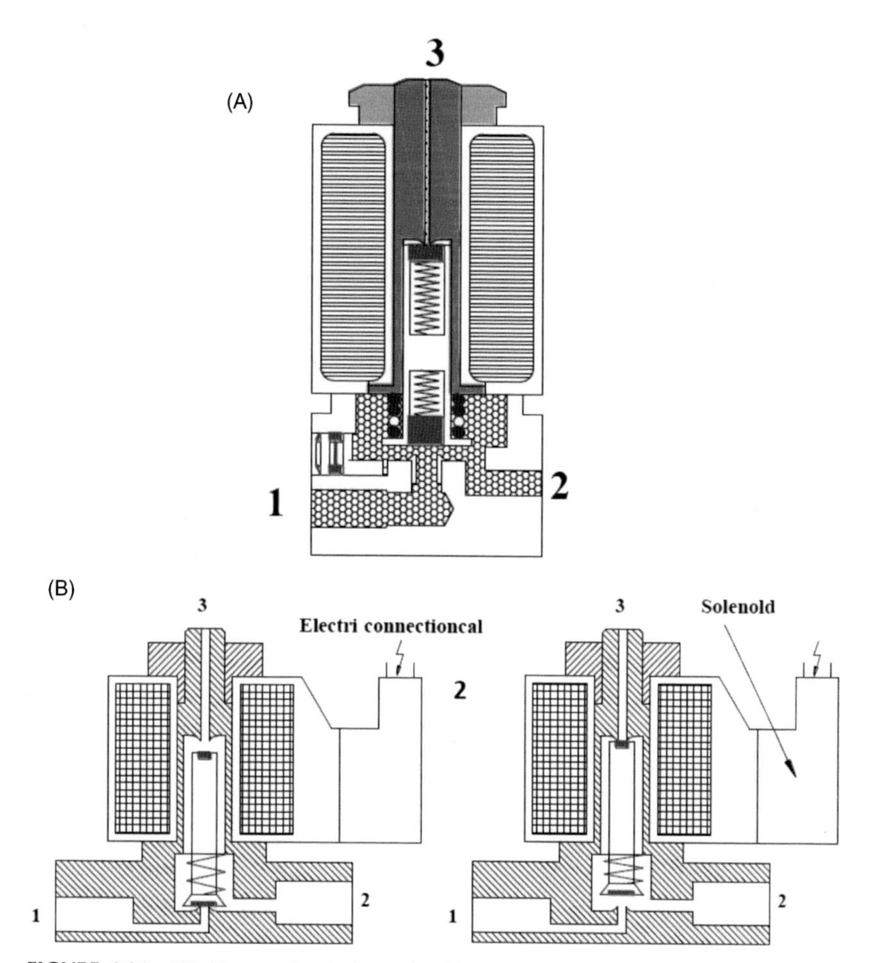

FIGURE 6.31 (A) Cross-sectional view of a 3/2 single solenoid valve. (B) Cross-sectional view of a 2/2 way solenoid operated valve.

6.3.6.2 5/2 Way individual solenoid valve, spring operated

The cross view of 5/2 way one solenoid in the regular and ON conditions are depicted in Fig. 6.32. In regular orientation, port 1 is linked to 2, and 4 is linked to 5, and 3 is closed. When the designed voltage is given to coil 14, the valve is operated by inner pilot valve. In this condition, port 1 is linked to 4, 2 is linked to 3, and 5 is closed. The valve back to the normal condition during voltage is removed. This kind of valves is commonly utilized as a last valve to control dual-activity cylinders.

The symbolization for the different solenoid/pilot operated valves Table 6.5.

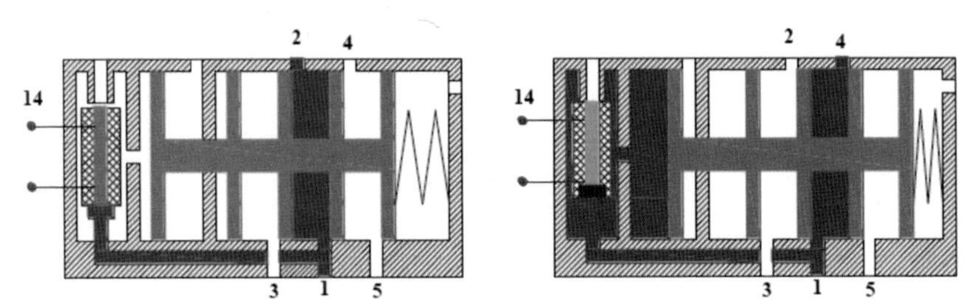

FIGURE 6.32 Sectional visual image of a 5/2 way solenoid valve.

TABLE 6.5 Symbols used for solenoid/pilot valves.

	3/2 way single solenoid valve (spring actuated)
	3/2 way pilot operated single solenoid valve (spring actuated)
	5/2 way single solenoid valve (spring actuated)
	5/2 way dual solenoid valve
	5/2 way piloted function, dual solenoid valve

6.3.7 Relays

A relay is an electromagnetic switch. It is applied for signal process. Relays are configured to resist heavy power surges and rough surroundings. When a voltage is given to the solenoid coil, an electromagnetic field forms and drives the armature to be forced to the coil core. The armature trigger the relay connection, by close or open action, based on the design. A return spring makes the armature to its first point, during the current to the coil is discontinued. Sectional perspective of a relay is depicted in Fig. 6.33.

A sizable figure of control contacts can be integrated in relays in opposideness to the instance of a push-button station. Relays are normally named as K1, K2, and K3, etc. and also have impinging capacity that is essentially a safety property in control circuits, interlocking debar coincident switching of certain coils.

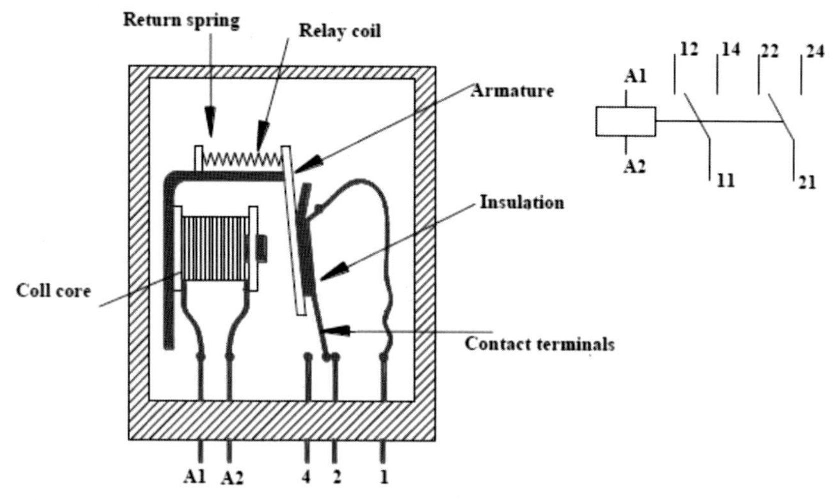

FIGURE 6.33 Cross-sectional view of a relay.

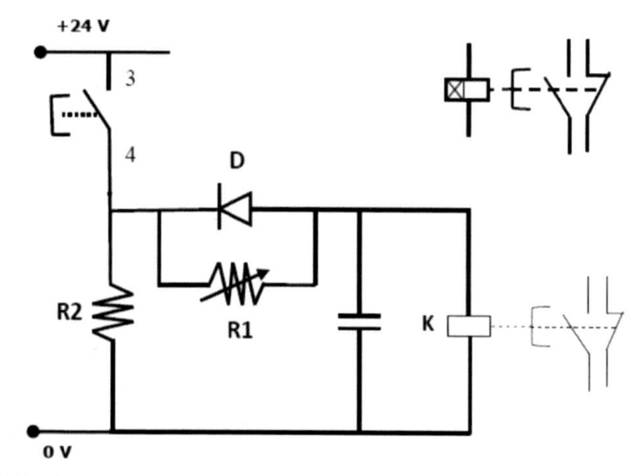

FIGURE 6.34 Structure and symbolization.

6.3.8 Timer/time delay relays

Timers are needful in control instrumentation to make the time break between work process. This is accomplishable by pausing the procedure of the connected control section with a timer. Majority of the timers we apply is electronically operated and are two variety of time relay

1. pull in delay (on-delay timer), and
2. drop −out delay (off-delay timer).

In the on-delay timer (Fig. 6.34) when push-button PB is ON, capacitor C is charged by potentiometer R1 as diode D is negative-biased. The time taken over to charge the capacitor, calculated on the potentiometer (R1) and capacitance (C). By changing the R, the requisite time delay could be obtained. While capacitor is charged fully, coil K is energized, and its contacts function, subsequently to the set time-lag. During push-button is OFF, the capacitor discharges quickly through a small resistance (R2) as the diode by passes resistor R1, and the contacts of relay (K) back to their regular position immediately.

In the off-delay timer, the contacts function immediately when the push-button is ON. The contacts coming back to the regular position after the set time-lag when the push-button is OFF. The structure and symbolization are in Fig. 6.34.

6.3.9 Temperature switch

Temperature switches in auto mode notes variations in temperature and open/close an electrical switch while a set temperature is attained. This switch may be wired either NO or NC. Temperature switches are applied to shield a liquid power system from high damages during a module like pumps/strainers/coolers, malfunctions.

6.3.10 Electronic sensors

Inductive, optical, and capacitive proximity switches are electronic sensors and usually have three electrical relays. One for line voltage, other for earth, and the lead for output signal.

In these sensors, no variable contact is switched, and as an alternative, the output is electrically coupled to supply voltage or with ground. The types of electronic sensors are with regard to the polarity of output voltage.

1. Positive switch sensor: In this output voltage is zero if no part is found in the nearness. The movement of machine element lead to switching of the output, giving out the voltage.
2. Negative switch sensors: In this output voltage is zero if no part is found in the nearness. The movement of machine element lead to switching of the output, giving out the 0 voltage.

6.3.10.1 Inductive sensors

Inductive sensors are used to sense the near by metal objects and it is working based on the principle of induction of magnetic field. It applies a coil or inductor to produce an elevated frequency magnetic field as depicted in Fig. 6.35. If a metal is there in the varying magnetic field, current will be produced in the object. This current sets up a fresh magnetic field which oppose the initial magnetic field. The final result is that it alters the

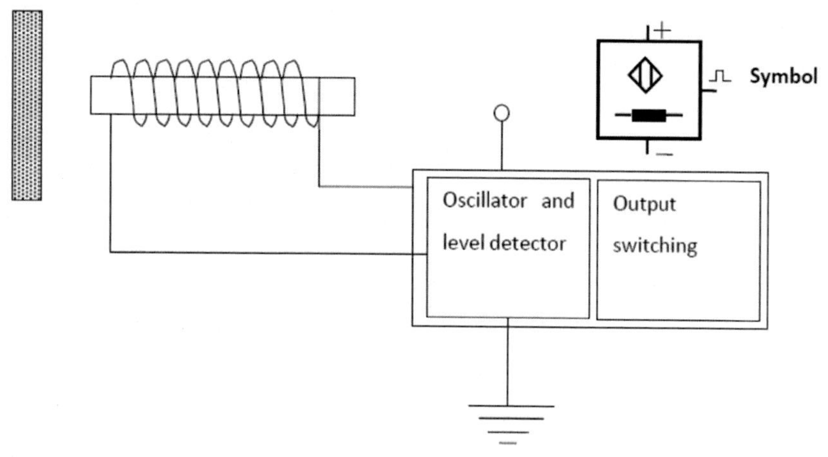

FIGURE 6.35 Inductive sensor.

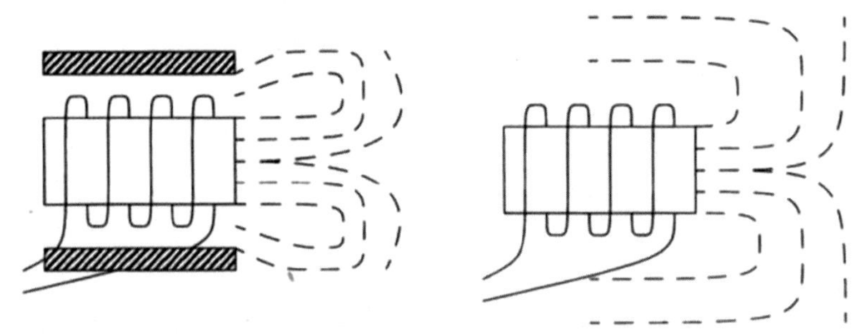

FIGURE 6.36 Shielded versus unshielded Inductive sensors.

inductance of the coil in the sensor. By quantifying the inductance, the sensor finds when a metal is near.

These will identify any metal, while finding several types of metal several sensors are applied. Further to metals, graphite is also detected. It is vital to note that it is accomplished by producing a high frequency field. If a target approach the field it will tempt eddy currents and it consumes power due resistive nature, hence energy is lost, and the signal value lowers. The detector checks the filed value, to find when it is in lower value to switch.

The sensors finds things nearer to it and the direction to the item can be random as in Fig. 6.36. The magnetic field of the sensors without shield, makes up a higher volume in the region of the head of the coil. By adding up a shield (a metal jacket inside of coil) the magnetic field is made small, and extra focused. Shields will frequently be obtainable for inductive sensors to give a good path and precision.

TABLE 6.6 Reduction factors for a range of materials.

Material	Reduction factor
Stainless steel	0.80–0.85
Nickel steel	0.70–0.90
Aluminum and brass	0.35–0.50
Copper	0.25–0.40

Advantage of proximity sensors are

1. self-contained, rugged, and extremely dependable;
2. extended service life;
3. reduced switching time; and
4. smaller and repair free.

Shortcomings of proximity sensors are

1. Similar to reed switches, they are not suitable in the conditions surrounded by magnetic fields (akin to resistive welding machine).

Uses of proximity sensors

1. Detection of end spot of linear actuators like cylinders and semi rotary actuators.
2. To find metal objects in conveyors. That is existence or nonexistence of work metal part in conveyor.
3. In press to find the ending position.
4. To find breakage of drill bit while drilling.

They are also used as feedback devices in speed measuring devices.

The sensing area of inductive sensors is related to the conductivity and permeability of metal device, and being there or not is detected. This area alters with material constituent of the indented item. The mild steel is taken for benchmark reference (Table 6.6). This is explained by the reduction factor, which is the sensing range of the inductive sensor, based on material composition, comparing with steel (FE 360) as the benchmark reference.

One more feature that influences the sensing range of inductive sensors is the diameter of sensing coil. A small sensor with a coil diameter of 18 mm has a characteristic range of 1 mm, while a high limit sensor with core diameter of 75 mm has detection limits up to 50 mm or higher.

6.3.10.2 Capacitive sensors

Capacitive sensors are able to detect most materials at distances up to a few centimeters.

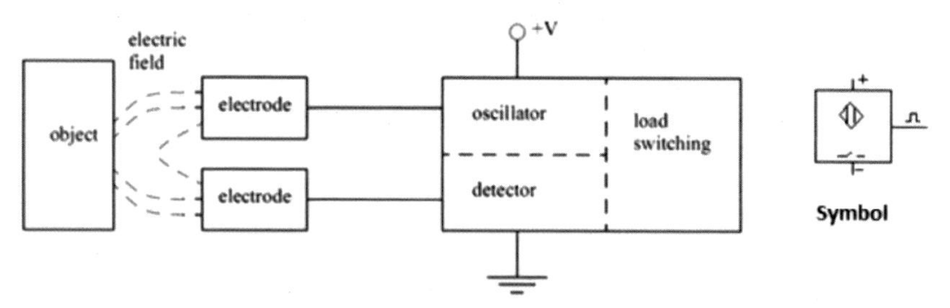

FIGURE 6.37 Capacitive sensors.

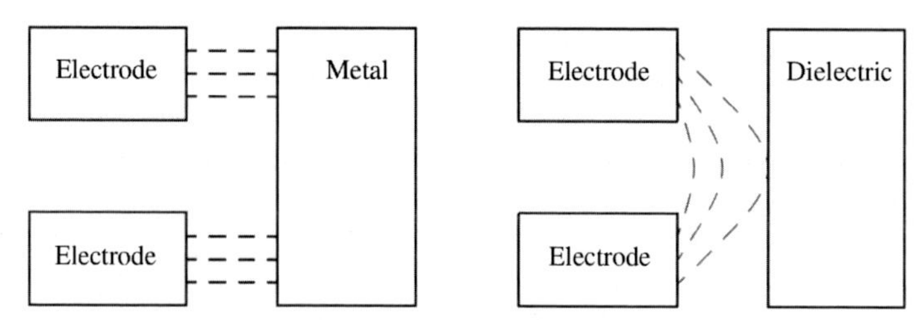

FIGURE 6.38 Capacitive sensors for metals and dielectrics.

$$\frac{\text{Areaofplates} \times \text{dielectricconstant}}{\text{Distancebetweenplates}} = \text{Capacitance}$$

The area of the plates and the distance between these are set in the sensor. But if specific material is brought near the sensor, the dielectric constant of the space around them can vary. An example of a capacitive sensor is shown in Fig. 6.37. The use of an oscillating field to determine the plates capacitance. If this changes beyond selected sensitivity the performance of the sensor is disabled.

The presence of any material near the electrodes may increase the capacitance for capacitive sensors. This will change the magnitude of the oscillating signal, and when this is large enough to determine proximity the detector will decide.

For insulators (such as plastics) that appear to have high dielectric coefficients, these sensors work well, thus increasing the power. But, they also function well enough for metals as the target's conductive materials appear as larger electrodes, thus increasing efficiency as shown in Fig. 6.38. The capacitance variations are generally in the order of pF in the whole.

Proximity sensor advantages are (1) their ability to react with wide variety of materials; (2) makes them widely used. We are likely to spot nonmetal objects; and (3) they can be used for level sensing and tracking in storage containers.

The disadvantages of the proximity sensors are (1) they are particularly susceptible to humid environment and (2) the sensor will be very sensitive to soil, oil, and other pollutants that could bind to the sensor without the compensator ring.

6.3.10.3 Optical proximity sensors

Light sensors have been used almost a century-photocells were first used on motion pictures for applications such as reading audio tracks. Yet current optical sensors are considerably more powerful. The optical sensors need a light source (emiter) as well as a detector. Emitters can use LEDs and laser diodes to create light beams in the visible and invisible spectrums. Detectors usually are built with photodiodes or phototransistors. The emitter and detector are designed to block or reflect a beam from an object while it is active. Fig. 6.39 displays a reference optical sensor.

The light beam is produced on the left in the figure, directed through a lens. On the detector side the beam with a second lens is centered on the detector. If the beam is interrupted the detector will show that there is an obstacle. The oscillating light wave is used to allow the sensor to filter out normal in-room light. At a fixed frequency the light from the emitter is switched ON and OFF. When the light is detected by the detector it scans to ensure it is at the same frequency. If light is emitted at the correct frequency then the beam will not be damaged. The oscillation frequency is within the KHz range, and is too quick to note. A side effect of the frequency approach is the possibility to use the sensors at greater distances with low power.

You can set an emitter to point directly to a detector, this is known as opposite mode. The component will be detected when the beam is split. As seen in Fig. 6.40, this sensor needs two separate parts. This design works well with opaque and transparent objects with the emitter and the detector segregated by distances of up to hundreds of feet.

Using the separate emitter and detector raises maintenance issues and requires synchronization. One alternative solution is to house the detector

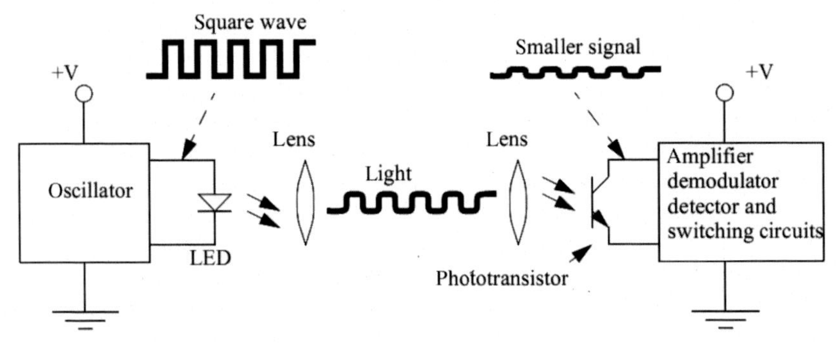

FIGURE 6.39 A basic optical sensor.

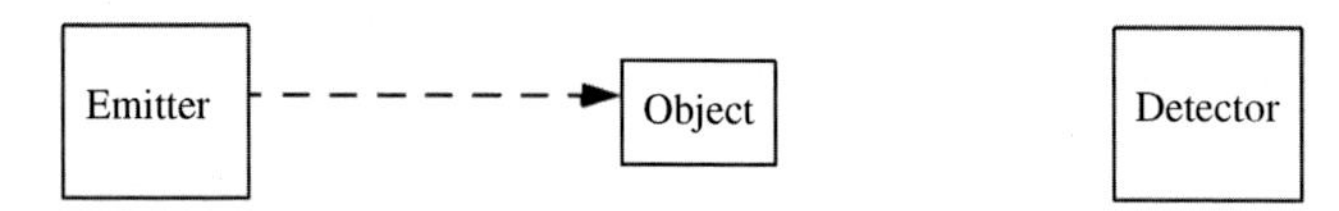

FIGURE 6.40 Opposed mode optical sensor.

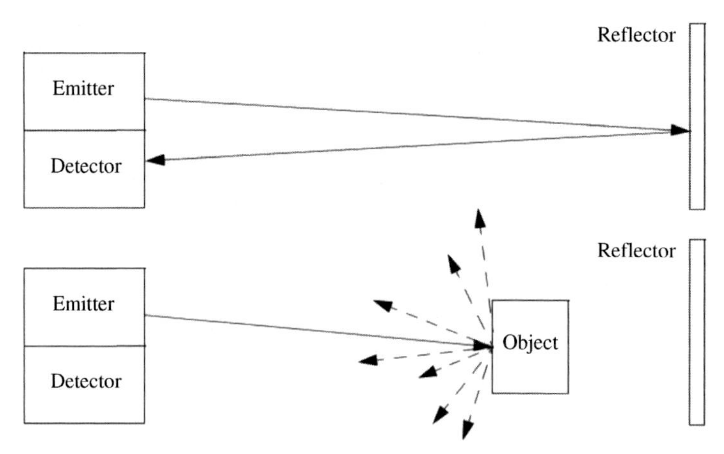

FIGURE 6.41 Emitter and detector in one unit.

and the emitter in one device. But, as seen in Fig. 6.41, this demands that light be reflecting back. Such sensors are ideal for bigger items up to a few feet away.

The reflector is built with 90° focused, polarizing screens. Unless the light is actually reflected back the light is not going through the mirror in front of detector. The reflector is equipped to rotate the light process by 90°, so it is now going through the screen in front of the detector.

The emitter sets out a beam of light in the figure. When the light returns much of the light beam is transferred to the detector from the reflector. The beam is no longer reflected back to the detector when an object blocks the beam between the emitter and the reflector, and the sensor remains operational. One possible issue with this sensor is that a good beam might return reflective artifacts. This problem is solved by polarizing the light at the emitter (with a mirror), then using the detector's polarized mirror. The reflector uses small cubic reflectors, and the polarity rotates by 90° as the light is reflected. The light will not rotate by 90° if the light is reflected off the surface. As seen in Fig. 6.42, then the polarizing filters on the emitter and detector are rotated by 90°.

The reflectors are relatively easy to align for retro reflectors but this approach still requires two installed components. A diffuse sensor is a single device not using a reflector, but using focused light as shown in Fig. 6.43.

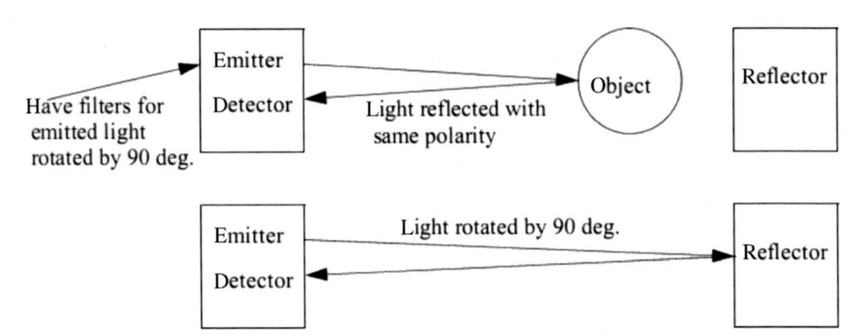

FIGURE 6.42 Polarized light in retro reflective sensors.

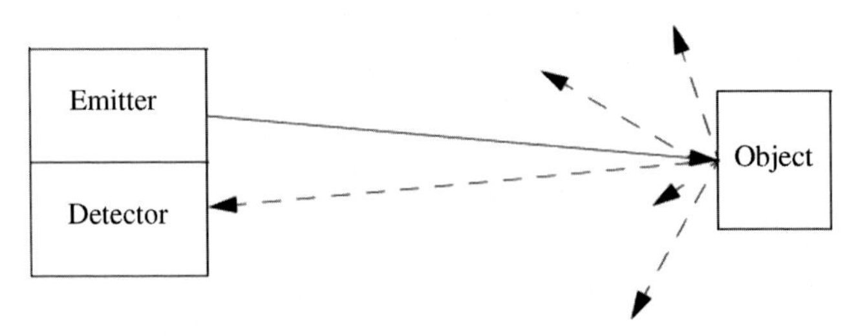

FIGURE 6.43 Diffuse sensor.

The light is dispersed with diffuse reflection. This reduces the amount of light given back. As a consequence the lenses need to amplify the sun.

6.3.10.4 *Diffuse sensors*

Diffuse sensors use light concentrated over a specified range, and for choosing a distance a sensitivity change is used. Such sensors are the most easy to mount but need well-controlled conditions. Under light and dark coloured objects diffuse sensors will have pick up issues.

The emitter and detector must be matched when using opposite-mode sensors so that the emitter beam and detector window match, as shown in Fig. 6.44. Emitter beams usually have a cone shape with a slight divergence angle (with a few degrees less). Detectors also have a detection volume molded to the tip. Therefore, when trying to align opposed mode sensor care is taken to point not only the emitter to the detector, but also the emitter detector. A thing to remember with this and other sensors is that the light intensity reduces over time, meaning that the sensors have a time limit.

If an object is smaller than the light beam width it cannot fully block the beam while it is in front, as seen in Fig. 6.45. That will generate detection difficulties, or probably fully stop detection. Using narrower beams, or wider objects, are solutions to this issue. To solve this problem, fiber optic cables

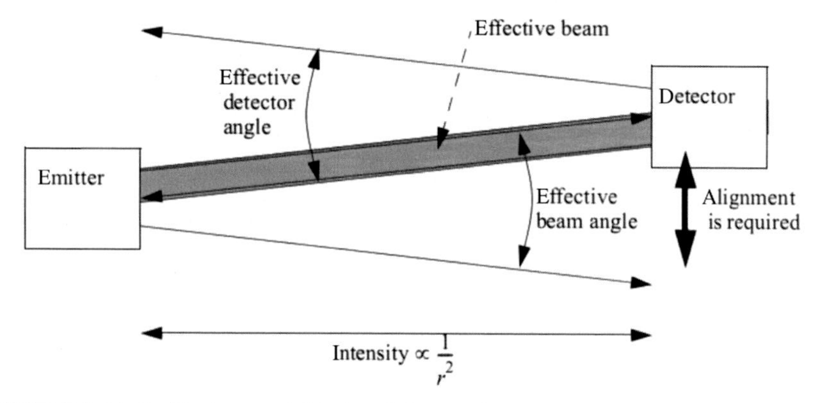

FIGURE 6.44 Emitter beam and detector window overlap.

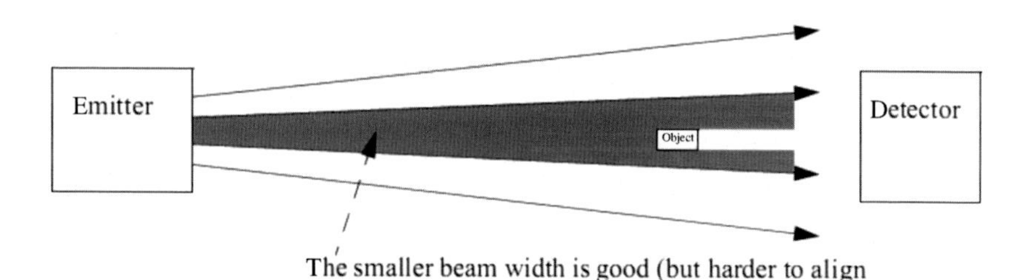

FIGURE 6.45 The relationship between beam width and object size.

can be used with an optical sensor in opposite mode, but the maximum safe distance is restricted to a few feet.

Separate sensors, as shown in Fig. 6.46, can detect reflective parts using reflex. The emitter and detector are placed so the light is returned to the detector when a reflective surface is in place. The light does not return when the surface is not present.

Certain types of optical sensors that often focus on a single point using converging beams instead of diverging (Fig. 6.47). The emitter beam is centered at a distance such that the light intensity at the focal distance is greater. The detector can look at the point from a different angle, so that the two emitter and detector center lines converge at the point of interest. If an object is present before or after the focal point the light reflected will not be detected by the detector. As shown in Fig. 6.48 this technique can also be used to detect several point ranges.

Many implementations do not require the use of full sized optic photosensors. Fiber optics can be used to keep the emitters and detectors isolated from the device. Some vendors also sell photosensors which separate the phototransistors and LEDs from the electronics.

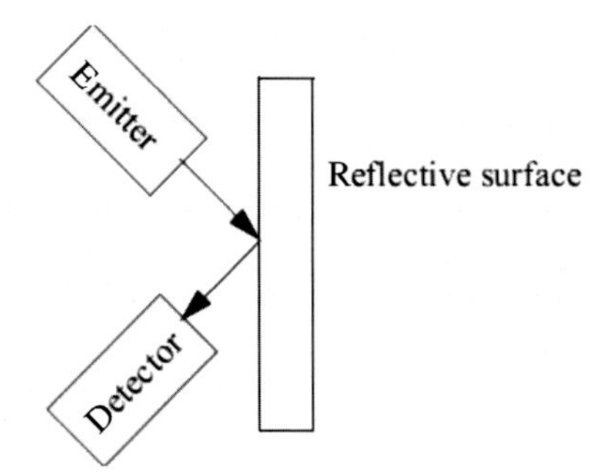

FIGURE 6.46 Separated sensors.

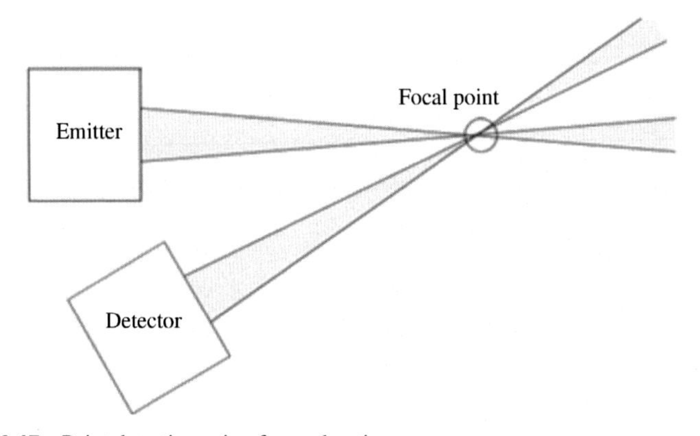

FIGURE 6.47 Point detection using focused optics.

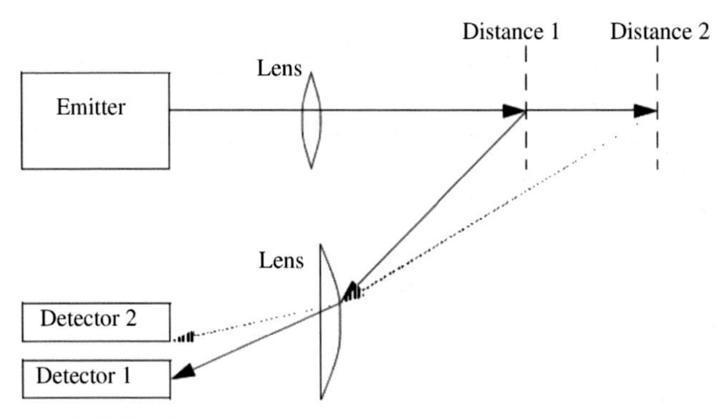

FIGURE 6.48 Multiple point detection using optics.

6.4 Electro pneumatics circuits

6.4.1 Control of system with timed response

Control systems to which a specific timing pattern is allocated must be fitted with electric time-lag relays. There are control systems that are solely influenced by time scanning or the mixture of route and time. Such time-lag relays, which are presently electronic time-lag relays, provide two specific forms of timed reaction. They are called time-lag relays with dynamic delay and deenergizing latency. Figs. 6.49 and 6.50 display time-lag relay with energizing delays and time-lag relay with deenergizing delays.

6.4.2 Control of double-acting cylinder with time delay (double-solenoid)

When pressing manual push-button PB1, relay K1 switches state and the relay usually open contact k1 is attached to solenoid coil Y1. As the usually open touch closes, the condition of the solenoid valve shifts, the cylinder moves to its final forward position where the limit switch S2 is actuated.

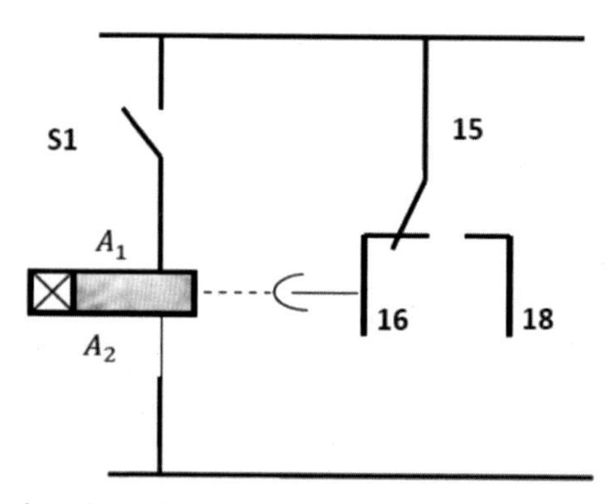

FIGURE 6.49 Time-lag relay with energizing delay.

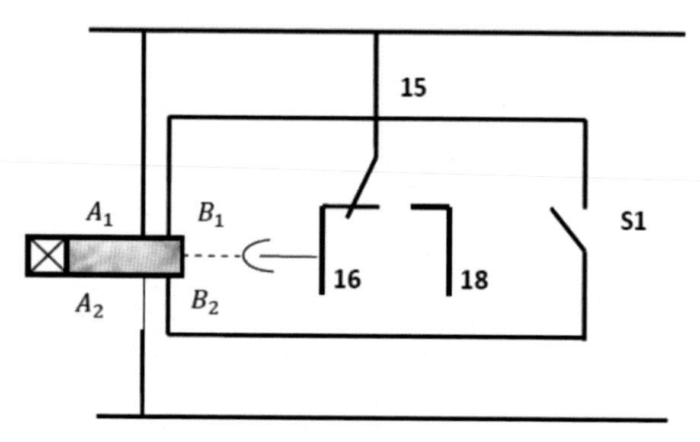

FIGURE 6.50 Time-lag relay with deenergizing delay.

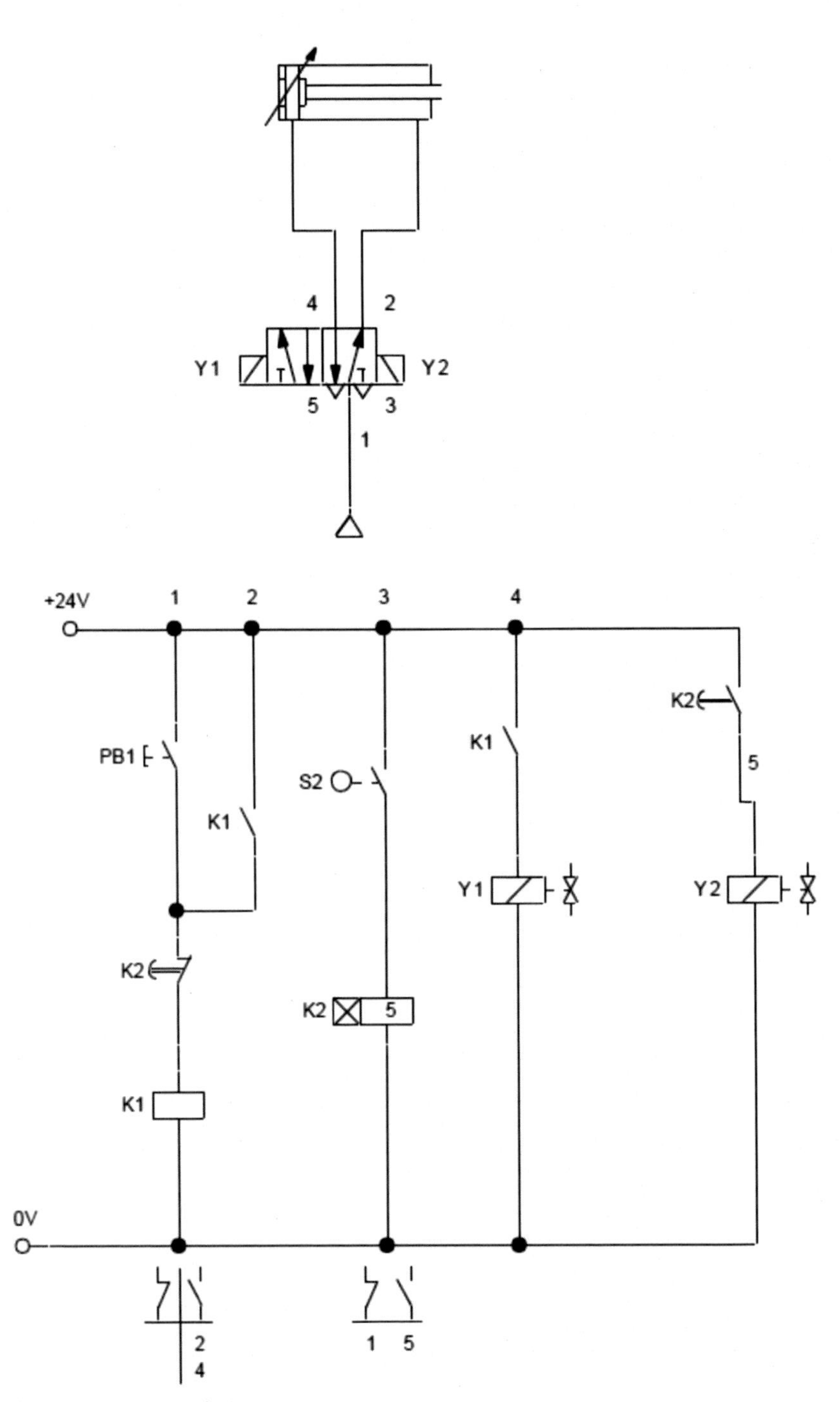

FIGURE 6.51 Control of double-acting cylinder with time delay.

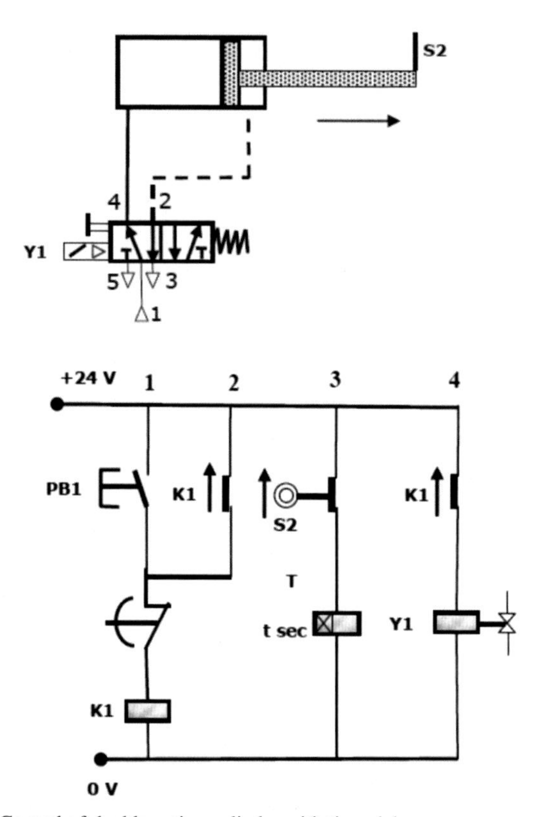

FIGURE 6.52 Control of double-acting cylinder with time delay.

This limit turn stats the time-lag relay K2 (with energy delay) After 5 seconds, the solenoid coil Y2 of the directional control valve is energized by the usually open time-lag relay touch. The valve turns over and the piston moves to its final rear position (Fig. 6.51A and B).

6.4.3 Control of double-acting cylinder using timer (single solenoid)

For the requisite memory feature, a latching circuit will be used. The circuit location is given in Fig. 6.52 when pushing button PB1 is pressed and then released. The cylinder stretches to its forward-end position and simultaneously actuates the limit turn S2. If the return motion is to be delayed, the delay timer is being used to obtain the time delay required. The time delay needed for this should be set on the timer. The Timer Coil T is operated by the S2 limit switch. The timer touch prevents the latching circuit after the specified delay thus triggering the cylinder return motion as seen in Fig. 6.52.

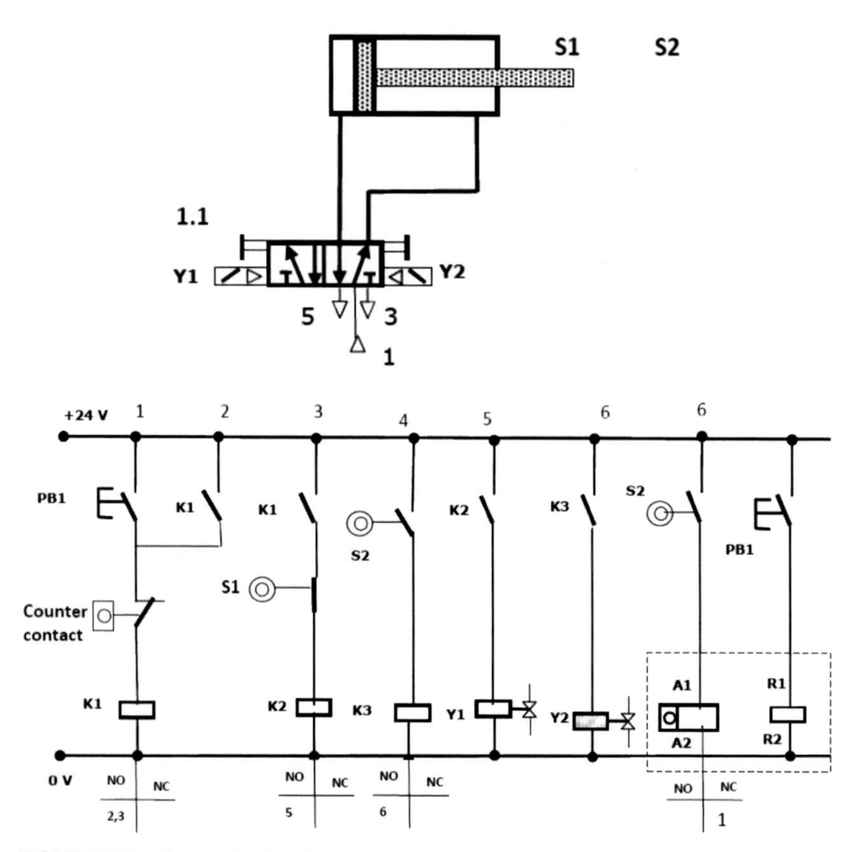

FIGURE 6.53 Control circuit using a timer.

6.4.4 Control of double-acting cylinder using electric counter with two end sensors

Pressing the PB1 push-button energizes the K1 coil in branch 1. The K1 is latched with PB1 in branch 2. Contact K1 in branch 3 energizes the coil K2 in branch 3, which in effect removes contact K2 in branch 5 allowing the solenoid coil Y1 in branch 5 to energize and shift the control valve in direction. Cylinder advances. In branch 7, when cylinder reaches the limit switch S2 it sends a signal pulse to counter coil (A1 andA2). After achieving a sufficient number of cycles (50 cycles), then counter contact C in branch 1 opens and de-energizes the stops of the K1 and the cylinder (Fig. 6.53).

6.4.5 Control of double-acting cylinder using pressure switch

Components shall be stamped with stamping tool. When two push buttons are pushed simultaneously, a double actuating lever is being used to push the

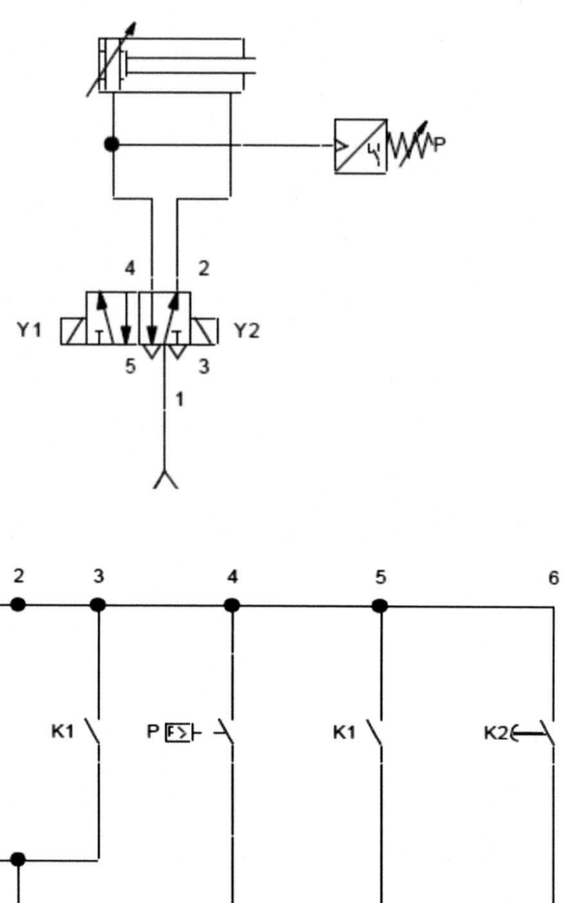

FIGURE 6.54 Control of double-acting cylinder using pressure switch.

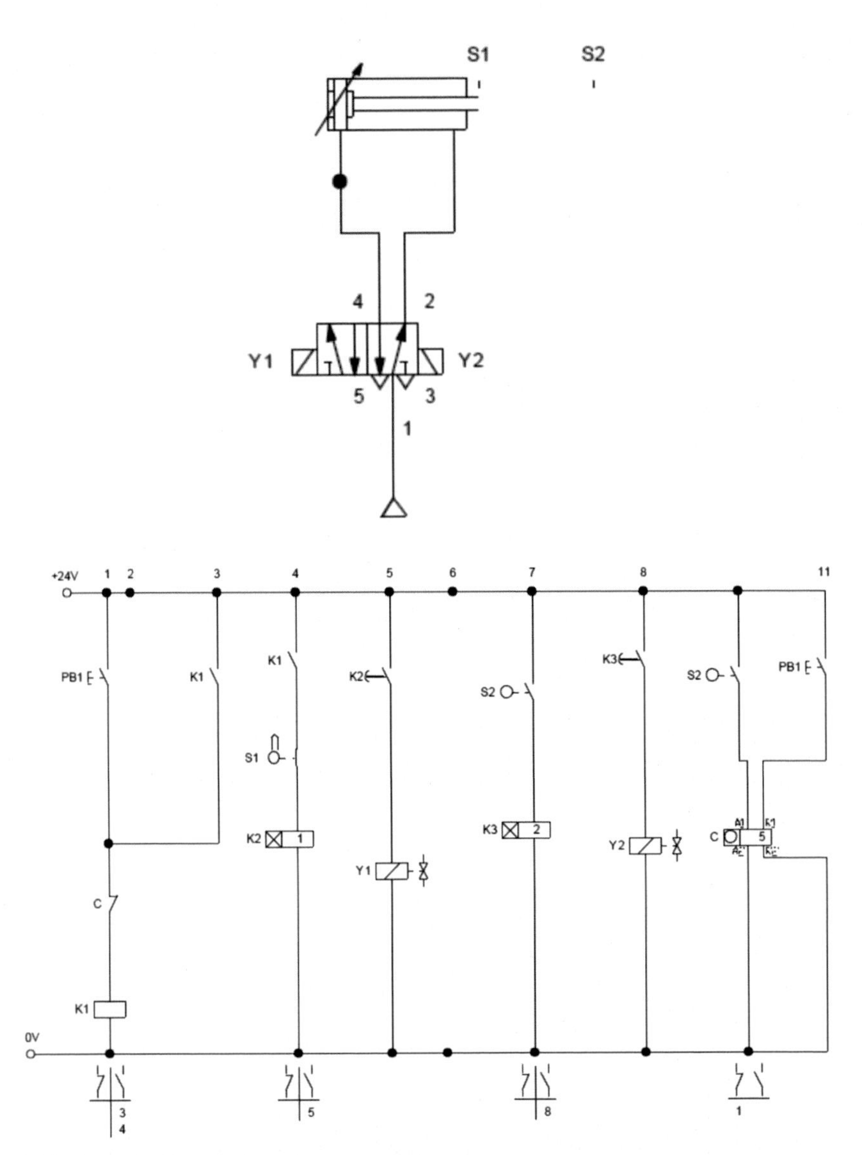

FIGURE 6.55 Control of double-acting cylinder using counter and delay.

die connected to a fixture down. The die is to returned to the initial location when a pressure transfer and one second delay exceeds adequate stamping pressure as sensed. To execute the control function for the stamping process, create an electro- control circuit. Fig. 6.54 indicates solution.

6.4.6 Control of double-acting cylinder using delay ON and OFF timer and counter

Components shall be stamped with stamping tool. After pressing the push-button, a double actuating cylinder is used to move the die connected to a fixture one second down. The die is to return to the initial location after achieving a correct two-second stamping time. After 5 cycles, the automatic process will stop. The counter should be reset to start. To execute the control function for the stamping process, create an electropneumatic control circuit. Fig. 6.55 indicates solution.

6.5 Summary

This chapter provides the basic structure of PLC and its applications, advantages, and disadvantages, and it applies the concepts of electrical ladder logic to various applications. Also, this chapter attracts the reader's interest with solved examples on pneumatic control circuit and ladder logic as well. The readers will be able to design and program basic PLC circuits for entry-level PLC applications after reading this chapter.

Chapter 7

Graphical Programming Using LabVIEW for Beginners

Chapter Outline

7.1 Introduction to LabVIEW and virtual instruments

LabVIEW is a graphical programming language that uses icons instead of lines of text to create applications. In contrast to text-based programming languages, where instructions determine program execution, LabVIEW uses dataflow programming, where the flow of data determines execution. In LabVIEW, one can build a user interface with a set of tools and objects. The user interface is known as the front panel. Then add code using graphical representations of functions to control the front panel objects. The block diagram contains this code. In some ways, the block diagram resembles a flowchart.

Software Tools for the Simulation of Electrical Systems. DOI: https://doi.org/10.1016/B978-0-12-819416-4.00007-7

LabVIEW programs are called virtual instruments, or VIs, because their appearance and operation imitate physical instruments, such as oscilloscopes and multimeters. Every VI uses functions that manipulate input from the user interface or other sources and display that information or move it to other files or other computers. A VI contains the following three components:

7.1.1 Front panel

Build the front panel with controls and indicators, which are the interactive input and output (I/O) terminals of the VI, respectively (Fig. 7.1). Controls are knobs, push buttons, dials, and other input devices. Indicators are graphs, LEDs, and other displays. Controls simulate instrument input devices and supply data to the block diagram of the VI. Indicators simulate instrument output devices and display data the block diagram acquires or generates.

Select Window and click show controls palette to display the controls palette, then select controls and indicators from the controls palette and place them on the front panel.

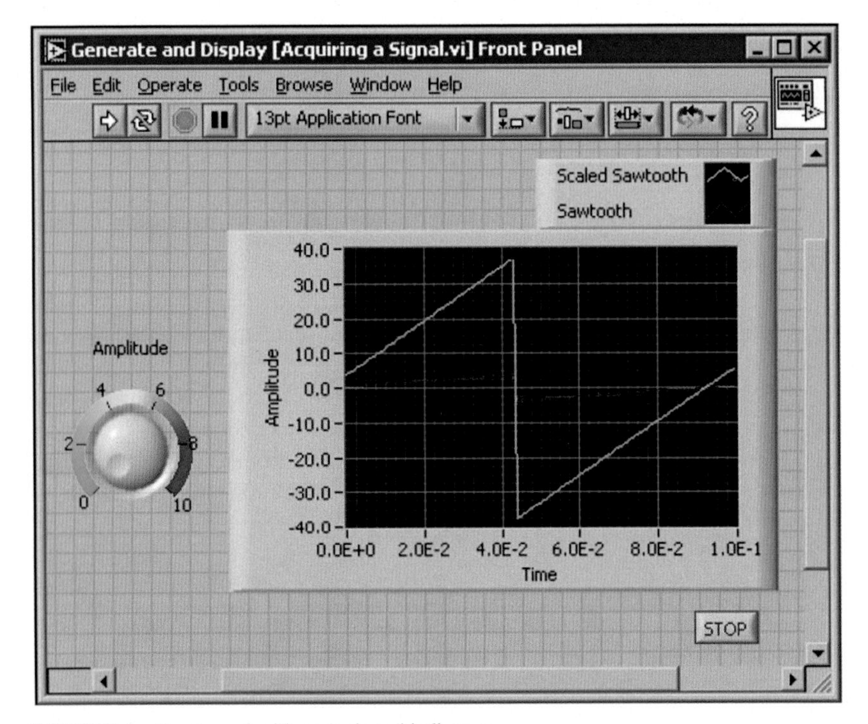

FIGURE 7.1 Front panel with controls and indicators.

7.1.2 Block diagram

After building the front panel, add code using graphical representations of functions to control the front panel objects. The block diagram contains this graphical source code. Front panel objects appear as terminals on the block diagram (Fig. 7.2).

The icon and connector panel identifies the VI so use the VI in another VI. The VI imported to any other VI is called sub-VI. A sub VI corresponds to a subroutine in text-based programming languages.

- *Terminals*

  The terminals represent the data type of the control or indicator. Configure front panel controls or indicators to appear as icon or data type terminals on the block diagram. By default, front panel objects appear as icon terminals. For example, a knob icon terminal shown on the left of work space, represents a knob on the front panel. The DBL at the bottom of the terminal represents a data type of double-precision, floating-point numeric control or indicator.
- *Nodes*

  Nodes are objects on the block diagram that have inputs and/or outputs and perform operations when VI runs. They are analogous to statements, operators, functions, and subroutines in text-based programming languages.
- *Wires*

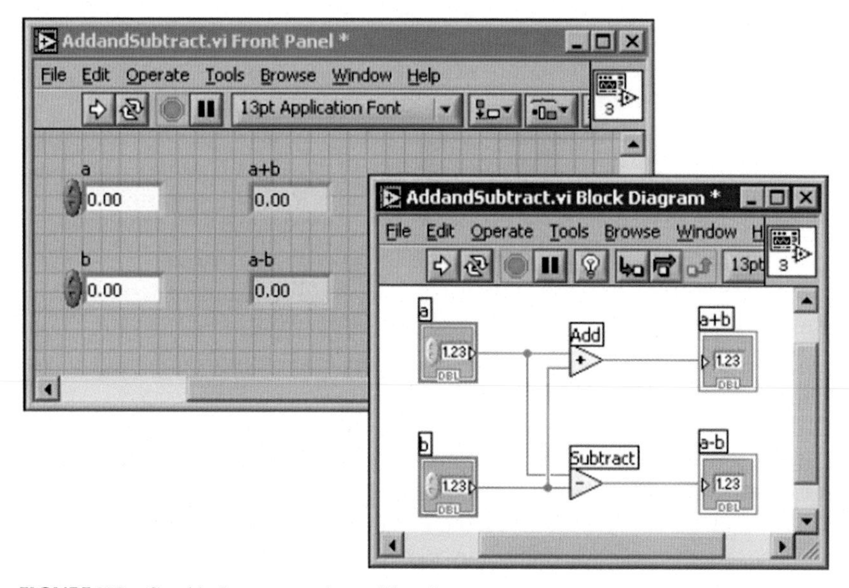

FIGURE 7.2 Graphical representations of functions.

Transfer data among block diagram objects through wires. Each wire has a single data source, but user can wire it to many VIs and functions that read the data. Wires are different colors, styles, and thicknesses, depending on their data types. A broken wire appears as a dashed black line with a red Xin the middle.

- *Structures*

Structures are graphical representations of the loops and case statements of text-based programming languages. Use structures on the block diagram to repeat blocks of code and to execute code conditionally or in a specific order.

7.1.3 Icon and connector pane

After building a VI front panel and block diagram, build the icon and connector pane. To use a VI as a sub-VI, need to build a connector pane. Every VI displays a connector pane, next to the VI icon in the upper right corner of the front panel window. An icon is a graphical representation of a VI. It can contain text, images, or a combination of both. If using a VI as a sub VI, the icon identifies the sub VI on the block diagram of the VI. Then double-click the icon to customize or edit it.

The connector pane is a set of terminals that correspond to the controls and indicators of that VI, similar to the parameter list of a function call in text-based programming languages. The connector pane defines the inputs and outputs wire to the VI so can use it as a sub VI. A connector pane receives data at its input terminals and passes the data to the block diagram code through the front panel controls and receives the results at its output terminals from the front panel indicators.

When the connector pane is viewed for the first time, a connector pattern is shown. Select a different pattern. The connector pane generally has one terminal for each control or indicator on the front panel. Up to 28 terminals can be assigned to a connector pane. If changes to the VI that would require a new input or output are anticipated, leave extra terminals unassigned.

7.1.4 Building the front panel

The front panel is the user interface of a VI. Generally, design the front panel first, then design the block diagram to perform tasks on the inputs and outputs created on the front panel.

Build the front panel with controls and indicators, which are the interactive I/O terminals of the VI, respectively. Controls are knobs, push buttons, dials, and other input devices. Indicators are graphs, LEDs, and other displays. Controls simulate instrument input devices and supply data to the block diagram of the VI. Indicators simulate instrument output devices and display data the block diagram acquires or generates.

Select Window >> Show Controls Palette to display the Controls palette, then select controls and indicators from the Controls palette and place them on the front panel.

- *Controls palette*

  The Controls palette is available only on the front panel. The Controls palette contains the controls and indicators used to create the front panel. The controls and indicators are located on subpalettes based on the types of controls and indicators.

  Select Window >> Show Controls Palette or right-click the front panel workspace to display the Controls palette. Place the Controls palette anywhere on the screen. LabVIEW retains the Controls palette position and size so when restarting LabVIEW, the palette appears in the same position and has the same size.
- *Functions palette*

  The Functions palette is available only on the block diagram. The Functions palette contains the VIs and functions used to build the block diagram. The VIs and functions are located on subpalettes based on the types of VIs and functions.

  Select Window >> Show Functions Palette or right-click the block diagram workspace to display the Functions palette. Place the Functions palette anywhere on the screen. LabVIEW retains the Functions palette position and size so when restarting LabVIEW, the palette appears in the same position and has the same size.
- *Navigating the controls and functions palettes*

When clicking a subpalette icon, the entire palette changes to the subpalette. Click an object on the palette to place the object on the cursor so it can be placed on the front panel or block diagram. Also right-click a VI icon on the palette and select Open VI from the shortcut menu to open the VI. Use the following buttons on the Controls and Functions palette toolbars to navigate the palettes, to configure the palettes, and to search for controls,

7.1.4.1 Virtual instruments, and functions

- *Up*—Takes up one level in the palette hierarchy. Click this button and hold the mouse button down to display a shortcut menu that lists each subpalette in the path to the current subpalette. Select a subpalette name in the shortcut menu to navigate to the subpalette.
- *Search*—Changes the palette to search mode so perform text-based searches to locate controls, VIs, or functions on the palettes. While a palette is in search mode, click the Return to Palette button to exit the search mode and return to the palette.
- *Options*—Displays the Controls/Functions Palettes page of the Options dialog box, in which one can select a palette view and a format for the palettes.

- *Restore Palette Size*—Resizes the palette to its default size. This button appears only if resizing the Controls or Functions palette.

The Tools palette is available on the front panel and the block diagram. A tool is a special operating mode of the mouse cursor. The cursor corresponds to the icon of the tool selected in the palette. Use the tools to operate and modify the front panel and block diagram objects. Select Window >> Show Tools Palette to display the Tools palette. Place the Tools palette anywhere on the screen. LabVIEW retains the Tools palette position so when restarting LabVIEW, the palette appears in the same position.

If automatic tool selection is enabled, move the cursor over objects on the front panel or block diagram and LabVIEW automatically selects the corresponding tool from the Tools palette. Disable automatic tool selection by clicking the Automatic Tool Selection button on the Tools palette, shown on the left. Press the <Shift-Tab> keys or click the Automatic Tool Selection button to enable automatic tool selection again. Disable automatic tool selection by manually selecting a tool on the Tools palette. Press the <Tab> key or click the Automatic Tool Selection button on the Tools palette to enable automatic tool selection again.

7.1.4.2 Customizing the controls and functions palettes

Customize the Control and Function palettes in the following ways:

- Add VIs and controls to the palettes.
- Set up different views for different users, hiding some VIs and functions to make LabVIEW easier to use for one user while providing the full palettes for another user.
- Rearrange the built-in palettes to make the VIs and functions use frequently more accessible.
- Convert a set of ActiveX controls into custom controls and add them to the palettes.
- Add toolsets to the palettes.

7.1.5 Data flow model

LabVIEW follows a dataflow model for running VIs. A block diagram node executes when it receives all required inputs. When a node executes, it produces output data and passes the data to the next node in the dataflow path. The movement of data through the nodes determines the execution order of the VIs and functions on the block diagram.

Visual Basic, C++, Java, and most other text-based programming languages follow a control flow model of program execution. In control flow, the sequential order of program elements determines the execution order of a program.

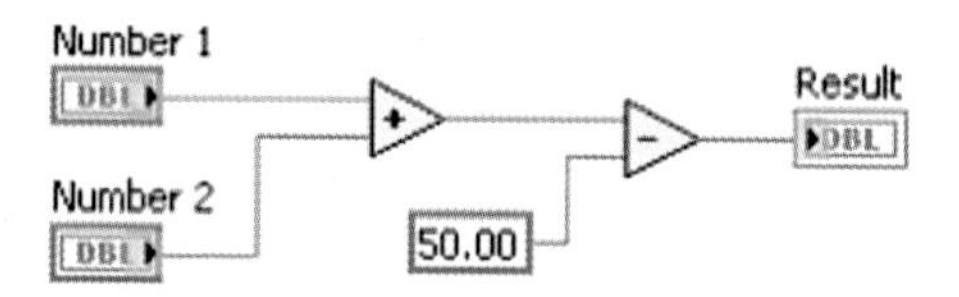

FIGURE 7.3 Dataflow programming example.

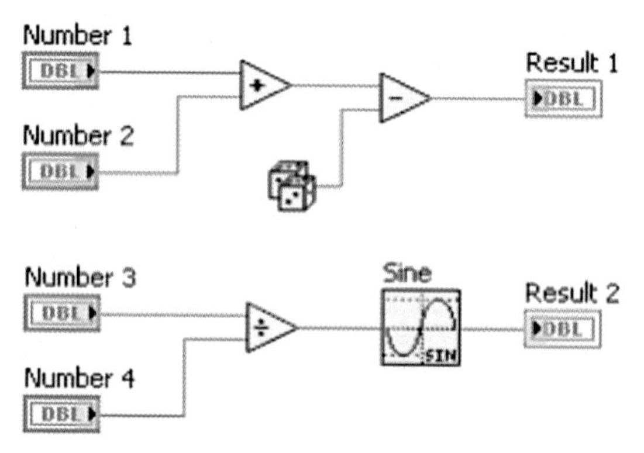

FIGURE 7.4 Dataflow example for multiple code segments.

For a dataflow programming example, consider a block diagram that adds two numbers and then subtracts 50.00 from the result of the addition, as shown in Fig. 7.3. In this case, the block diagram executes from left to right, not because the objects are placed in that order, but because the Subtract function cannot execute until the Add function finishes executing and passes the data to the Subtract function. Remember that a node executes only when data is available at all of its input terminals and supplies data to the output terminals only when the node finishes execution.

In Fig. 7.4, consider which code segment would execute first—the Add, Random Number, or Divide function. The user cannot predict,. because inputs to the Add and Divide functions are available at the same time, and the Random Number function has no inputs. In a situation where one code segment must execute before another, and no data dependency exists between the functions, use other programming methods, such as sequence structures or error clusters, to force the order of execution.

7.1.5.1 Wires

Transfer data among block diagram objects through wires. In Figs. 7.3 and 7.4, wires connect the control and indicator terminals to the Add and Subtract functionWires are different colors, styles, and thicknesses, depending on their data types. Each wire has a single data source, but user can wire it to many VIs and functions that read the data.

TABLE 7.1 Common wire types.

Wire type	Scalar	1D array	2D array	Color
Numeric				Orange (floating-point), blue (integer)
Boolean				Green
String				Pink

A broken wire appears as a dashed black line with a red X in the middle, as shown above. Broken wires occur for a variety of reasons, such as when one tries to wire two objects with incompatible data types. Table 7.1 shows the most common wire types.

In LabVIEW, one can use wires to connect multiple terminals together to pass data in a VI. And one must connect the wires to inputs and outputs that are compatible with the data that is transferred with the wire. For example, one cannot wire an array output to a numeric input. In addition the direction of the wires must be correct. It must connect the wires to only one input and at least one output. Also one cannot wire two indicators together. The components that determine wiring compatibility include the data type of the control and/or the indicator and the data type of the terminal. For example, if a switch has a green border, wire a switch to any input with a green label on an Express VI. If a knob has an orange border, wire a knob to any input with an orange label. However, one cannot wire an orange knob to an input with a green label. Notice the wires are the same color as the terminal.

7.1.5.2 Automatically wiring objects

When moving a selected object close to other objects on the block diagram, LabVIEW draws temporary wires to show valid connections. When releasing the mouse button to place the object on the block diagram, LabVIEW automatically connects the wires. Also it can automatically wire objects already on the block diagram. LabVIEW connects the terminals that best match and does not connect the terminals that do not match.

By default automatic wiring is enabled when selecting an object from the Functions palette or when copying an object already on the block diagram by pressing the <Ctrl> key and dragging the object. Automatic wiring is disabled by default when using the Positioning tool to move an object already on the block diagram. Adjust the automatic wiring settings by selecting Tools then Options and selecting Block Diagram from the Category list.

7.1.5.3 Manually wiring objects

When passing the Wiring tool over a terminal, a tip strip appears with the name of the terminal. In addition, the terminal blinks in the Context Help window and on the icon to help verify wiring to the correct terminal. To wire objects together, pass the Wiring tool over the first terminal, click, pass the cursor over the second terminal, and click again. After wiring, right-click the wire and select Clean up wire from the shortcut menu to have LabVIEW automatically choose a path for the wire. If there are broken wires to remove, press <Ctrl-B> to delete all the broken wires on the block diagram.

7.1.6 Programming concepts of virtual instrument

7.1.6.1 String data type

A string is a sequence of displayable or nondisplayable ASCII characters. Strings provide a platform-independent format for information and data. Some of the more common applications of strings include the following:

- Creating simple text messages.
- Controlling instruments by sending text commands to the instrument and returning data values in the form of either ASCII or binary strings, which then convert to numeric values.
- Storing numeric data to disk. To store numeric data in an ASCII file, first convert numeric data to strings before writing the data to a disk file.
- Instructing or prompting the user with dialog boxes.

On the front panel, strings appear as tables, text entry boxes, and labels. LabVIEW includes built-in VIs and functions used to manipulate strings, including formatting strings, parsing strings, and other editing. LabVIEW represents string data with the color pink shown in below.

7.1.6.2 Numeric data type

LabVIEW represents numeric data as floating-point numbers, fixed-point numbers, integers, unsigned integers, and complex numbers. Double and Single precision as well as Complex numeric data is represented with the color orange in LabVIEW. All Integer numeric data is represented with the color blue.

7.1.6.3 Boolean data type

LabVIEW stores Boolean data as 8-bit values. A Boolean value is used in LabVIEW to represent a 0 or 1, or a TRUE or FALSE. If the 8-bit value is zero, the Boolean value is FALSE. Any nonzero value represents TRUE. Common applications for Boolean data include representing digital data and serving as a front panel control that acts as a switch that has a mechanical action often used to control an execution structure such as a Case structure. A Boolean control is typically used as the conditional statement to exit a While Loop. In LabVIEW, the color green represents Boolean data.

7.1.6.4 Dynamic data type

- Most Express VIs accept and/or return the dynamic data type, which appears as a dark blue terminal. Using the Convert to Dynamic Data and Convert from Dynamic Data VIs, convert floating-point numeric or Boolean data of the following data types:
- 1D array of waveforms
- 1D array of scalars
- 1D array of scalars—most recent value
- 1D array of scalars—single channel
- 2D array of scalars—columns are channels
- 2D array of scalars—rows are channels
- Single scalar
- Single waveform

Wire the dynamic data type to an indicator that can best present the data. Indicators include a graph, chart, or numeric, or Boolean indicator. However, because dynamic data undergoes an automatic conversion to match the indicator to which it is wired, Express VIs can slow down the block diagram execution speed.

The dynamic data type is for use with Express VIs. Most other VIs and functions that are shipped with LabVIEW do not accept this data type. In order to use a built-in VI or function to analyze or process the data the dynamic data type includes, convert the dynamic data type.

7.1.6.5 *Arrays*

Sometimes it is beneficial to group related data. Use arrays and clusters to group related data in LabVIEW. Arrays combine data points of the same data type into one data structure, and clusters combine data points of multiple data types into one data structure.

An array consists of elements and dimensions. Elements are the data points that make up the array. A dimension is the length, height, or depth of an array. An array can have one or more dimensions and as many as (2^{31})—1 elements per dimension, memory permitting.

Build arrays of numeric, Boolean, path, string, waveform, and cluster data types. Consider using arrays when working with a collection of similar data points and performing repetitive computations. Arrays are ideal for storing data collected from waveforms or data generated in loops, where each iteration of a loop produces one element of the array.

Fig. 7.5 shows an example of an array of numeric data. The first element shown in the array (3.00) is at index 1, and the second element (1.00) is at index 2. The element at index 0 is not shown in this image because element 1 is selected in the index display. The element selected in the index display always refers to the element shown in the upper left corner of the element display.

Creating array controls and indicators

Create an array control or indicator on the front panel by adding an array shell to the front panel, as shown in Fig. 7.6. And dragging a data object or element, such as a numeric or string control, into the array shell.

If attempting to drag an invalid control or indicator into the array shell, one is unable to place the control or indicator in the array shell. One must

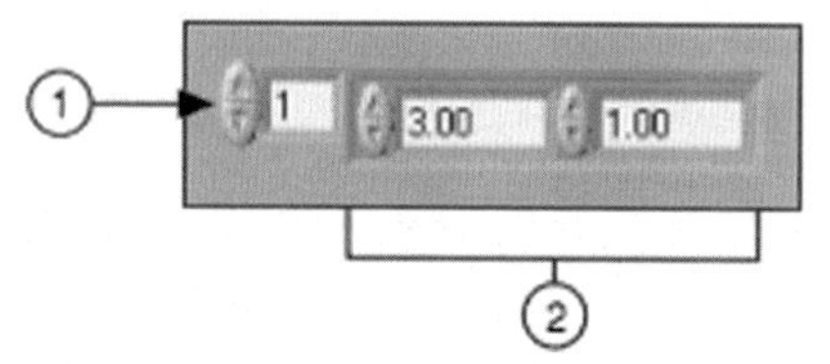

FIGURE 7.5 Array control of numerics.

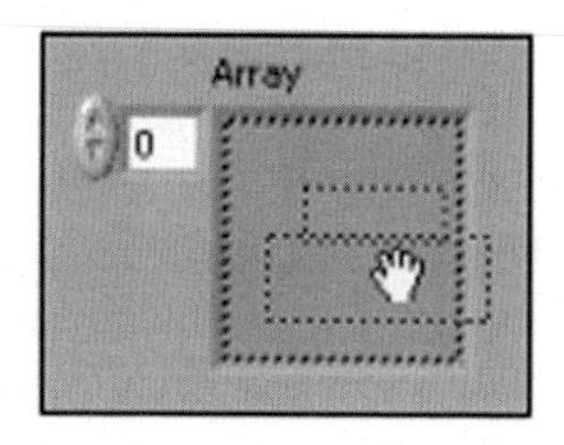

FIGURE 7.6 Placing a numeric control in an array shell.

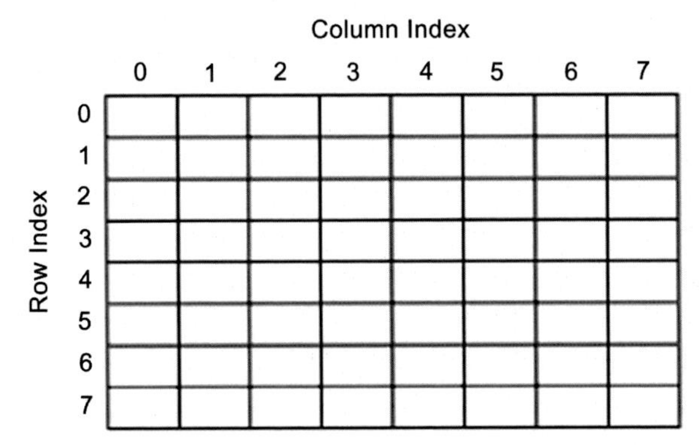

FIGURE 7.7 2D array.

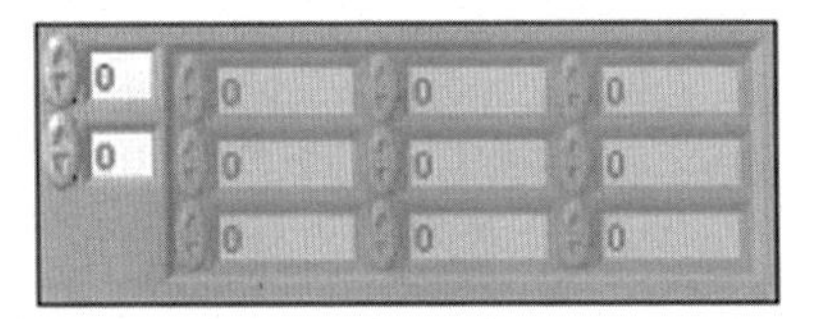

FIGURE 7.8 2D uninitialized array.

insert an object in the array shell before using the array on the block diagram. Otherwise, the array terminal appears black with an empty bracket and has no data type associated with it.

Two-dimensional arrays

The previous examples use 1D arrays. A 2D array stores elements in a grid. It requires a column index and a row index to locate an element, both of which are zero-based. Fig. 7.7 shows an 8 column by 8 row 2D array, which contains $8 \times 8 = 64$ elements.

To add a multidimensional array to the front panel, right-click the index display and select Add Dimension from the shortcut menu. Also, one can resize the index display to have as many dimensions as required.

Initializing arrays

Initialize an array or leave it uninitialized. When an array is initialized, the number of elements in each dimension and the contents of each element are defined. An uninitialized array contains a fixed number of dimensions but no elements. Fig. 7.8 shows an uninitialized 2D array control. Notice that the elements are all dimmed. This indicates that the array is uninitialized.

In a 2D array, after initializing an element, any uninitialized element in that column and in previous columns are initialized and populated with the default value for the data type. In Fig. 7.9, a value of 4 was entered in

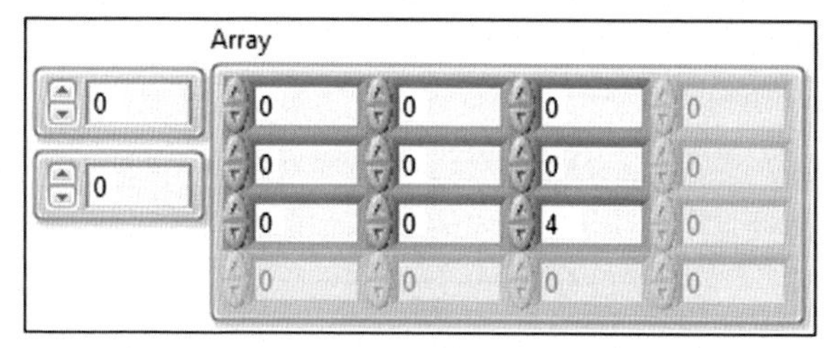

FIGURE 7.9 An initialized 2D array with nine elements.

column 2, of the 0-based array. The previous elements in column 0, 1, and 2 are initialized to 0, which is the default value for the numeric data type.

Creating array constants

To create an array constant on the block diagram, select an array constant on the Functions palette, place the array shell on the block diagram, and place a string constant, numeric constant, Boolean constant, or cluster constant in the array shell. Use an array constant to store constant data or as a basis for comparison with another array.

Auto-indexing array inputs

Wire an array to or from a For Loop or While Loop, link each iteration of the loop to an element in that array by enabling auto-indexing. The tunnel image changes from a solid square to the image to indicate auto-indexing. Right-click the tunnel and select Enable Indexing or Disable Indexing from the shortcut menu to toggle the state of the tunnel.

Array inputs

If enabling auto-indexing on an array wired to a For Loop input terminal, LabVIEW sets the count terminal to the array size so you do not need to wire the count terminal. Because For Loops process arrays one element at a time, LabVIEW enables auto-indexing by default for every array wire to a For Loop. Disable auto-indexing if you do not need to process arrays one element at a time.

In Fig. 7.10 the For Loop executes a number of times equal to the number of elements in the array. Normally, if the count terminal of the For Loop is not wired, the run arrow is broken. However, in this case the run arrow is not broken.

If enabling auto-indexing for more than one tunnel or if wiring the count terminal, the actual number of iterations becomes the smaller of the choices. For example, if two auto-indexed arrays enter the loop, with 10 and 20

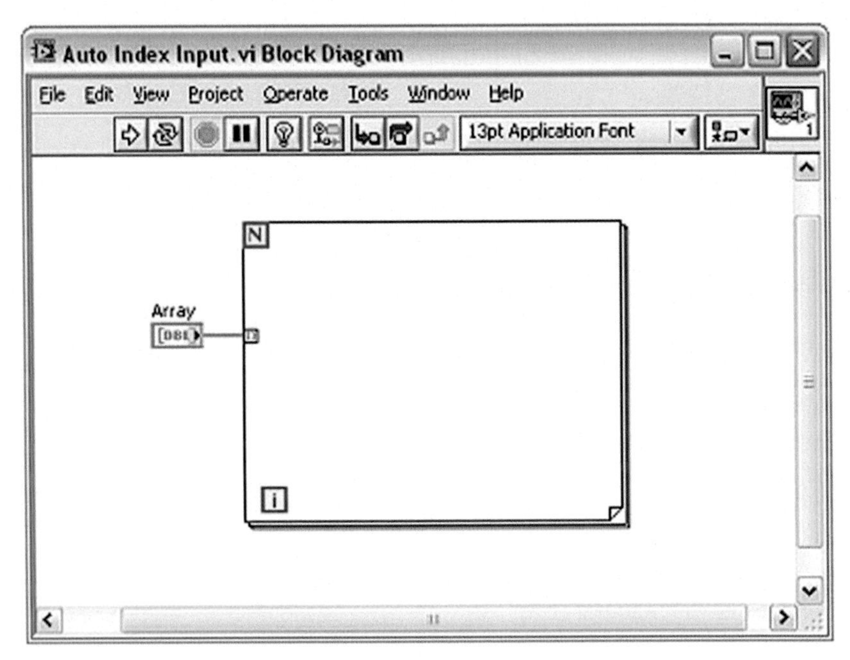

FIGURE 7.10 Array used to set for loop count.

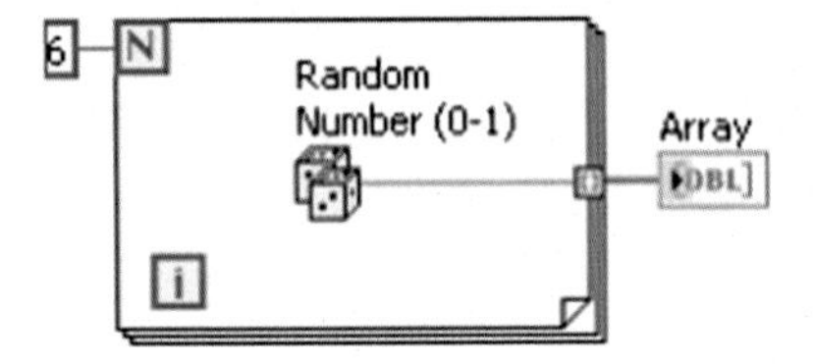

FIGURE 7.11 Auto-indexed output.

elements respectively, and you wire a value of 15 to the count terminal, the loop still executes only 10 times, indexing all elements of the first array but only the first 10 elements of the second array.

Array outputs

When auto-indexing an array output tunnel, the output array receives a new element from every iteration of the loop. Therefore auto-indexed output arrays are always equal in size to the number of iterations (Fig. 7.11).

The wire from the output tunnel to the array indicator becomes thicker as it changes to an array at the loop border, and the output tunnel contains square brackets representing an array.

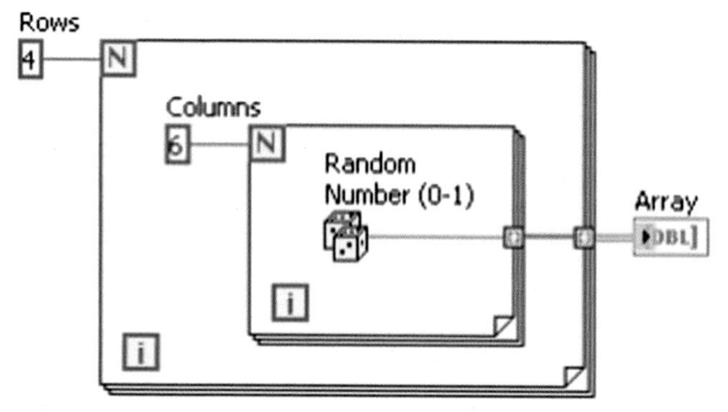

FIGURE 7.12 Creating a 2D array.

Right-click the tunnel at the loop border and select Enable Indexing or Disable Indexing from the shortcut menu to enable or disable auto-indexing. Auto-indexing for While Loops is disabled by default. For example, disable auto-indexing if you need only the last value passed out of the tunnel.

Creating two-dimensional arrays

Use two For Loops, nested one inside the other, to create a 2D array. The outer For Loop creates the row elements, and the inner For Loop creates the column elements (Fig. 7.12).

7.1.6.6 Clusters

Clusters group data elements of mixed types. An example of a cluster is the LabVIEW error cluster, which combines a Boolean value, a numeric value, and a string. A cluster is similar to a record or a struct in text-based programming languages.

Bundling several data elements into clusters eliminates wire clutter on the block diagram and reduces the number of connector pane terminals that sub VIs need. The connector pane has, at most, 28 terminals. If the front panel contains more than 28 controls and indicators that want to pass to another VI, group some of them into a cluster and assign the cluster to a terminal on the connector pane.

Most clusters on the block diagram have a pink wire pattern and data type terminal. Error clusters have a dark yellow wire pattern and data type terminal. Clusters of numeric values, sometimes referred to as points, have a brown wire pattern and data type terminal. Wire brown numeric clusters to Numeric functions, such as Add or Square Root, to perform the same operation simultaneously on all elements of the cluster.

Order of cluster elements

Although cluster and array elements are both ordered, one must unbundle all cluster elements at once using the Unbundle function. Use the Unbundle by Name function to unbundle cluster elements by name. If using the Unbundle by Name function, each cluster element must have a label. Clusters also differ from arrays in that they are a fixed size. Like an array, a cluster is either a control or an indicator. A cluster cannot contain a mixture of controls and indicators.

Creating cluster controls and indicators

Create a cluster control or indicator on the front panel by adding a cluster shell to the front panel, as shown in the following front panel, and dragging a data object or element, which can be a numeric, Boolean, string, path, refnum, array, or cluster control or indicator, into the cluster shell. Resize the cluster shell by dragging the cursor while placing the cluster shell (Fig. 7.13).

Fig. 7.14 is an example of a cluster containing three controls: a string, a Boolean switch, and a numeric.

Creating cluster constants

To create a cluster constant on the block diagram, select a cluster constant on the Functions palette, place the cluster shell on the block diagram, and place a string constant, numeric constant, Boolean constant, or cluster constant in the cluster shell. Use a cluster constant to store constant data or as a basis for comparison with another cluster.

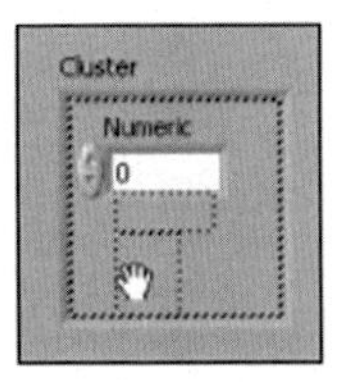

FIGURE 7.13 Creation of a cluster control.

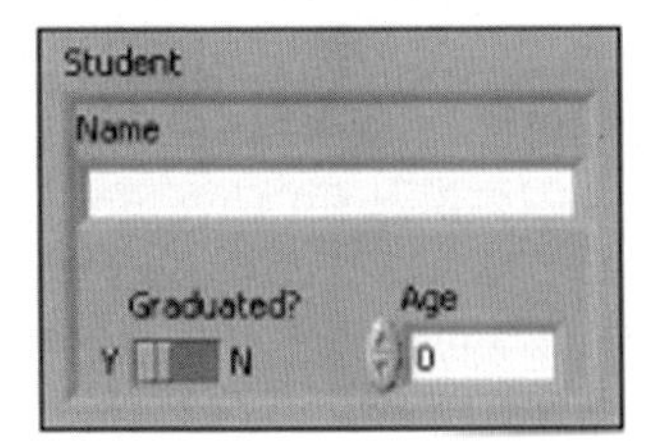

FIGURE 7.14 Cluster control example.

If you have a cluster control or indicator on the front panel window then to create a cluster constant containing the same elements on the block diagram, either drag that cluster from the front panel window to the block diagram or right-click the cluster on the front panel window and select Create and click Constant from the shortcut menu.

Using cluster functions

Use the Cluster functions to create and manipulate clusters, for example, to perform tasks similar to the following:

- Extract individual data elements from a cluster.
- Add individual data elements to a cluster.
- Break a cluster out into its individual data elements.

Use the Bundle function to assemble a cluster, use the Bundle function and Bundle by Name function to modify a cluster, and use the Unbundle function and the Unbundle by Name function to disassemble clusters.

Also place the Bundle, Bundle by Name, Unbundle, and Unbundle by Name functions on the block diagram by right-clicking a cluster terminal on the block diagram and selecting the Cluster, Class & Variant Palette from the shortcut menu. The Bundle and Unbundle functions automatically contain the correct number of terminals. The Bundle by Name and Unbundle by Name functions appear with the first element in the cluster. Use the Positioning tool to resize the Bundle by Name and Unbundle by Name functions to show the other elements of the cluster.

Assembling clusters

Use the Bundle function to assemble a cluster from individual elements or to change the values of individual elements in an existing cluster without having to specify new values for all elements (Fig. 7.15). Use the Positioning tool to resize the function or right-click an element input and select Add Input from the shortcut menu.

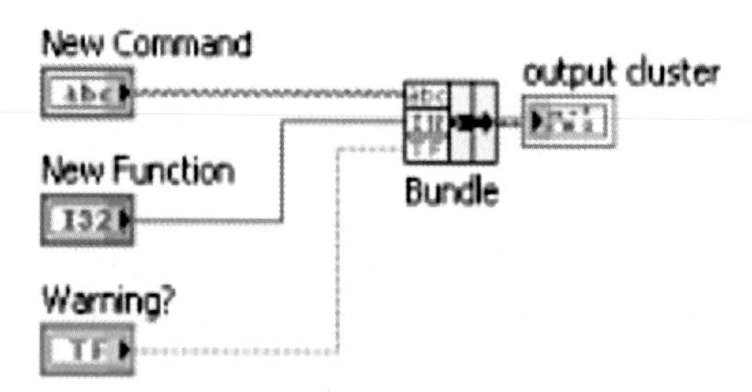

FIGURE 7.15 Assembling a cluster on the block diagram.

Modifying a cluster

If wiring the cluster input, wire only the elements you want to change. For example, the Input Cluster shown in Fig. 7.16 contains three controls.

If you know the cluster order, use the Bundle function to change the Command value by wiring the elements shown in Fig. 7.16.

Also use the Bundle by Name function to replace or access labeled elements of an existing cluster. The Bundle by Name function works like the Bundle function, but instead of referencing cluster elements by their cluster order, it references them by their owned labels. Access only elements with owned labels. The number of inputs does not need to match the number of elements in an output cluster.

Use the Operating tool to click an input terminal and select an element from the pull-down menu. Also, right-click the input and select the element from the Select Item shortcut menu.

In Fig. 7.17, use the Bundle by Name function to update the values of Command and Function with the values of New Command and New Function.

Use the Bundle by Name function for data structures that might change during development. If adding a new element to the cluster or modifying its order, you do not need to rewire the Bundle by Name function because the names are still valid.

Disassembling clusters

Use the Unbundle function to split a cluster into its individual elements. Use the Unbundle by Name function to return the cluster elements whose names are specified. The number of output terminals does not depend on the number of elements in the input cluster.

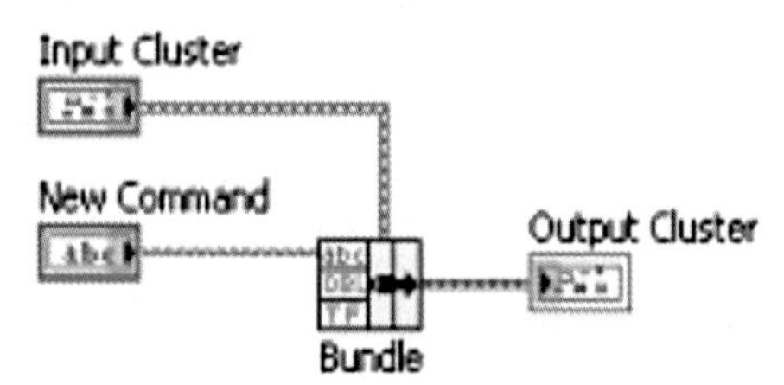

FIGURE 7.16 Bundle used to modify a cluster.

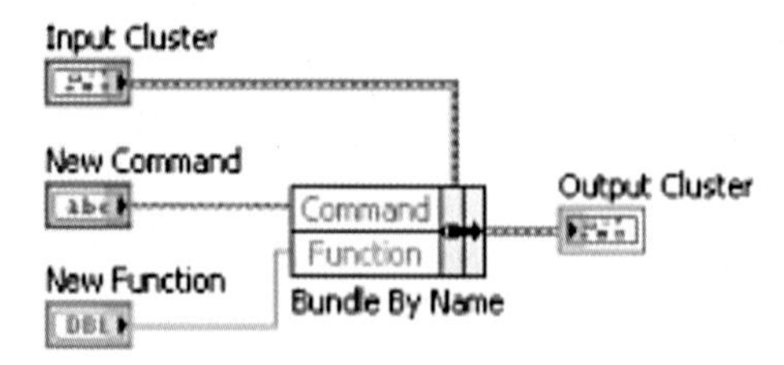

FIGURE 7.17 Bundle by name used to modify a cluster.

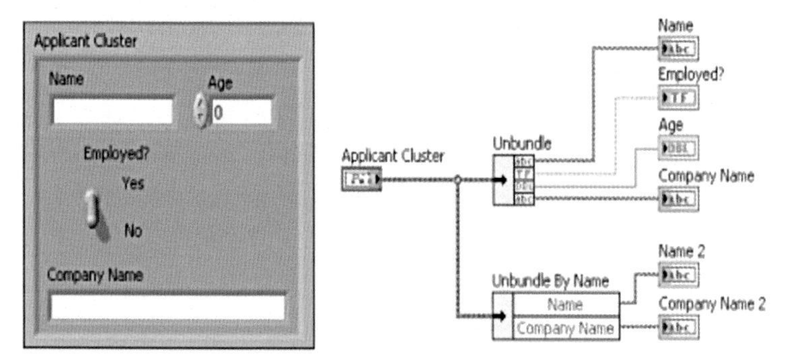

FIGURE 7.18 Unbundle and unbundle by name.

Use the Operating tool to click on an output terminal and select an element from the pull-down menu. Also, right-click the output terminal and select the element from the Select Item shortcut menu.

For example, if using the Unbundle function with the cluster in Fig. 7.18, it has four output terminals that correspond to the four controls in the cluster. You must know the cluster order so associate the correct Boolean terminal of the unbundled cluster with the corresponding switch in the cluster. In this example, the elements are ordered from top to bottom starting with element 0. If using the Unbundle by Name function, there are an arbitrary number of output terminals and access individual elements by name in any order.

Enums

An enum (enumerated control, constant or indicator) is a combination of data types. An enum represents a pair of values, a string and a numeric, where the enum can be one of a list of values. For example, if creating an enum type called Month, the possible value pairs for a Month variable are January-0, February-1, and so on through to December-11. Fig. 7.19 shows an example of these data pairs in the Properties dialog box for an enumerated control. This is directly accessed by right-clicking the enum control and selecting Edit Items.

Enums are useful because it is easier to manipulate numbers on the block diagram than strings. Fig. 7.20 shows the Month enumerated control, the selection of a data pair in the enumerated control, and the corresponding block diagram terminal.

7.1.7 Running and debugging virtual instruments

Running a VI executes the operation for which the VI is designed. The Run button shows as a solid white arrow on the toolbar, shown as follows. The solid white arrow also indicates the use of the VI as a sub VI if creating a connector pane for the VI.

FIGURE 7.19 Properties for the month enumerated control.

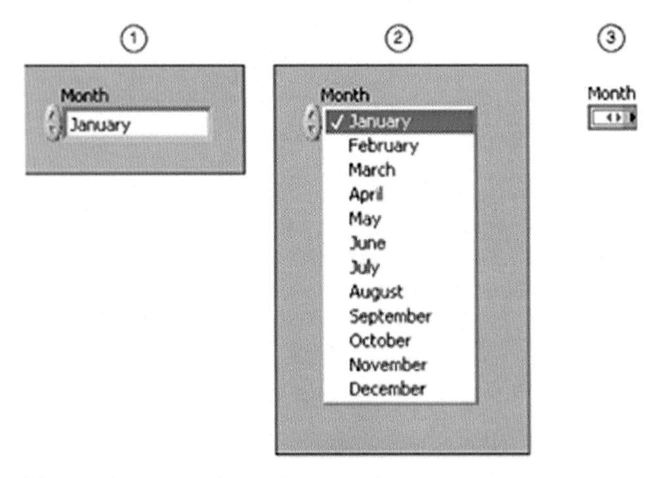

FIGURE 7.20 Month enumerated control.
(1) Front Panel Control. (2) Selecting an Item. (3) Block Diagram Terminal.

A VI runs when the *Run* or *Run Continuously* buttons or the single-stepping buttons on the block diagram toolbar are clicked. While the VI runs, the *Run* button changes to a darkened arrow, shown as follows, to indicate that the VI is running. One cannot edit a VI while the VI runs.

Clicking the Run button runs the VI once. The VI stops when the VI completes its data flow. Clicking the Run Continuously button, shown as follows, runs the VI continuously until it is stopped manually.

If a VI does not run, it is a broken, or nonexecutable, VI. The Run button appears broken, shown as follows, when the VI created or edited contains errors. If the button still appears broken when the wiring of the block diagram is finished then the VI is broken and cannot run.

7.1.7.1 Finding causes for broken virtual instruments

Warnings do not prevent a VI from running. They are designed to help avoid potential problems in VIs. Errors, however, can break a VI. Any errors must be resolved before running the VI.

Click the broken *Run* button or select *View* and *Error List* to find out why a VI is broken. The *Error list* window lists all the errors. The *Items with errors* section lists the names of all items in the memory, such as VIs and project libraries that have errors. If two or more items have the same name, this section shows the specific application instance for each item.

The *Errors and warnings* section lists the errors and warnings for the VI to select in the *Items with errors* section. Click the *Show Error* button or double-click the error description to highlight the area on the block diagram or front panel that contains the error. The toolbar includes the *Warning* button, shown as follows, if a VI includes a warning then a checkmark is placed in the *Show Warnings* checkbox in the *Error list* window.

7.1.7.2 Common causes of broken virtual instruments

The following list contains common reasons why a VI is broken while editing it:

The block diagram contains a broken wire because of a mismatch of data types or a loose, unconnected end.
A required block diagram terminal is unwired.
A sub VI is broken or its connector pane is edited after placing its icon on the block diagram of the VI.

7.1.7.3 Debugging techniques

If a VI is not broken, but unexpected data is received, use several techniques to identify and correct problems with the VI or the block diagram data flow.

7.1.7.4 Execution highlighting

View an animation of the execution of the block diagram by clicking the Highlight Execution button, shown as follows. Execution highlighting shows the movement of data on the block diagram from one node to another using bubbles that move along the wires. Use execution highlighting in conjunction with single-stepping to see how data values move from node to node through a VI.

If the error out cluster reports an error, the error value appears next to the error out with a red border. If no error occurs, OK appears next to the error out with a green border.

7.1.7.5 Single-stepping

Single-step through a VI to view each action of the VI on the block diagram as the VI runs. The single-stepping buttons, shown as follows, affect execution only in a VI or sub VI in single-step mode.

Step Into Step Over Step Out

Enter single-step mode by clicking the Step Over or Step Into button on the block diagram toolbar. Move the cursor over the Step Over, Step Into, or Step Out button to view a tip strip that describes the next step if that button is clicked. Single-step through sub VIs or run them normally.

If single-stepping through a VI with an execution highlighted, an execution glyph, shown as follows, appears on the icons of the sub VIs that are currently running.

7.1.7.6 Probe tool

Use a generic probe to view the data that passes through a wire. Right-click a wire and select *Custom Probe* then *Generic Probe* from the shortcut menu to use the generic probe.

7.1.7.7 Breakpoints

Use the Breakpoint tool, shown as follows, to place a breakpoint on a VI, node, or wire on the block diagram and pause execution at that location. When setting a breakpoint on a wire, execution pauses after data passes through the wire. Place a breakpoint on the block diagram to pause execution after all nodes on the block diagram execute.

When a VI pauses at a breakpoint, LabVIEW brings the block diagram to the front and uses a marquee to highlight the node or wire that contains the breakpoint. When moving the cursor over an existing breakpoint, the black area of the Breakpoint tool cursor appears white.

When reaching a breakpoint during execution, the VI pauses and the Pause button appears red. Take the following actions:

- Single-step through execution using the single-stepping buttons.
- Probe wires to check intermediate values.
- Change values of front panel controls.
- Click the Pause button to continue running to the next breakpoint or until the VI finishes running.

LabVIEW saves breakpoints with a VI, but they are active only when run the VI. View all breakpoints by selecting *Operate* and *Breakpoints* and clicking the *Find button*.

7.1.7.8 Handling errors

Without a mechanism to check for errors, on can know only that the VI does not work properly. Error checking tells why and where errors occur. When performing any kind of I/O, consider the possibility that errors might occur. Almost all I/O functions return error information. Include error checking in VIs, especially for I/O operations (file, serial, instrumentation, data acquisition, and communication), and provide a mechanism to handle errors appropriately.

By default, LabVIEW automatically handles any error when a VI runs by suspending execution, highlighting the sub VI or function where the error occurred, and displaying an error dialog box.

To disable automatic error handling for the current VI, select *File* and *VI Properties* and select *Execution* from the category pull-down menu. To disable automatic error handling for any new, blank VIs, select *Tools* and *Options* then select Block Diagram from the Category list. To disable automatic error handling for a sub VI or function within a VI, wire its error out parameter to the error in parameter of another sub VI or function or to an error out indicator.

Choose other error handling methods. For example, if an I/O VI on the block diagram times out, one might not want the entire application to stop and display an error dialog box. Also one might want the VI to retry for a certain period of time. In LabVIEW, make these error handling decisions on the block diagram of the VI.

Use the LabVIEW error handling VIs and functions on the Dialog & User Interface palette and the error in and error out parameters of most VIs and functions to manage errors. For example, if LabVIEW encounters an error, display the error message in different kinds of dialog boxes.

Use error handling in conjunction with the debugging tools to find and manage errors. VIs and functions return errors in one of two ways with numeric error codes or with an error cluster. Typically, functions use numeric error codes, and VIs use an error cluster, usually with error inputs and outputs.

Error handling in LabVIEW follows the dataflow model. Just as data values flow through a VI, so can error information. Wire the error information from the beginning of the VI to the end. Include an error handler VI at the end of the VI to determine if the VI ran without errors. Use the error in and error out clusters in each VI build to pass the error information through the VI. The error clusters are flow-through parameters.

As the VI runs, LabVIEW tests for errors at each execution node. If LabVIEW does not find any errors, the node executes normally. If LabVIEW detects an error, the node passes the error to the next node without executing that part of the code. The next node does the same thing, and so on. At the end of the execution flow, LabVIEW reports the error.

7.1.7.9 Error clusters

The error in and error out clusters include the following components of information:

- Status is a Boolean value that reports TRUE if an error occurred.
- Code is a 32-bit signed integer that identifies the error numerically. A nonzero error code coupled with a status of FALSE signals a warning rather than an error.
- Source is a string that identifies where the error occurred.

Some VIs, functions, and structures that accept Boolean data also recognize an error cluster. For example, wire an error cluster to the Boolean inputs of the Select, Quit LabVIEW, or Stop functions. If an error occurs, the error cluster passes a TRUE value to the function.

7.1.7.10 Using while loops for error handling

Wire an error cluster to the conditional terminal of a While Loop to stop the iteration of the While Loop. When wiring the error cluster to the conditional terminal, only the TRUE or FALSE value of the status parameter of the error cluster is passed to the terminal. When an error occurs, the While Loop stops. When an error cluster is wired to the conditional terminal, the shortcut menu items stop if True and continue if True changes to Stop on Error and Continue while Error.

7.1.7.11 Using case structures for error handling

When wiring an error cluster to the selector terminal of a Case structure, the case selector label displays two cases—Error and No Error—and the border of the Case structure changes color—red for Error and green for No Error. If an error occurs, the Case structure executes the Error subdiagram. Use the Sub VI with Error Handling template VI to create a VI with a Case structure for error handling.

7.1.8 Graphs and charts

After acquiring or generating data, use a graph or chart to display data in a graphical form. Graphs and charts differ in the way they display and update data. VIs with a graph usually collect the data in an array and then plot the data to the graph. This process is similar to a spreadsheet that first stores the data then generates a plot of it. When the data is plotted, the graph discards the previous data and displays only the new data. Typically use a graph with fast processes that acquire data continuously.

In contrast, a chart appends new data points to those points already in the display to create a history. On a chart, see the current reading or measurement in context with data previously acquired. When more data points are added than can be displayed on the chart, the chart scrolls so that new points are added to the right side of the chart while old points disappear to the left. Typically use a chart with slow processes in which only a few data points per second are added to the plot.

7.1.8.1 Types of graphs and charts

LabVIEW includes the following types of graphs and charts:

- Waveform graphs and charts display data typically acquired at a constant rate.
- XY graphs display data acquired at a nonconstant rate and data for multi-valued functions.
- Intensity graphs and charts display 3D data on a 2D plot by using color to display the values of the third dimension.
- Digital waveform graphs display data as pulses or groups of digital lines.
- (Windows) 3D graphs display 3D data on a 3D plot in an ActiveX object on the front panel.

7.1.8.2 Waveform graphs

The waveform graph displays one or more plots of evenly sampled measurements. The waveform graph plots only single-valued functions, as in $y = f(x)$, with points evenly distributed along the x-axis, such as acquired time-varying waveforms. The following figure shows an example of a waveform graph (Fig. 7.21).

The waveform graph can display plots containing any number of points. The graph also accepts several data types, which minimizes the extent to which the data must be manipulated before being displayed.

7.1.8.3 Waveform charts

The waveform chart is a special type of numeric indicator that displays one or more plots of data typically acquired at a constant rate. The following figure shows an example of a waveform chart (Fig. 7.22).

The waveform chart maintains a history of data, or buffer, from previous updates. Right-click the chart and select Chart History Length from the shortcut menu to configure the buffer. The default chart history length for a waveform chart is 1024 data points. The frequency at which data is sent to the chart determines how often the chart redraws.

7.1.8.4 Waveform data type

The waveform data type carries the data, start time, and delta t of a waveform. Create a waveform using the Build Waveform function. Many of the VIs and functions used to acquire or analyze waveforms accept and return waveform data by default. When wire waveform data is sent to a waveform graph or chart, the graph or chart automatically plots a waveform based on

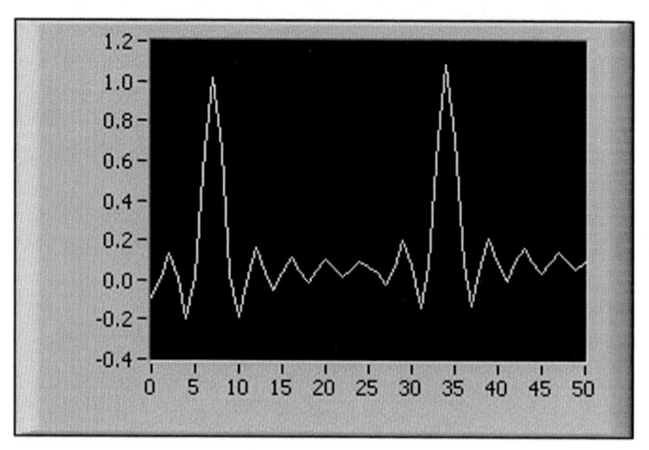

FIGURE 7.21 Waveform graph.

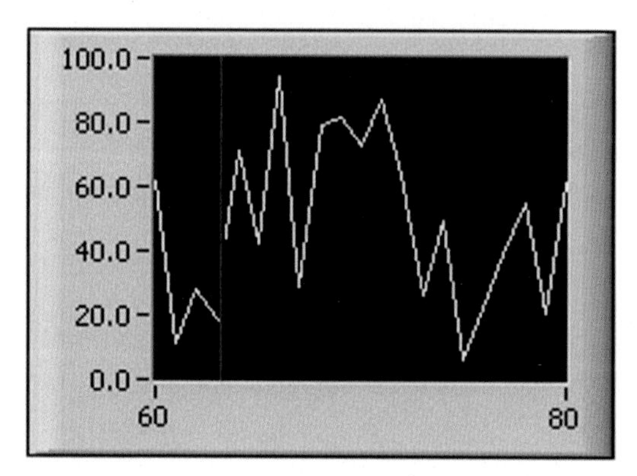

FIGURE 7.22 Waveform chart.

the data, start time, and delta x of the waveform. When a wire array of waveform data is sent to a waveform graph or chart, the graph or chart automatically plots all waveforms.

7.1.8.5 XY graphs

The XY graph is a general purpose, Cartesian graphing object that plots multivalued functions, such as circular shapes or waveforms with a varying time base. The XY graph displays any set of points, evenly sampled or not (Fig. 7.23).

Also, Nyquist planes, Nichols planes, *S* planes, and *Z* planes can be displayed on the XY graph. Lines and labels on these planes are the same color as the Cartesian lines, and cannot modify the plane label font. The following figure shows an example of an XY graph.

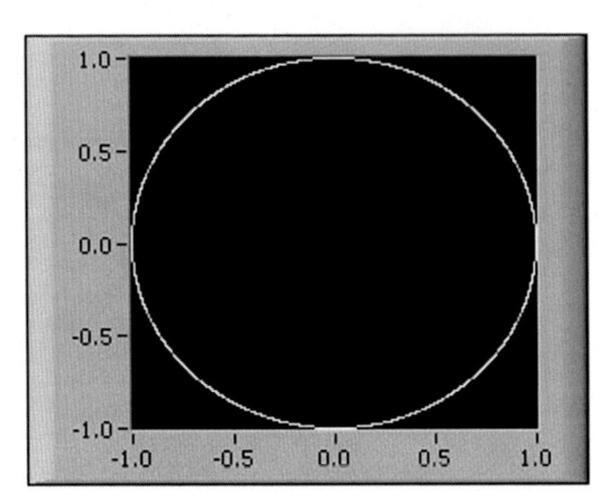

FIGURE 7.23 XY graph.

Input array

Column = y

Row = x	0	1	2
0	50	50	13
1	45	61	10
2	6	13	5

Color map definition

Array element = z	Color
5	blue
6	purple
10	lt red
13	dk red
45	orange
50	yellow
61	green

Resulting plot

3			
	dk red	lt red	blue
2			
	yellow	green	dk red
1			
	yellow	orange	purple
0	1	2	3

FIGURE 7.24 Intensity graph.

The XY graph can display plots containing any number of points. The XY graph also accepts several data types, which minimizes the extent to which data must be manipulated before being displayed.

7.1.8.6 Intensity graphs and charts

Use the intensity graph and chart to display 3D data on a 2D plot by placing blocks of color on a Cartesian plane (Fig. 7.24.). For example, use an intensity graph or chart to display patterned data, such as temperature patterns and terrain, where the magnitude represents altitude. The intensity graph and chart accept a 3D array of numbers. Each number in the array represents a specific color. The indexes of the elements in the 2D array set the plot locations for the colors. The following figure shows the concept of the intensity chart operation.

The rows of the data pass into the display as new columns on the graph or chart. If you want rows to appear as rows on the display, wire a 2D array data type to the graph or chart, right-click the graph or chart, and select Transpose Array from the shortcut menu. The array indexes correspond to the lower left vertex of the block of color. The block of color has a unit area, which is the area between the two points, as defined by the array indexes. The intensity graph or chart can display up to 256 discrete colors.

Intensity charts

After plotting a block of data on an intensity chart, the origin of the Cartesian plane shifts to the right of the last data block (Fig. 7.25). When the chart processes new data, the new data values appear to the right of the old data values. When a chart display is full, the oldest data values scroll off the left side of the chart. This behavior is similar to the behavior of a strip chart. The following figure shows an example of an intensity chart.

The intensity chart shares many of the optional parts of the waveform chart, including the scale legend and graph palette, which can be shown or hidden by right-clicking the chart and selecting *Visible Items* from the shortcut menu. In addition, because the intensity chart includes color as a third dimension, a scale similar to a color ramp control defines the range and mappings of values to colors.

Intensity graphs

The intensity graph works the same as the intensity chart, except it does not retain previous data values and does not include update modes. Each time

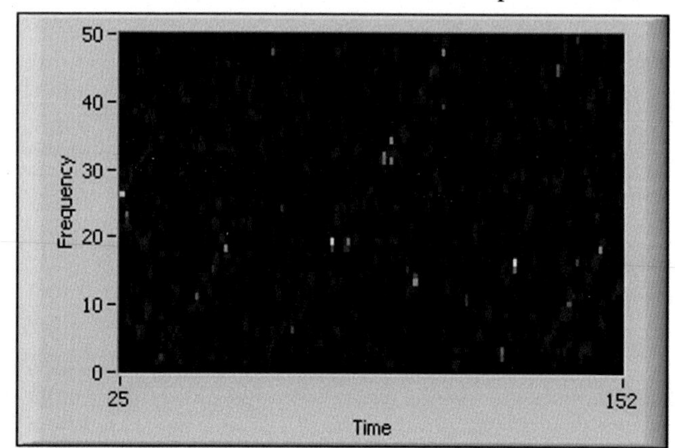

FIGURE 7.25 Intensity charts.

new data values pass to an intensity graph, the new data values replace old data values. Like other graphs, the intensity graph can have cursors. Each cursor displays the x, y, and z values for a specified point on the graph.

Using color mapping with intensity graphs and charts

An intensity graph or chart uses color to display 3D data on a 2D plot. When setting the color mapping for an intensity graph or chart, configure the color scale of the graph or chart. The color scale consists of at least two arbitrary markers, each with a numeric value and a corresponding display color. The colors displayed on an intensity graph or chart correspond to the numeric values associated with the specified colors. Color mapping is useful for visually indicating data ranges, such as when plot data exceeds a threshold value. Set the color mapping interactively for the intensity graph and chart the same way to define the colors for a color ramp numeric control.

7.1.8.7 Digital waveform graphs

Use the digital waveform graph to display digital data, especially when work with timing diagrams or logic analyzers. The digital waveform graph accepts the digital waveform data type, the digital data type, and an array of those data types as an input. By default, the digital waveform graph collapses digital buses, so the graph plots digital data on a single plot. If wiring an array of digital data, the digital waveform graph plots each element of the array as a different plot in the order of the array.

The digital waveform graph in the following front panel plots digital data on a single plot. The VI converts the numbers in the *Numbers* array to digital data and displays the binary representations of the numbers in the *Binary Representations* digital data indicator. In the digital graph, the number 0 appears without a top line to symbolize that all the bit values are zero. The number 255 appears without a bottom line to symbolize that all the bit values are 1.

Right-click the y-scale and select Expand Digital Buses from the shortcut menu to plot each sample of digital data. Each plot represents a different bit in the digital pattern. The digital waveform graph in the following front panel displays the six numbers in the numbers array (Fig. 7.26).

The binary representations digital indicator displays the binary representations of the numbers. Each column in the table represents a bit. For example, the number 89 requires 7 bits of memory (the 0 in column 7 indicates an unused bit). Point 3 on the digital waveform graph plots the 7 bits necessary to represent the number 89 and a value of 0 to represent the unused eighth bit on plot 7 (Fig. 7.27). The following VI converts an array of numbers to digital data and uses the Build Waveform function to assemble the start time, delta t, and the numbers entered in a digital data control and to display the digital data (Fig. 7.28).

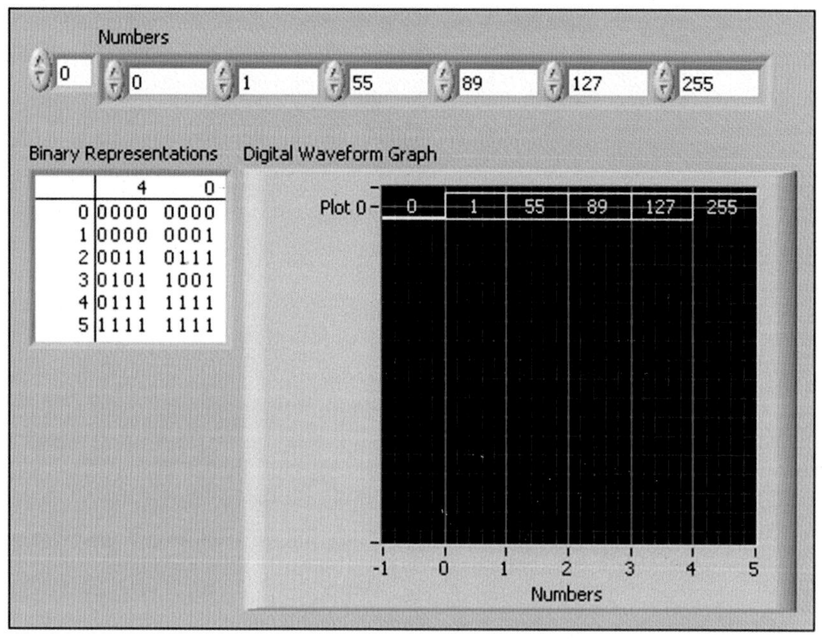

FIGURE 7.26 Digital waveform graph.

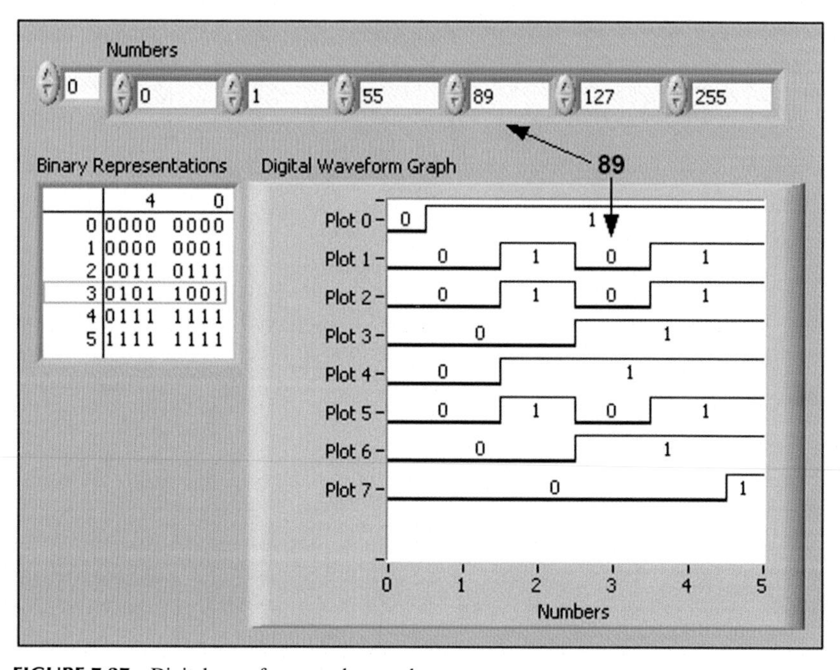

FIGURE 7.27 Digital waveform graph—numbers array.

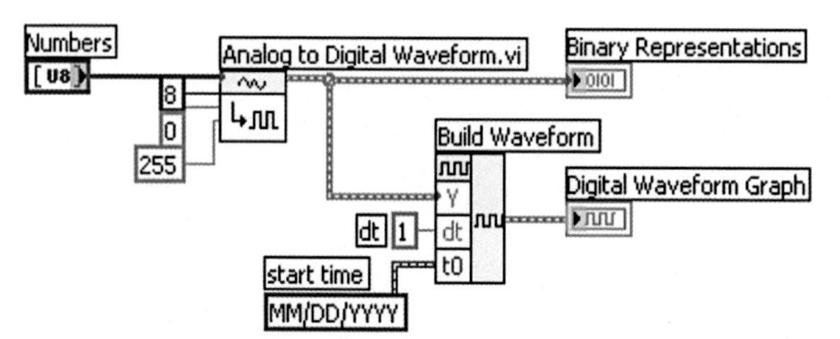

FIGURE 7.28 Digital waveform graph—binary representations.

Digital waveform data type

The digital waveform data type carries start time, delta x, the data, and the attributes of a digital waveform. Use the Build Waveform function to create a digital waveform. When wiring digital waveform data to the digital waveform graph, the graph automatically plots a waveform based on the timing information and data of the digital waveform. Wire digital waveform data to a digital data indicator to view the samples and signals of a digital waveform.

7.1.8.8 3D graphs

For many real-world data sets, such as the temperature distribution on a surface, joint time-frequency analysis, and the motion of an airplane, there is a need to visualize data in three dimensions. 3D graphs visualize three-dimensional data and you can alter the way that data appears by modifying the 3D graph properties.

LabVIEW includes the following types of 3D graphs:

- *3D Surface Graph*—draws a surface in 3D space.
- *3D Parametric Surface Graph*—draws a parametric surface in 3D space.
- *3D Curve Graph*—draws a line in 3D space.

Use the 3D graphs in conjunction with the 3D Graph VIs to plot curves and surfaces. A curve contains individual points on the graph, each point having an x, y, and z coordinate. The VI then connects these points with a line. A curve is ideal for visualizing the path of a moving object, such as the flight path of an airplane. The following figure shows an example of a 3D curve graph (Fig. 7.29).

A surface plot uses x, y, and z data to plot points on the graph. The surface plot then connects these points, forming a three-dimensional surface view of the data. For example, use a surface plot for terrain mapping. The following figure shows examples of a 3D surface graph and a 3D parametric surface graph (Figs. 7.30 and 7.31).

The 3D graphs use ActiveX technology and VIs that handle 3D representation. When selecting a 3D graph, LabVIEW places an ActiveX container

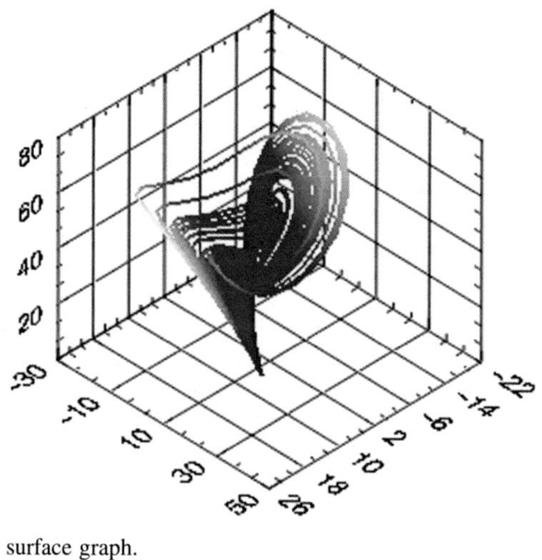

FIGURE 7.29 3D surface graph.

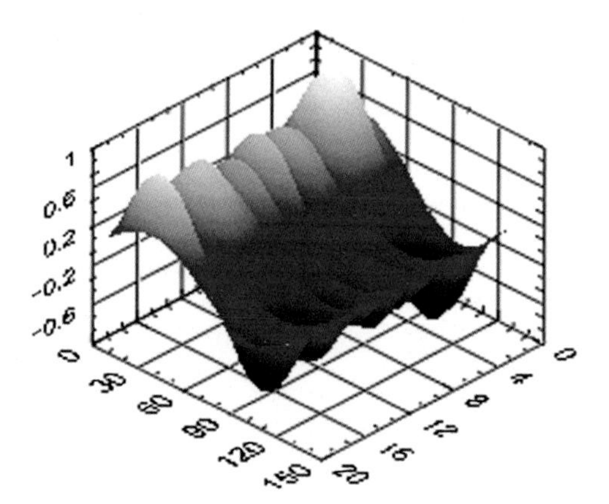

FIGURE 7.30 3D parametric surface graph.

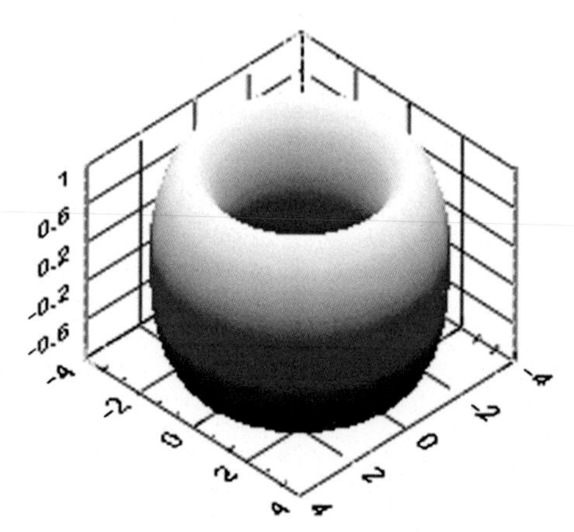

FIGURE 7.31 3D curve graph.

on the front panel that contains a 3D graph control. LabVIEW also places a reference to the 3D graph control on the block diagram. LabVIEW wires this reference to one of the three 3D Graph VIs.

7.2 LabVIEW examples

7.2.1 Introduction to basic operations

Step 1

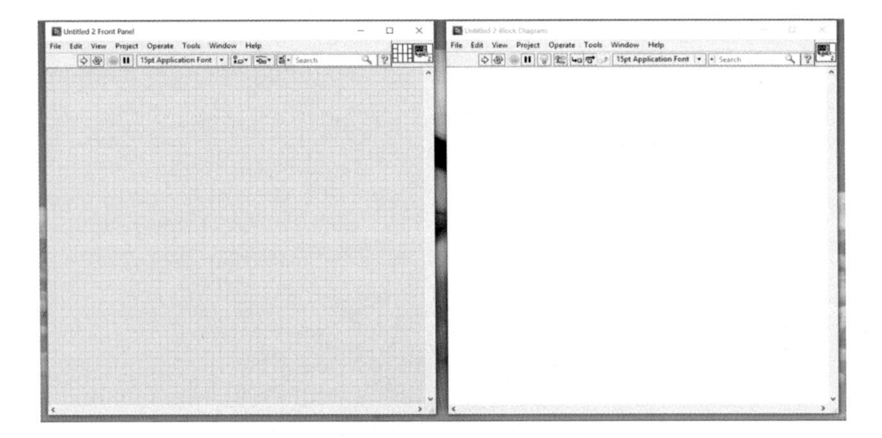

- Open the LabVIEW Software, it contains two separate windows.
- Left Window is the Output Verify Window.
- And Right Window is the Worksheet Window. All Works should be complete in this column.

Step 2

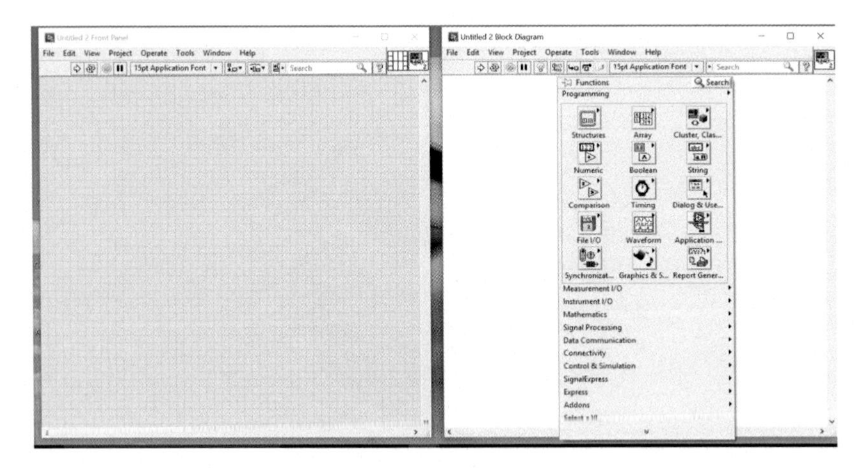

- Go to the Worksheet window. And Right click.
- Like above Function Blocks will open. That is the Main component in the LabVIEW.

Step 3

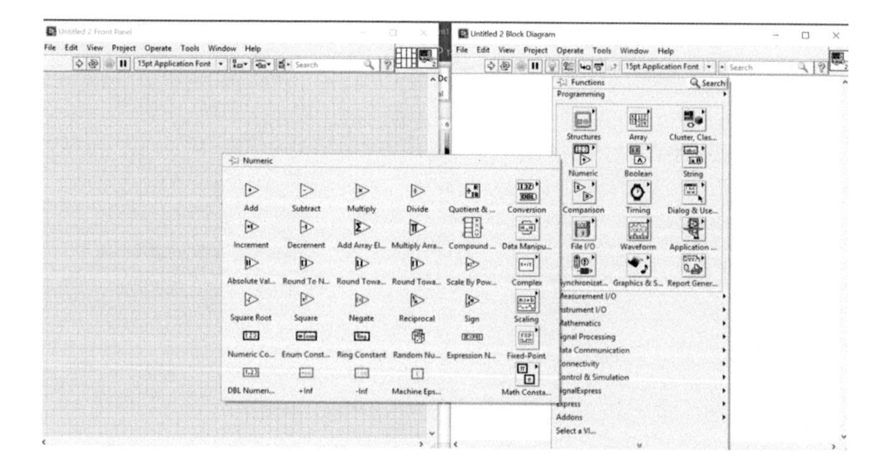

- Then Most of the Projects and Lab Experiments done using this Mathematical Option.
- That is available in Numeric Option.

Step 4

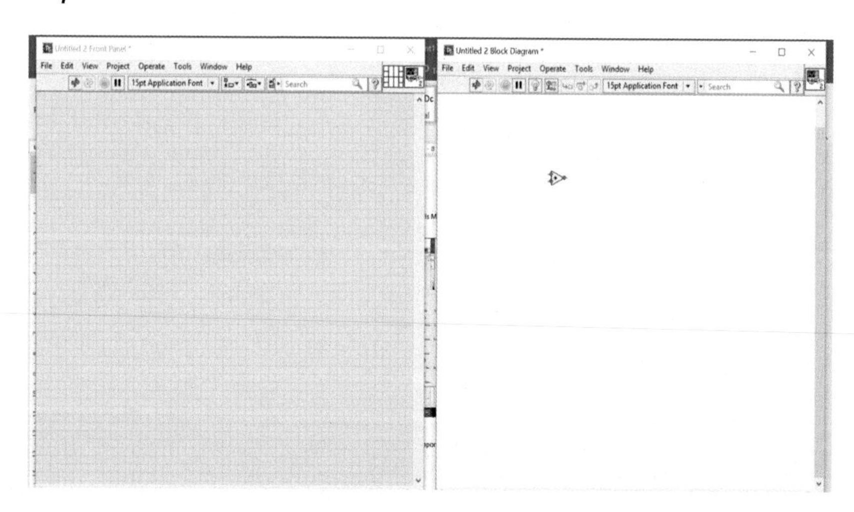

- Drag our Required Mathematical Component to the Worksheet Window. And Every Component has its Input indicators and Output Indicators in their blocks.
- Block Indicators are fixed on the basis of Mathematical Logic.

Step 5

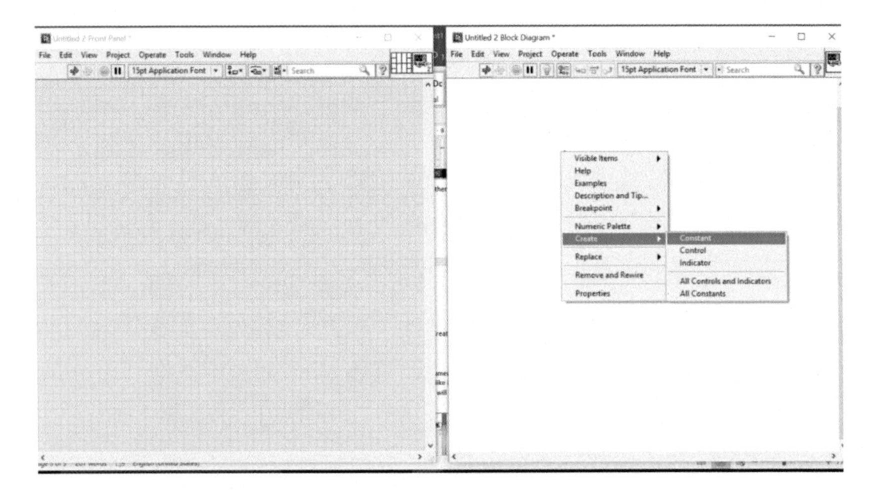

- For Constant Input, Right Click Create Constant.
- Enter the Specific Numeric Value in the Constant Block.

Step 6

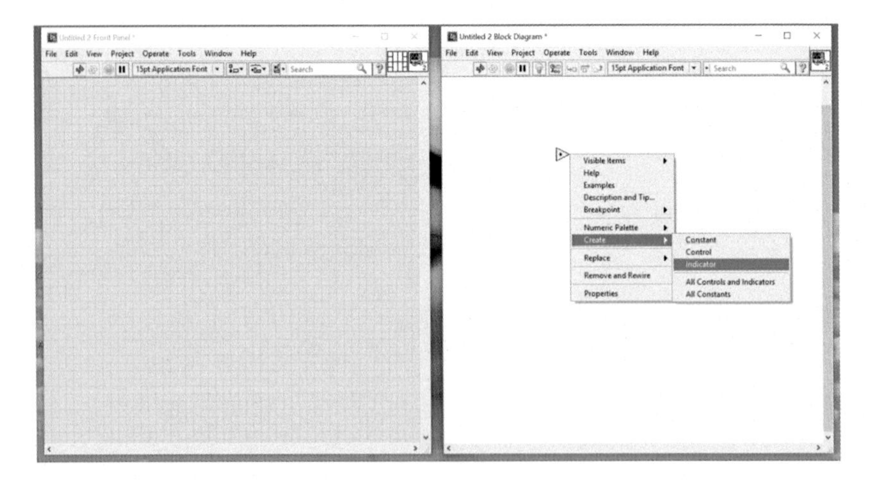

- Mathematical Output Indicators require Numeric Output Indicators.
- And Condition Operations require some True or False indicators like Bulb Blinking.
- And we can place the Output indicators in Worksheets. Automatically it will Copy in the Output Window. The image below shows everything.

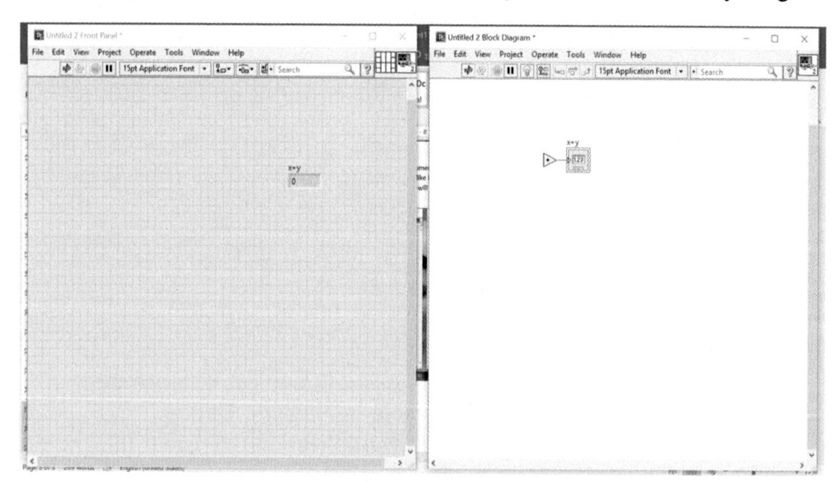

Step 7 Comparison operation's bulb output

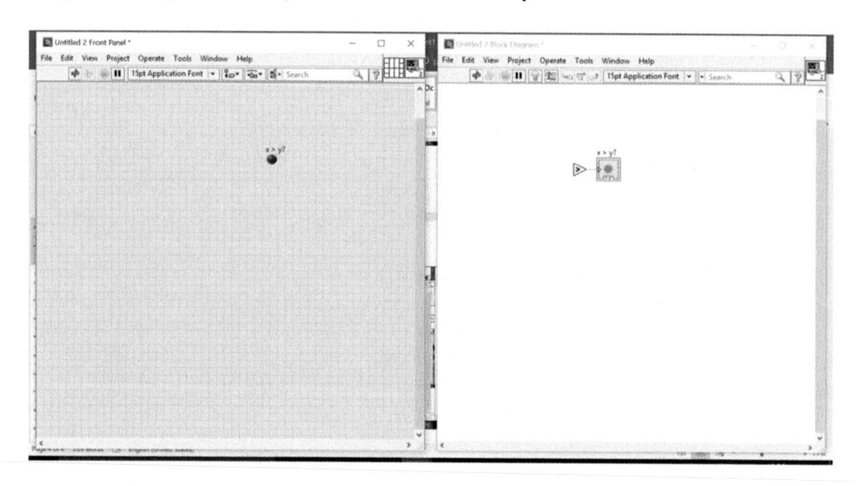

7.2.2 Basic mathematical operations

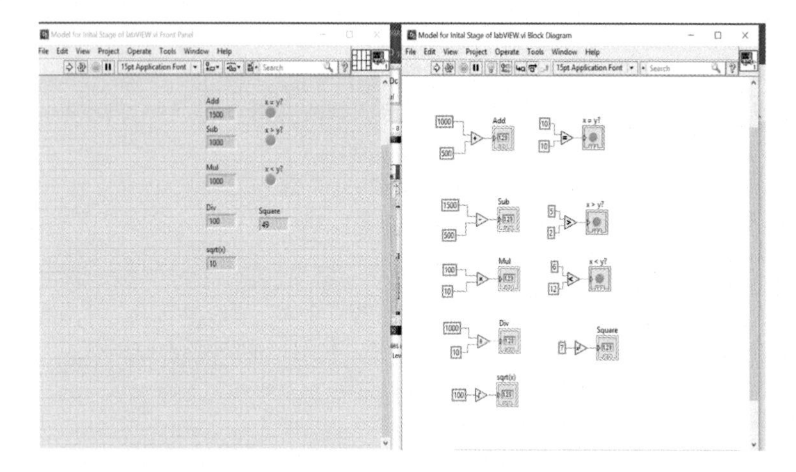

- All Mathematical Operations can complete in LabVIEW Software.
- All Mathematical Components available in Numeric Option.
- For example, Addition, Subtraction, Multiplication, Division, Integration, Differentiation, Square Root, Square, Comparison Process, etc.

7.2.3 Loop functions

7.2.3.1 For loop

Odd or even numbers formulation using for loop

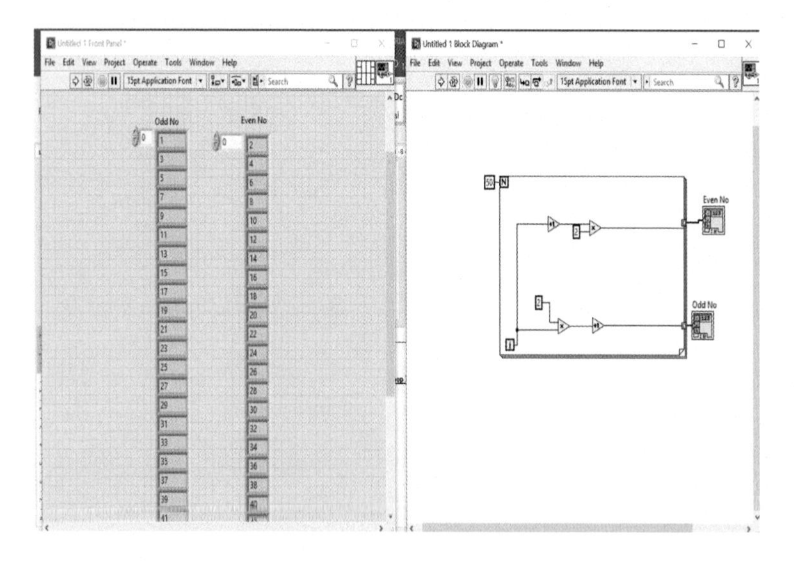

- Take For loop in Structure block and arrange the components as per the above Fig.
- This program gives the set of Odd and Even numbers in the given range.

Find out if the given number is odd or even

Odd

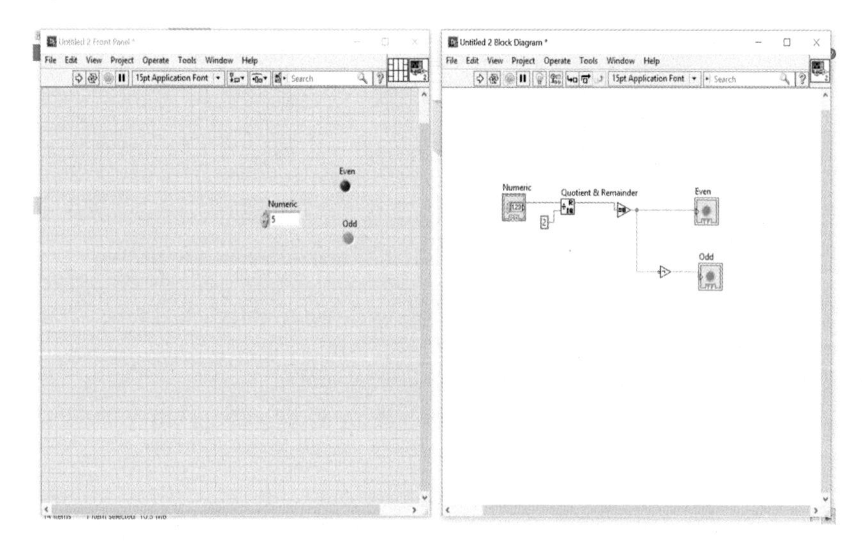

Even

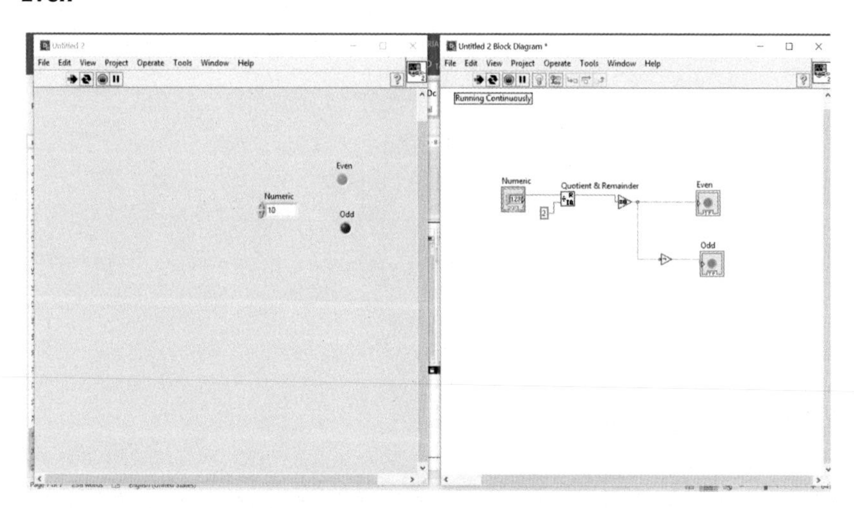

- Also find out the Odd and Even numbers without using For loop Structure.
- Blocks are made as per the above figure. If the given number is Odd, then the Odd LED will blink. And if the given number is Even, then the Even LED will blink.

7.2.3.2 While loop

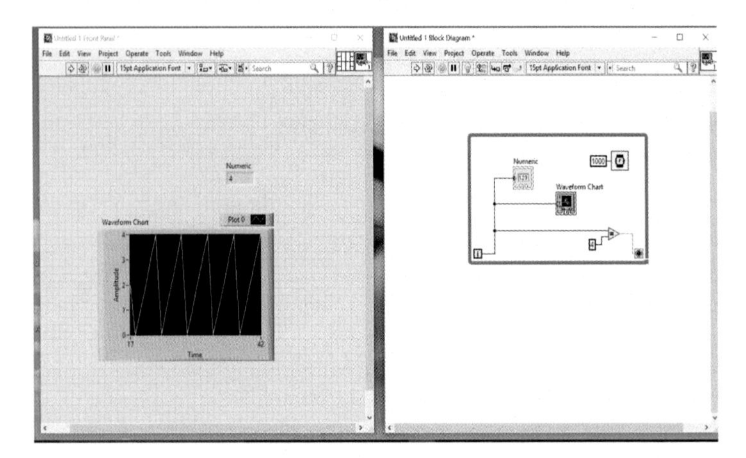

- Next loop structure is While loop. Go to Structure block and take While loop block.
- And While loop structure is the Conditional loop.
- A simple numerical condition is constructed in the above Fig. The range is in between 0 to 4. The loop increments do not go beyond the 4.

7.2.4 Traffic light program using LabVIEW

7.2.4.1 Red signal

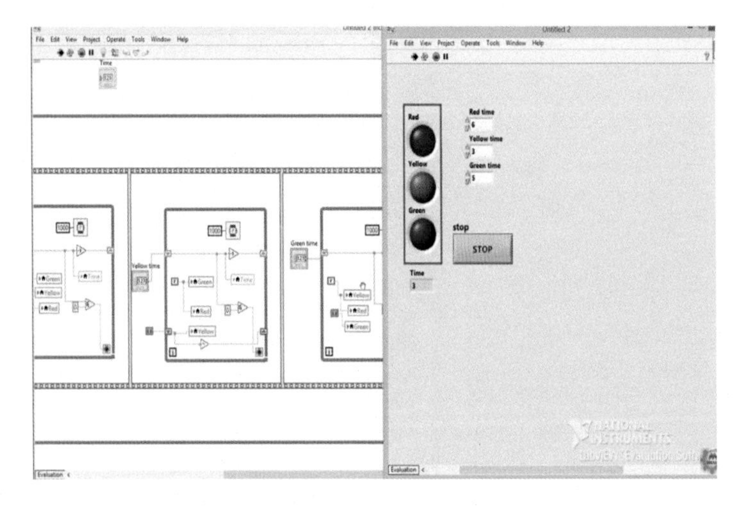

7.2.4.2 Yellow signal

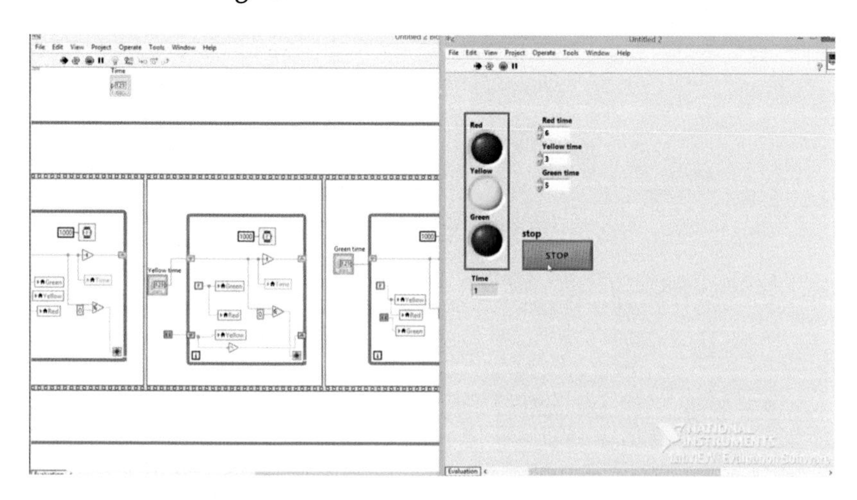

7.2.4.3 Green signal

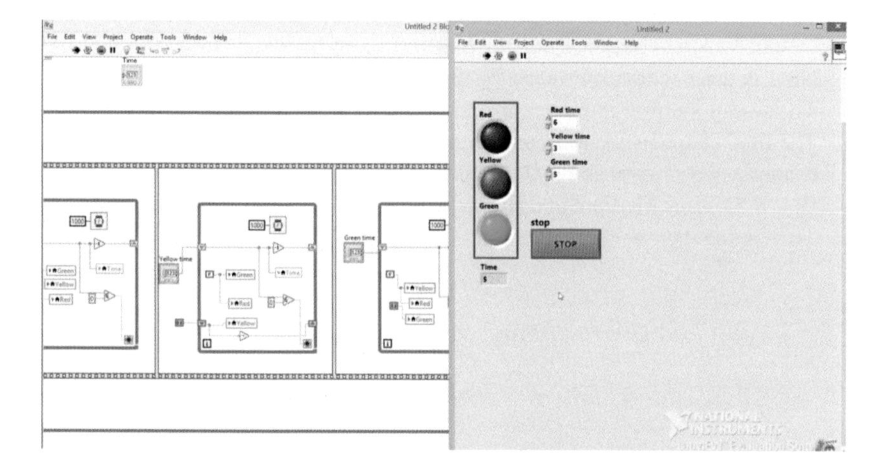

- This is the Traffic Light Program in LabVIEW Software.
- Using the Timer Concept and Formula Node Module, we can achieve this concept.
- Initially arrange the Lamps using Boolean functions and timer values of each signal.
- While loop is used to restrict the time period within the limit. And Formula Node is used to arrange sequential operation.

7.2.5 Water level indicator

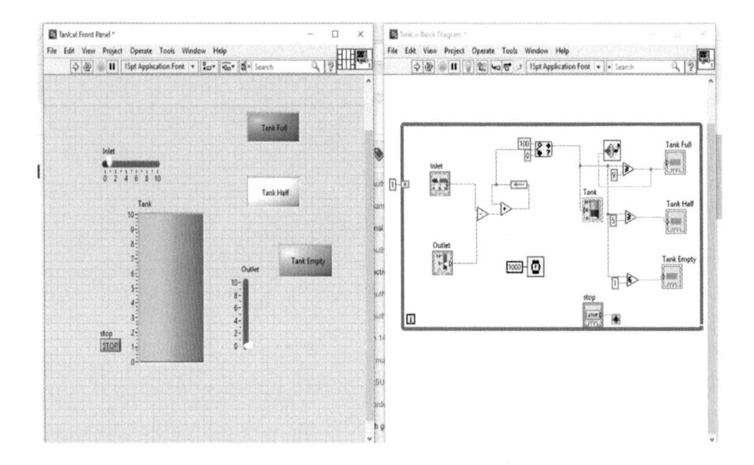

- In Worksheet window, Inlet and Outlet indicate Tank Inlet and Outlet water levels.
- The figure Tank level is zero. Adjust the Inlet level and Outlet level in Output Verify Window. Then the Subtraction of Inlet and Outlet takes place.
- Next is the Feedback Process of the Inlet and Outlet water entries.
- The watch symbol indicates the Time interval of the Process.
- The final elements are the Tank High, Medium, and Low-level indicators. For example, water level above 9 means it will be indicate by the Tank High bulb. Below 5 means Medium and Below 1 means Tank Empty bulb will blink.
- The figures below clearly explain the water level indicator process.

7.2.5.1 Tank empty (ordinary white bulb changed into red bulb)

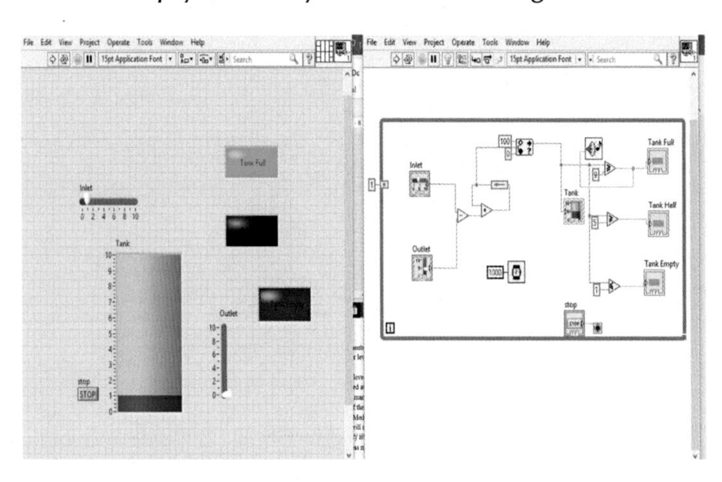

7.2.5.2 Tank medium (in medium, black bulb changed into white bulb)

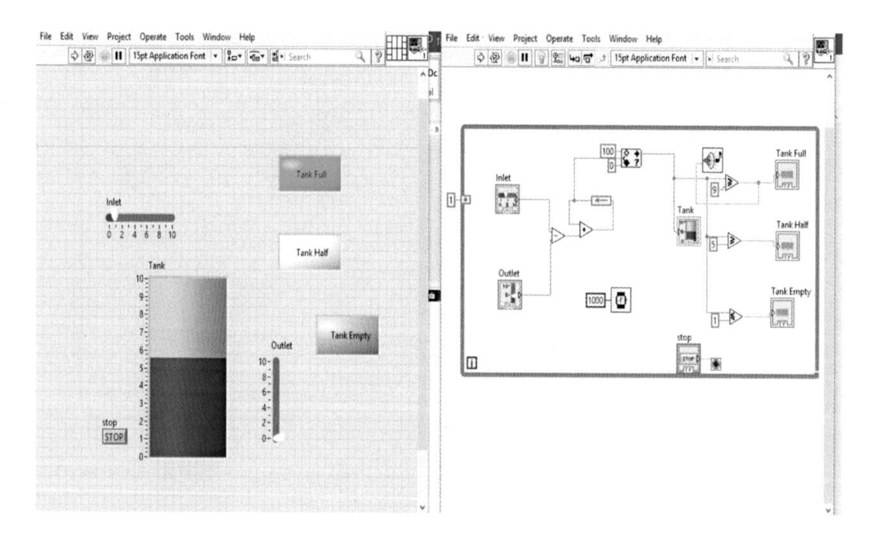

7.2.5.3 Tank full (at full condition green lamp changed into blue lamp)

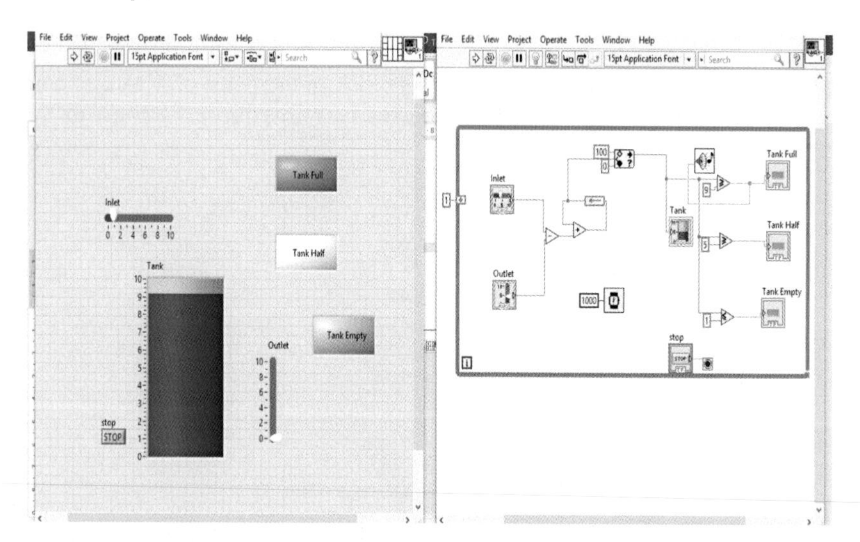

- At Full level, on a laptop or PC an alarm will ring because of the Alarm block in the LabVIEW Function.

7.2.6 Temperature indicator

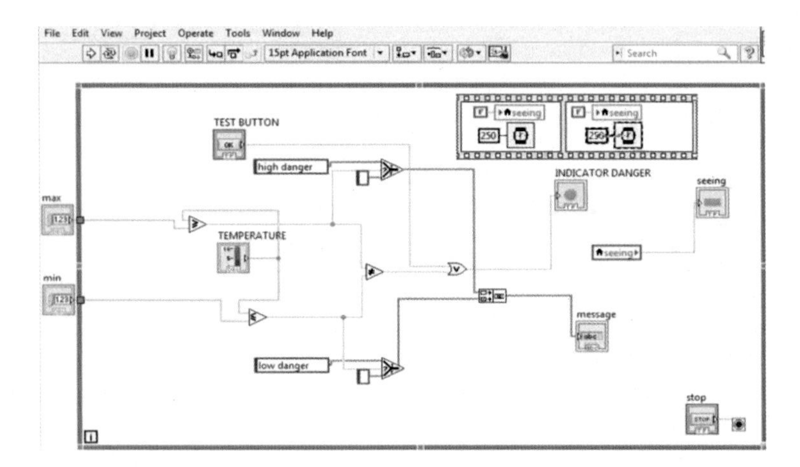

- This is the program for the temperature indication process at Dangerous level and Normal level.
- Initially add Max and Min Values of Temperature blocks, and Comparison blocks for Max and Min values.
- Then an Automatic Selection block is added. This is used to select the highest and lowest ranges of the Temperature values. It should initiate the Temperature Range values in this block.
- In Advance, we can connect the DAQ block in the place of Max and Min Values Block. The DAQ block receives real-time process values to the LabVIEW software.

7.2.6.1 Temperature at critical value

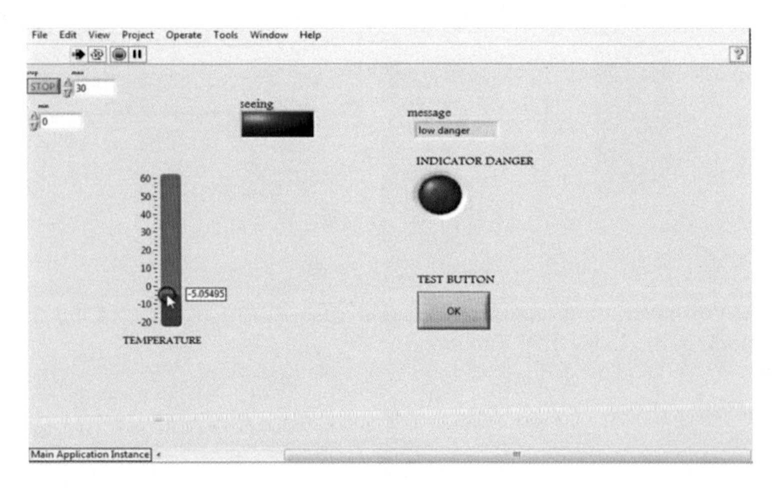

7.2.6.2 Temperature at normal condition

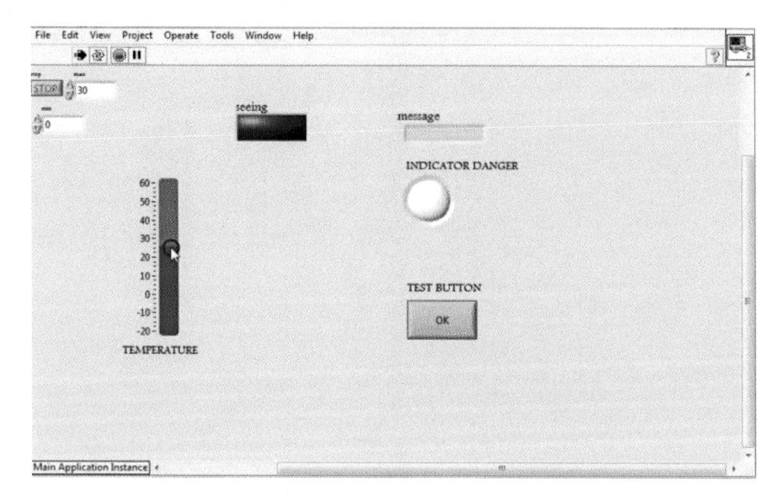

7.2.7 Simple calculator in LabVIEW

- It is a simple calculator program in LabVIEW. This example explains basic calculator operations within the single case structure.
- Case structure is a structure which can contains more than one operations.
- Enum block is used to connect the various applications and case structures.
- Initially we should add multiple operations in the case structure through the Enum block.
- We can adjust the required operations in the Enum block in the Output window. It automatically changes the case structure and gives the Output of the required operation.

7.2.7.1 Addition

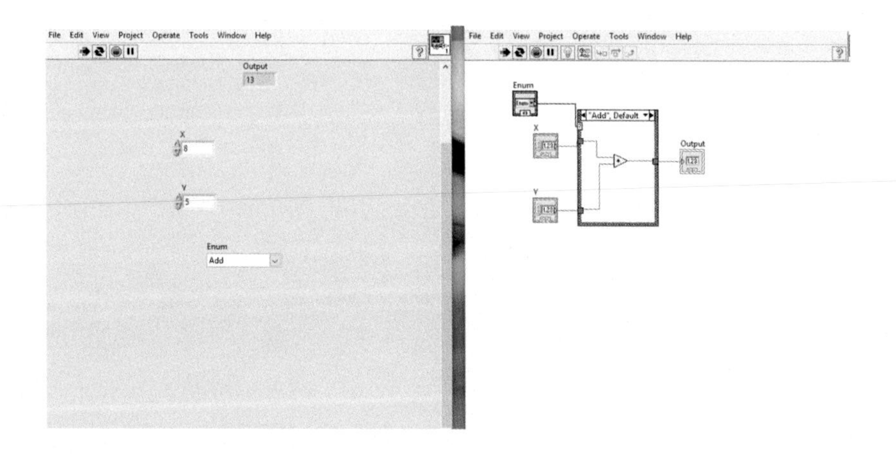

7.2.7.2 Subtraction

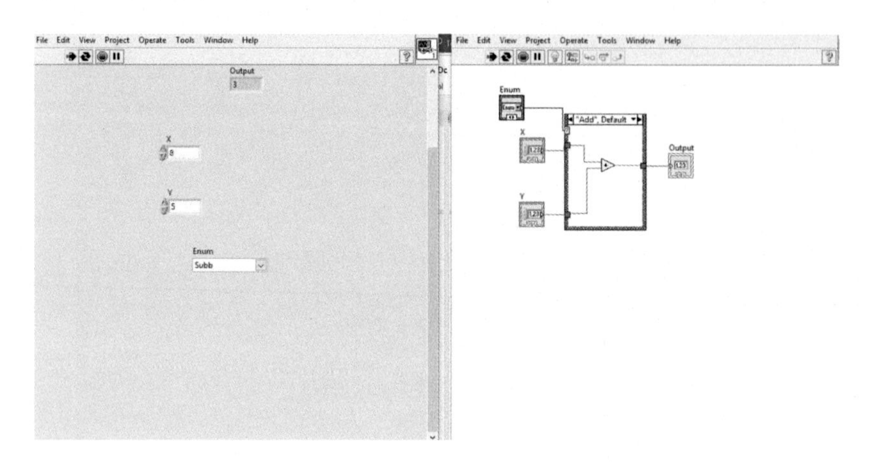

7.2.7.3 Multiplication

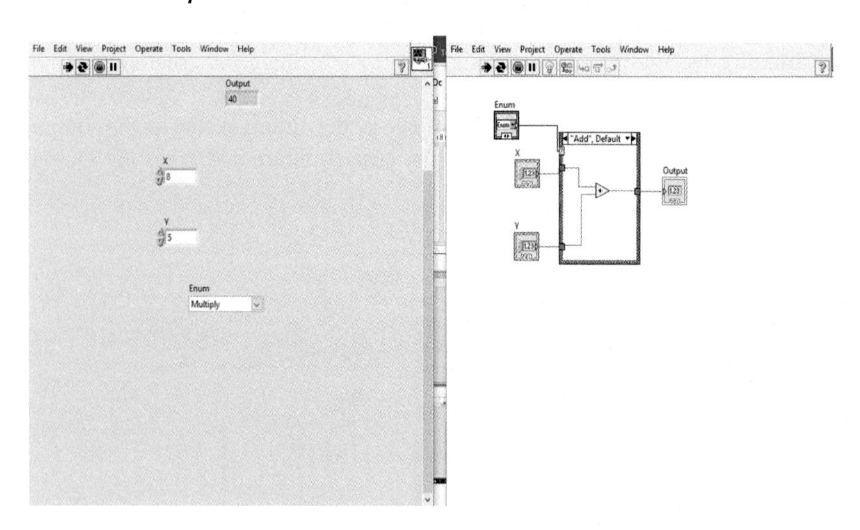

7.2.7.4 Division

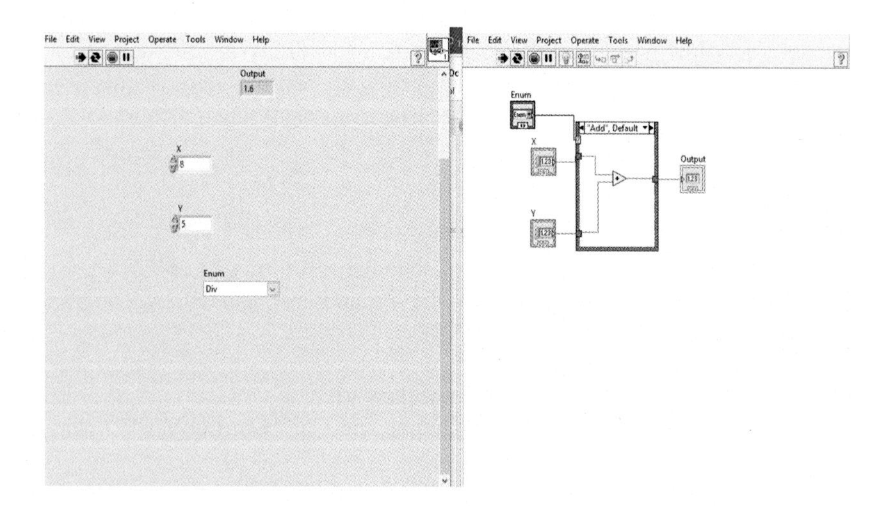

- Additionally, in LabVIEW, we can see the running process of the specified operation with the use of the Highlight Execution option.

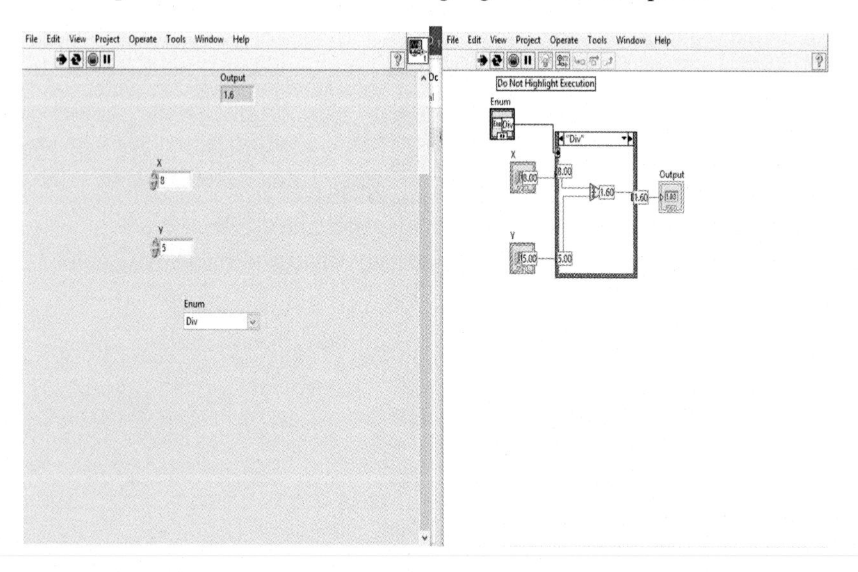

7.3 Summary

This LabVIEW chapter gives you the chance to explore the LabVIEW environment, dataflow programming, and common LabVIEW development techniques in a hands-on format. It also helps to develop data acquisition, instrument control, data-logging, and measurement analysis applications.

7.4 Review questions

1. What is meant by LabVIEW? And state the full form of LabVIEW.
2. What is the difference between the Front Panel and Block Diagram Panel in LabVIEW?
3. What is meant by Enum?
4. What are the common reasons for broken VIs?
5. What is the difference between Graphs and Charts in LabVIEW? State their types.
6. What is meant by Highlight Execution in LabVIEW?
7. Should we implement the PID concept in LabVIEW like in Matlab? Explain.
8. State the uses of Formula Node in LabVIEW.
9. What is meant by sub VI?
10. Which device is used to acquire data from an external hardware device to LabVIEW Software?
11. What is meant by Break Point in LabVIEW?
12. Explain the color coding of the wires.
13. Give the function of Probe Tool in LabVIEW.
14. Explain the difference between Matlab and LabVIEW.
15. State the difference between an Intensity Chart and Intensity Graph?

Chapter 8

Introduction to Power Systems Computer-Aided Design

Chapter Outline

8.1 Introduction

The world-renowned Power Systems Computer-Aided Design (PSCAD) is a coercive and supple visual communication user program that works with the EMTDC. PSCAD helps users pictorially generate a circuit, trial a simulation, examine the outcome, and carry off the data in entirely incorporated, graphical means.

PSCAD envisages a collection of tailor-made programs and tested models for simulated parameters, with a wide range of simple passive control

Software Tools for the Simulation of Electrical Systems. DOI: https://doi.org/10.1016/B978-0-12-819416-4.00008-9

functions and elements, and analyzable models such as electrical machines, complete FACTS device applications, circuits of transmission lines over head transmission (OHT), and underground (UG) cables. In case a needed model is not available, PSCAD renders ways for creating needed models. For instance, required models can be developed to receive a required module through the composition of available models or by constructing essential models starting from the initial stage in a negotiable design facility.

Some models commonly saved in PSCAD master library include the following:

- Elements such as inductors, capacitors, and resistors.
- Electrical equipment such as electromagnetically coupled windings, transformers, switchgears, breakers, AC/DC equipment, governors, stabilizers, inertial models, and excitation equipment.
- Frequency-influenced power networks for OH transmission and cables.
- Current and voltage sources.
- Protective relays, meter, and measuring functions.
- Electronic devices such as thyristors, diodes, GTOs, and FACTS power electronic circuits of high voltage direct current (HVDC), SVC, and other controllers.
- All control functions for analog and digital, including DC and AC controls.
- Wind energy turbines and governors.

8.1.1 Studies carried out using Power Systems Computer-Aided Design

AC network studies on contingent conditions comprising motors and generators, excitation systems, turbine speed governors, turbine models, transformers, transmission network cables, and overhead lines and its connected loads include the following:

- power system protection for the coordination of relays;
- equipment magnetic saturation and impulse testing on transformers, and also for insulation coordination for transformers, lightening arrestors, and breakers;
- power system networks, inclusive of machines, transmission lines, and HVDC systems;
- an evaluation of filter design and harmonic analysis;
- studies on subsynchronous resonance of power networks with machines, transmission system, and HVDC systems;
- design of control systems and effects of FACTS devices like VSC, STATCOM, and cycloconverters with controller design;
- probe into new control circuit concepts;

- faults due to lightning strikes in power network and any flaw on breaker operations;
- studies on fast and steep front;
- naval vessels design, which are propelled on electric power; and
- checking pulsing implications on wind turbine, diesel engines, and also power system networks.

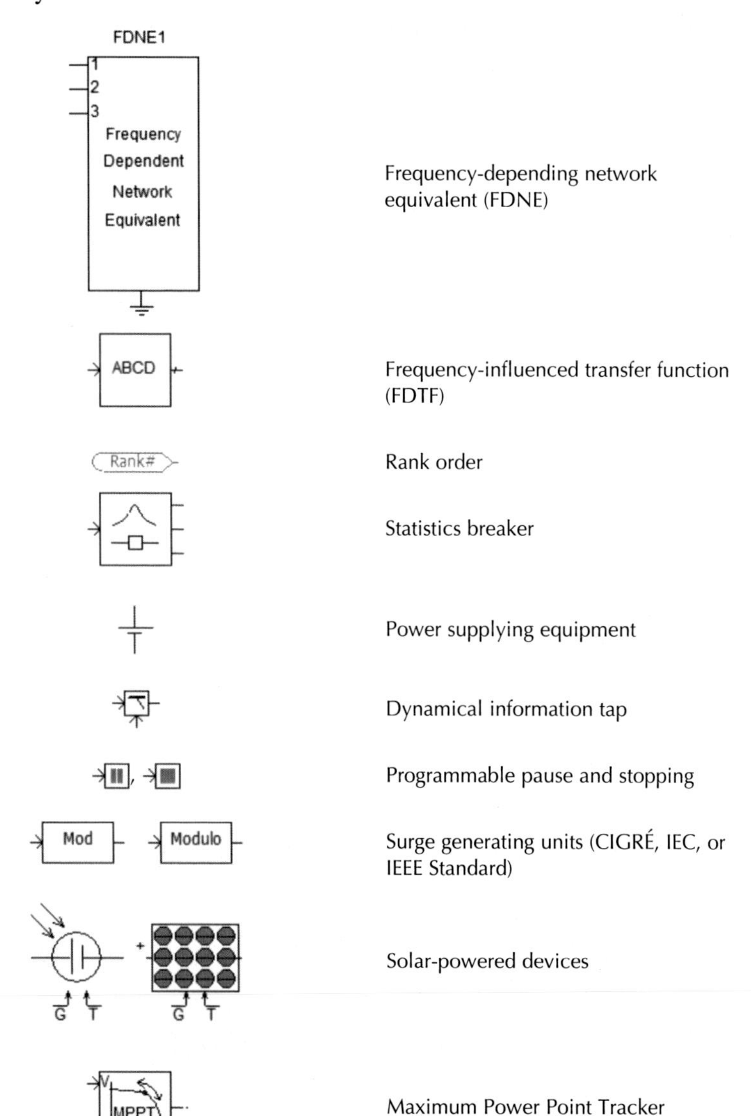

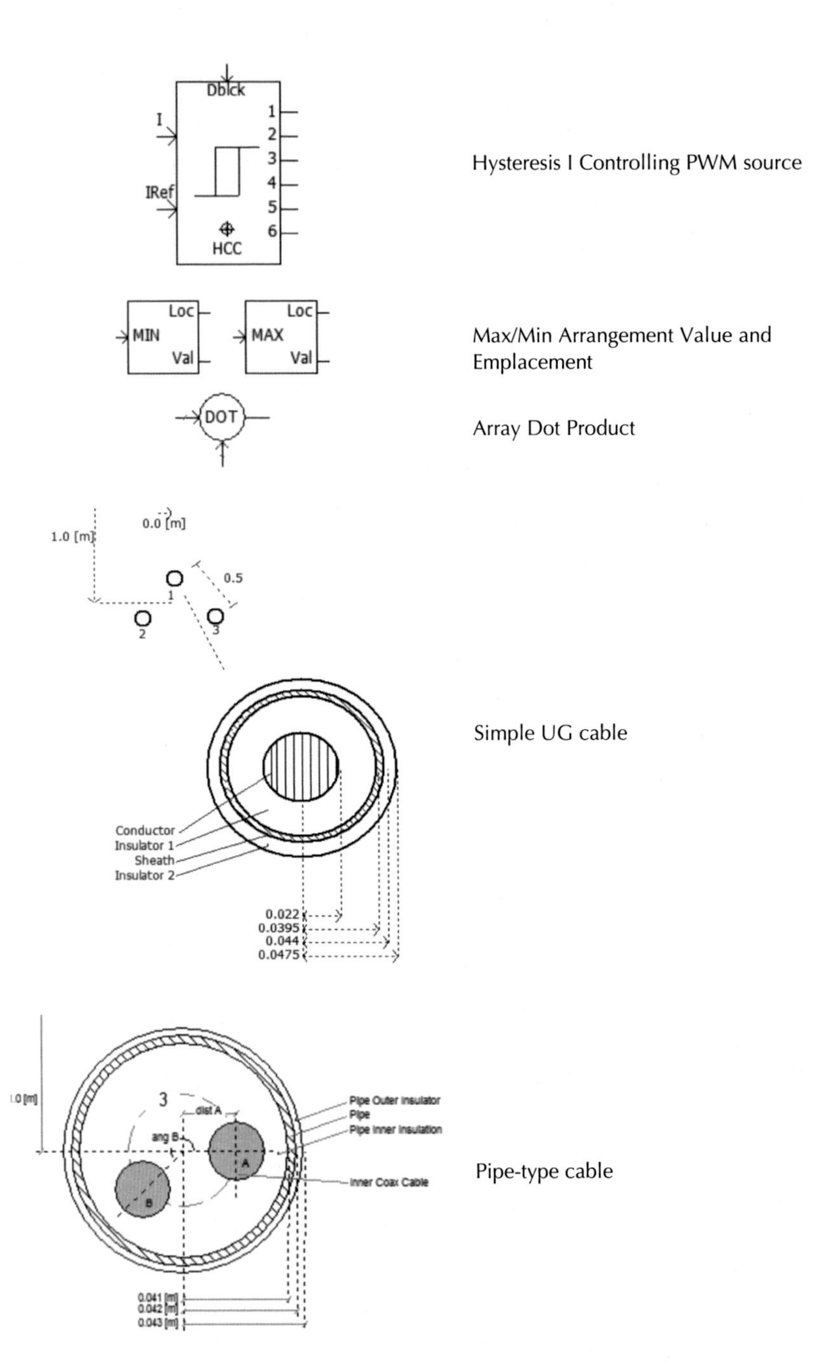
Dblck
I
IRef
1
2
3
4
5
6
HCC
Hysteresis I Controlling PWM source
MIN
Loc
Val
MAX
Loc
Val
Max/Min Arrangement Value and Emplacement
DOT
Array Dot Product
1.0 [m]
0.0 [m]
0.5
1
2
3
Conductor
Insulator 1
Sheath
Insulator 2
0.022
0.0395
0.044
0.0475
Simple UG cable
3
dist A
ang B
A
B
Pipe Outer Insulator
Pipe
Pipe Inner Insulation
Inner Coax Cable
0.041 [m]
0.042 [m]
0.043 [m]
Pipe-type cable

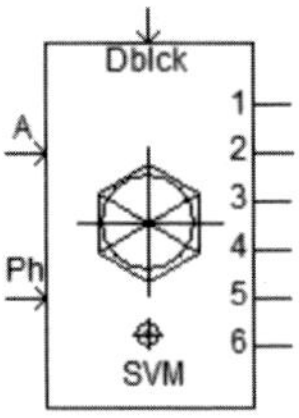

Space-vector modulation (SVM)

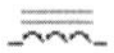

Saturating reactor

Spark gap

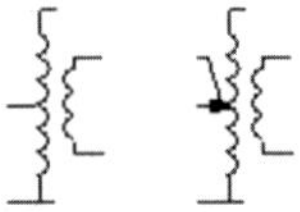

Singleφ/three Winding Auto Transformer

3-Y-Y Auto Transformer with Tertiary

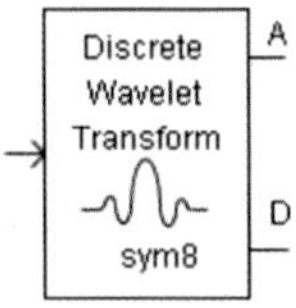

Discrete Wavelet Transform (DWT)

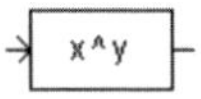

X Power Y

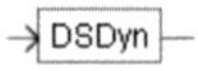

Force to dynamics subroutine

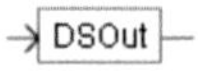

Force to dynamics subroutine

Runtime Configurable Passive Branch

XY Array

Variable Series Impedance Branch

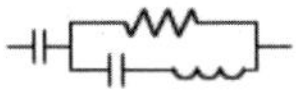

C-Type Filter

Multiple Run Additional Recording

8.1.2 Opening Power Systems Computer-Aided Design

When PSCAD has been installed in the system, to start with, go → Start

Then to →| All Programs |→ PSCAD | →PSCADX4 | →PSCAD <Version> in the Windows start menu for opening the application program.

8.1.2.1 Schematic tabs and ribbon control bar

The most vital modules of PSCAD are ribbon control bars and schematic tabs.

Ribbon Control Bar: The envisaged area under the application directly at title bar is the ribbon control bar. Almost all usefulness of the application accessible with PSCAD is here.

Schematic Tabs: The schematic tab is portrayed for all projects in the workspace and is depicted by a schematic tab, which displays name of project, path, and number call of each module with a canvas presently in view.

8.1.2.2 Working space and Windows for message

At the left-hand top area of environment, we can see a dock window named as the workspace window. When it is not available, it can be seen by going to the ribbon control bar, then → Select Workspace by clicking in View tab and further in the Panes drop list button.

The workspace give a full outlook of all libraries inclusive of all case study projects already loaded, it is used to carry out a wide range of performance, such as navigating or to access files.

There is one more dock window directly beneath the workspace or at the definition editor, which is used to the build messages pane. When it is not available to be seen, ribbon control bar can be used; hence, click the View tab access Messages in Panes drop list button.

8.1.2.3 Opening a case project

To begin with, the simple of example of problem module, for the lesson. This effect helps to make sure that PSCAD and FORTRAN compilers are being applied equally, and whether installed precisely. It can be learned to generate a problem module from initial condition in tutorial to create a fresh project.

A supplementary project the exists in the list of workspace window is *vdiv* (Single Phase Voltage Divider). This appears straight beneath the main library project. The key scheme canvas for a project will open by itself on schematic window tab. On observing assembly of circuit for voltage divider as in the drawing, is located in top left spot of key page at the project that has been opened. The plots located directly in the right side of circuit.

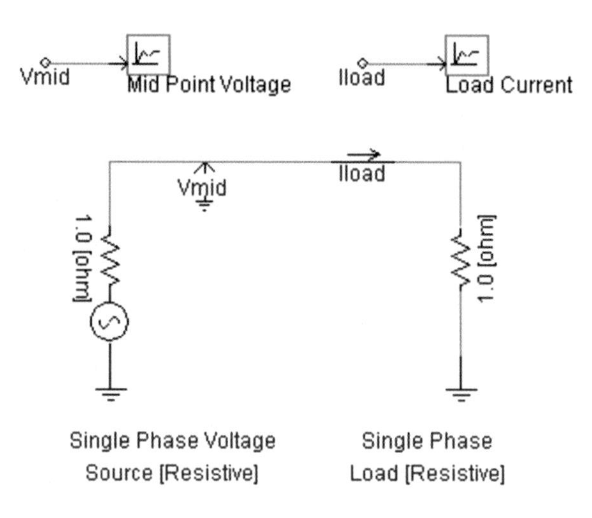

This circuit envisages $1-\varphi$ resistance voltage source connected with resistive load. Since the values of the source and load resistances are equal, the value of the voltage exits at load terminal is 50% that of the voltage at source resistance. This voltage is quantified by a voltmeter named as Vmid linked to the node in between the load and source. The value of I of circuit is $E/(R_S+R_L)=I_{\text{Load}}$

The graphs and plots contain the voltage levels exists at the midpoint of the circuit, the current which flows in the circuit while the project simulation is carried out.

8.1.3 Carrying out simulation of circuits

An easy calculation to find out value of load current and midpoint voltage is done before running the simulation. The source component is opened by double-clicking to view its parameter; it is to be noted that the source voltage value is 70.71 kV which is RMS of 100 kV peak. This dialog is closed with "Cancel" button in lower side of dialog and by left-clicking in empty place of page in order to de-select source component. The midpoint voltage is 50 kV peak for 100 kV source voltages, and the load current is 50 kA peaks. Now it can be simulated, and it can be actually verified for waveforms of current and voltage.

While the button is pushed, PSCAD goes with various levels of dealing out the circuit prior to initiating the simulation. The messages can be observed in the status bar in the lower side window of the PSCAD that is carried in a range of stages of procedure. It is impossible to read these with respect to how quickly the computer is processed.

By observing the graphs as the simulation carried out in near the lower-right corner of the environment, a message is observed in percentage, it represents the percentage of the whole simulation extent. At the right side the current simulation period is observed, which varies with the simulation. Again, with respect to how quickly the computer is processes, the simulation is completed instantly:

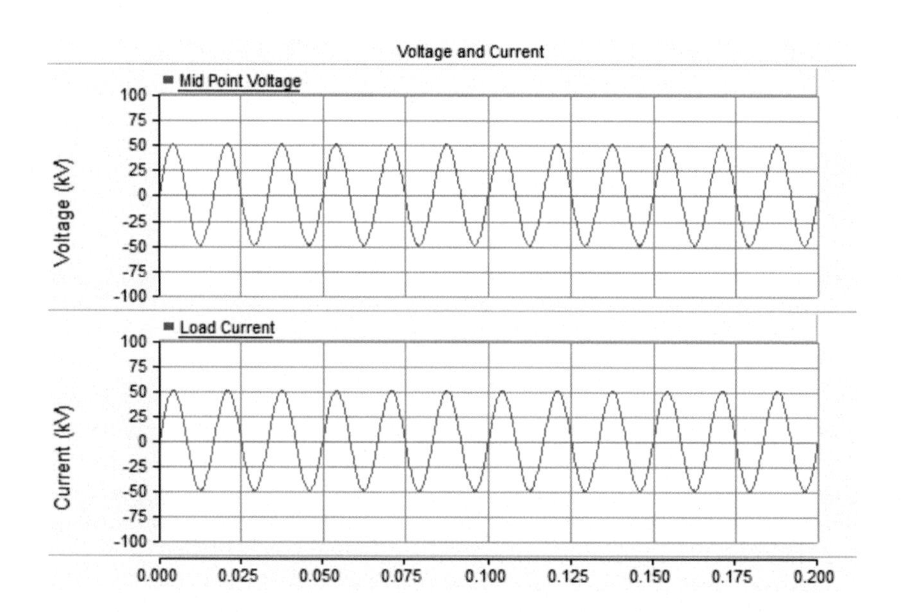

8.1.4 Printing the circuit

To get the output in print form of the circuit together with the graph which has been simulated, by select any of Print Page or Print Preview Page options gives the printed output.

If the Print Page is chosen, its dialog will be displayed. The content in dialog of Print Page depends what print is needed; clicking the OK button will print the content.

8.2 Control and Plotting online

8.2.1 Frames in Graph

A unique object during runtime container applied to encapsulate overlaying or poly graphs is the "graph frame" and can be located anywhere on the canvas schema. After graph frame is included, it can move forward to include many unlimited graph numbers as wanted by the user.

The application of graph frames is solely used in the curve plotting against the time. The horizontal axis of the graph frame is always representing the simulation time period. If it is required to draw a curve as a function with a variable, observing the *XY* plots will reveal the results.

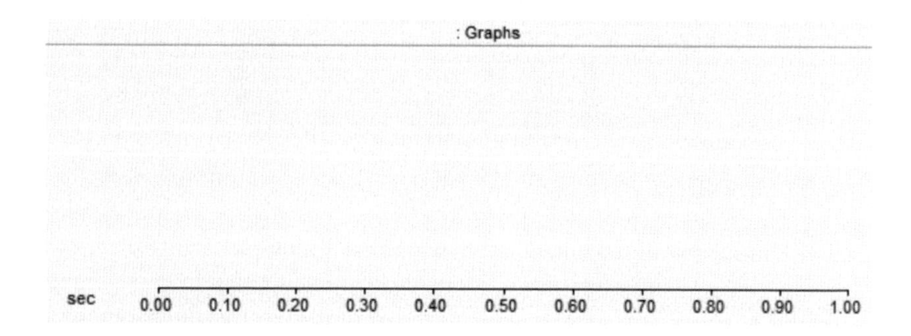

Addition of Graph Frame: The required project is opened in the schematic perspective. In the blank portion, right-click on the page and go to Component Addition I Graph Frame. Also, this can be pulled up by selecting the component tab from the graph frame button, which is in the ribbon control bar.

Resizing and repositioning graph framing: To resize the graph, left-click in title bar. To reposition the frame of the graph, move the mouse pointer in the title bar by left-clicking and holding. The frame is dragged and placed in the suitable place of choice, and then releasing the button will find the frame in the new position. By moving the mouse pointer in the title bar and then left-clicking, the frame can be selected and resized.

Frames for Cut and Copy: On the graph frame title bar, right-click and choose either Cut or Copy Frame correspondingly. After the graph frame is selected and either cut or copied, the frame can be copied in other positions in simulation project.

Frame Pasting: Frame pasting is carried by cutting or copying the frame as said in the earlier method and then right-clicking in the blank space in schematic vie and can be pasted any number of times.

Frame Property Adjustment: To approach the Graph Frame Attributes dialog, left double-click the plot title bar, or right-click over the title bar.

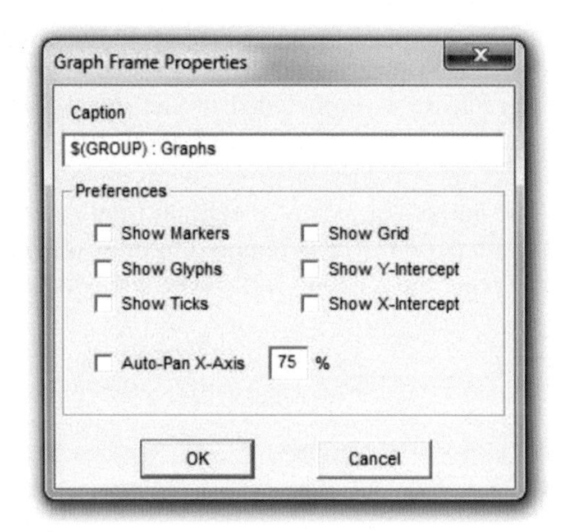

Items accessible in the dialog are explained as follows:

Caption Header for graph frame entered (this text appears in graph frame at title bar). The option text appears as bit cryptically: The dollar group $ syntax is applied in the naming rule for abstraction objects at workspace. For further substance in syntax, observe the Group of Runtime Objects. The instant syntax $ finds its usefulness in displaying name of module and event number of ease in transferal with reports.

Preferences:

Show

Markers: Choosing the choice to exhibit the *X* and *O* markers of every graph.

Glyphs: Choosing the choice to exhibit glyph symbols on every curve of frame.

Ticks: Choosing this choice to exhibit ticks along the *Y*-intercepts in every graph.

Grid: Choosing this choice to exhibit the grid in every graph.

Y-Intercept: Choosing this choice to exhibit the *Y*-Intercept at every graph.

X-Intercept: Choosing this choice to exhibit the *X*-intercept sat every graph.

Auto-Pan X-coordinate: This is the way to modify, the Pan act. The input field directly beside this check box acquire an information correspond to the proportion of the presently seen graph window (or aperture).

8.2.1.1 Altering X-axis attributes

To approach the *X*-coordinate attribute dialog box, either "double-click" the frame's *X*-coordinate, or horizontal coordinate, and choose attribute of coordinates.

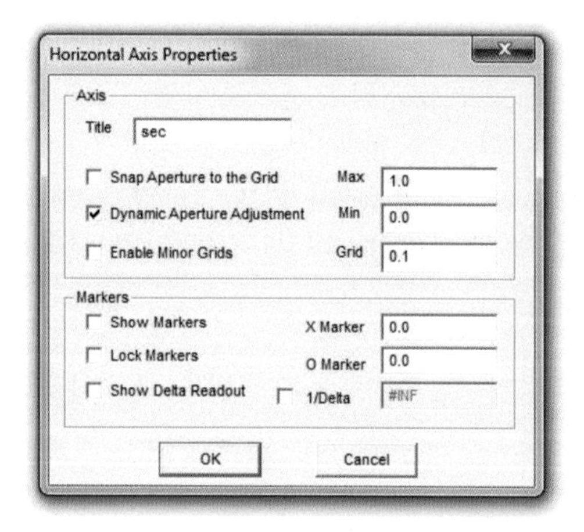

Axis:

Title: *X*-coordinate title for, text appears at lower-left corner in frame, direct side of *X*-coordinate.

Snap Aperture to Grid: By entering this while needed for dynamic aperture fitting, its view snaps to a major grid while scrolling.

Adjusting Dynamic Aperture: By entering we can able to do dynamic aperture adjustment *X*-coordinate scrolling is possible.

Enable Minor Grids: By entering the minor grid, ticks appear in the *X*-coordinate, and it always display the midpoint in between major grid points; these are unlabeled.

Max: By entering we can access instance of the observed extent of duration.

Min: By entering we can access the minimal instance of the observed extent of duration.

Grid: By entering we can access the duration in between coordinates of major grid point and are named in frame *X*-coordinate.

8.2.1.2 Marker

Show Marker: By entering this will access to exhibit the *X and O* marker.

Show Delta Readout:

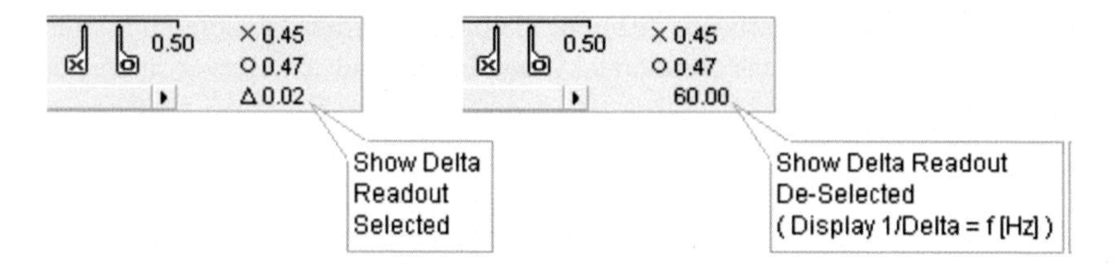

X Marker: By entering this we get (in seconds) to position *X* marker.

O Marker: By entering this we get (in seconds) to position *O* marker.

1/Delta: By entering this we get we can access the frequency 1/*D* between markers.

8.3 Overlay

8.3.1 Poly graphs

An individual graph exhibits many curves, and these curves are founded on the identical *Y*-axis. An illustration of the graph frame, which shows overlay graph in apex, and a polygraph beneath it is shown subsequently.

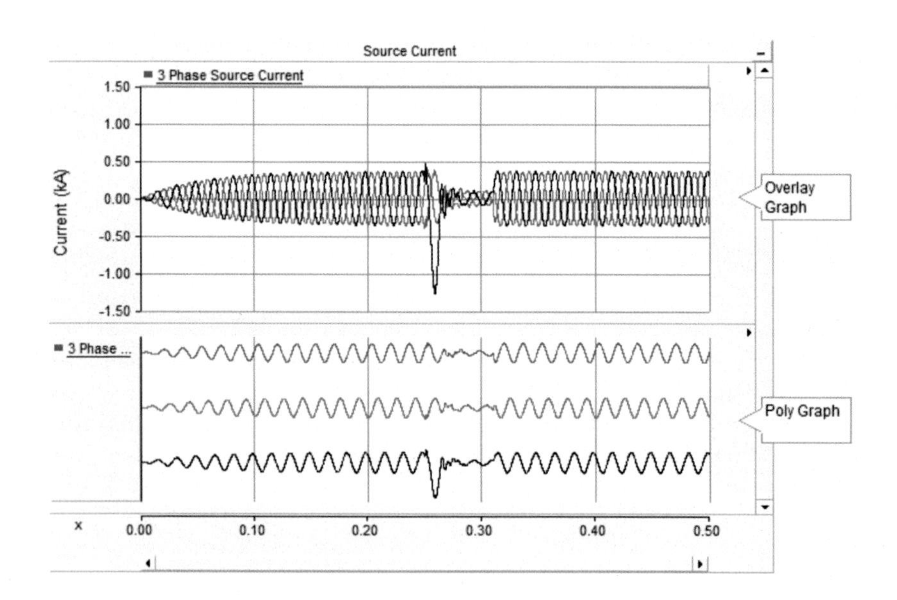

8.3.1.1 Add graphs with graph frame

Graph frames can be accommodated by individually or in aggregated graphs. To add a single or number of graphs, frame title bar and select and Add overlay graph (Analog) or Add Poly Graph (Analogue/Digital) option can be chosen. It can be directly selected by overlay graph directly by selecting the insert key with the mouse in the graph frame.

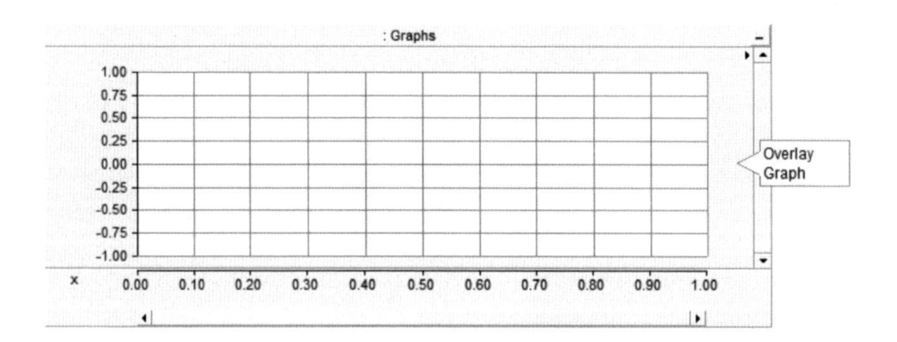

Changing the order of graph: When the aggregate of the graph is created with a frame, the order of the graphs can be rearranged by right-clicking on the graph, which is to be rearranged and choose any of the method as under by moving the graph:

Up
Down
Top
Bottom

Cut or Copy of Graph: Right-click on the graph, which is to be cut (removed) or to be copied, and choose it accordingly. After that, cut or copy it can be pasted in the desired graph.

Paste the Graph in required area: Right-click in the graph frame title by selecting paste; the cut or copied item can be pasted any number of times.

Copy Data in Clipboard: If the simulation result yields curve data, copy the entire or a component part of the entire data in the clipboard by selecting Copy Data in the clipboard; after that, select any one of pop-up menus as follows:

"All": copy entire curve information addressable.

"Visible Area": copy all curve information viewable at graph.

"Between Markers": copy all curve information that exits in between markers. It is to be noted that "Show Markers" should be chosen in Axis attribute dialog.

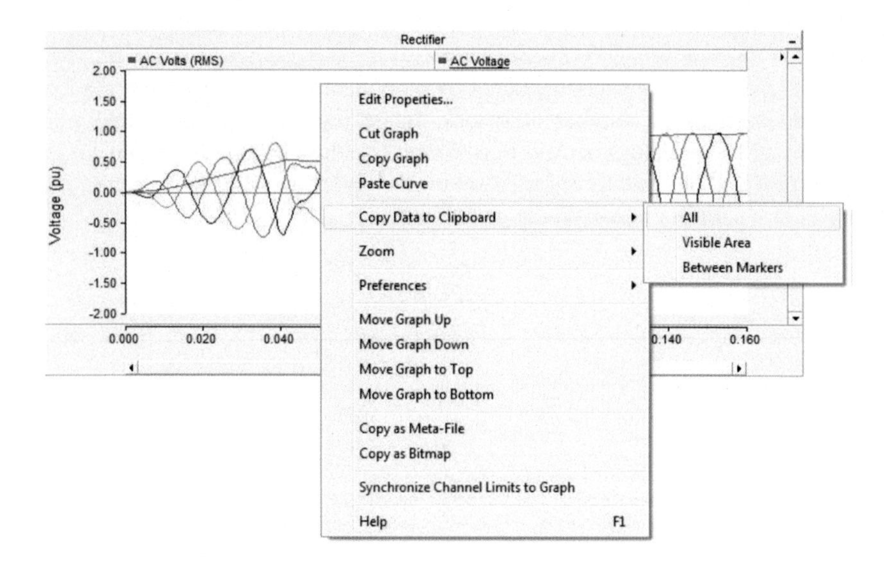

8.3.2 Overlaying graphs

Overlaying graphs are a popular and well-known kind of plot tool in PSCAD and exhibit calculated information with respect to instance, and any number of curves can be put up in one graph.

8.3.2.1 Altering overlay graph attribute

By selecting Properties, pop-up the overlay graph dialog option window, which appears as given under:

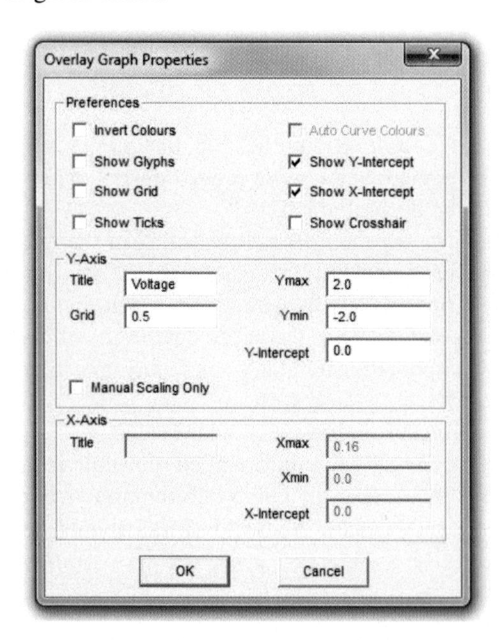

Preferences:

Invert Colors: Choosing this fetches a black color at rear of the graph.

Show Glyphs: Choosing this fetches glyph sign in every curve of graph.

Show Grid: Choosing this fetches grid line for the $X-Y$-coordinates main grid.

Show Ticks: Choosing this fetches main grid tick mark in Y-axis cutting mark.

Auto Curve Colors: Choosing this fetches for using self-loading color of curve at graph.

Show Y-Intercept: Choosing this fetches Y-coordinate interacting mark.

The Y-intercept line will be tuned applying Y-intercept field and detailed as follows.

Show X-Intercept: Choosing this fetches X-intercept line and is ever at zero instances, and not able to be varied at overlay graphs.

Show Cross-Hair: Choosing this fetches cross hairs condition.

Y-Coordinate:

Title: Giving textual matter for showing as graph header (placed on left part in graphical record).

Grid: Indicates the Y-coordinate grid measure. To visual image of Y-coordinate grids, choose choice *Show Grid* as explained in the preceding section.

Y min: Indicate marginal Y-coordinate showing the boundary in chart.

Y max: Maximal Y-coordinate showing the boundary in chart.

Y-Intercept: Indicates Y-coordinate placement of Y-intercepting line, and is observable if *Show Y-Intercept* is chosen (as indicated higher up).

Manual Scaling Only: Choosing this fetches for locking Y-coordinate boundaries given in

Y min/Max The Y-coordinate will remain nonfunctional with any later zoom functions.

8.3.3 Poly graphs

These are utilized generally to exhibit the drawn trend of functions in a "stacked" appearance. Every curve is contained inside its display area, and is stacked one over another. These can be selected over a modular overlaying graph in case it is needed to have a snapshot of many time functional trends in a single diagram, and also for using it for making the digitized manner for creating the logical transformational diagrams.

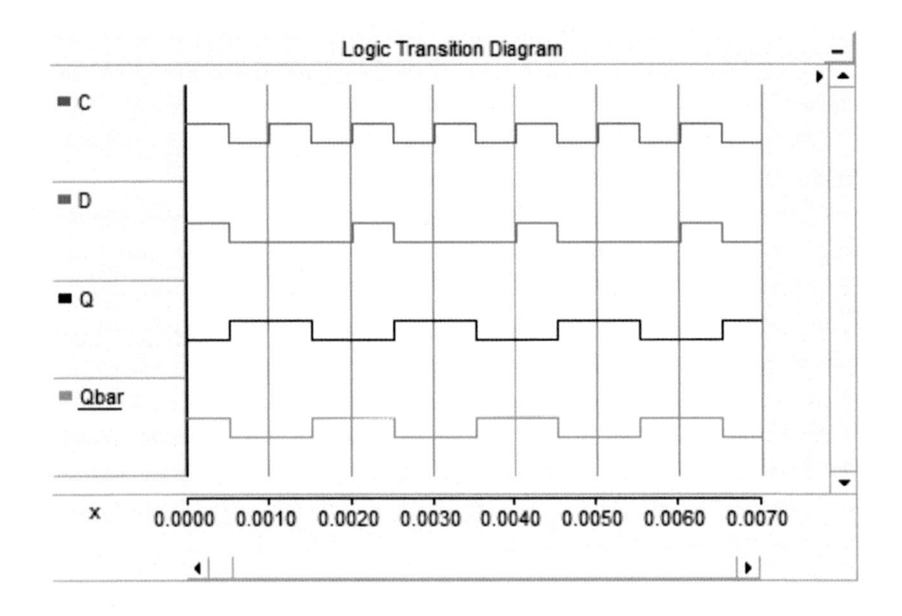

8.3.3.1 *Changing poly graph attributes*

Polygraph is chosen by clicking two times or by right-click over graph, and with Edit options will display Poly Graph attributes window:

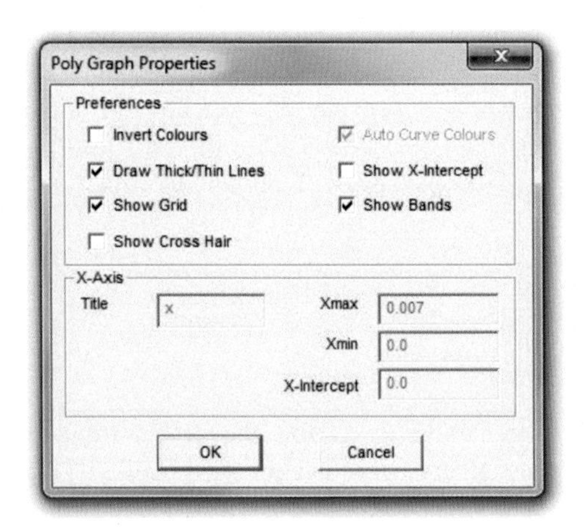

The respective factors which can be changed are explained as follows:
Preferences:

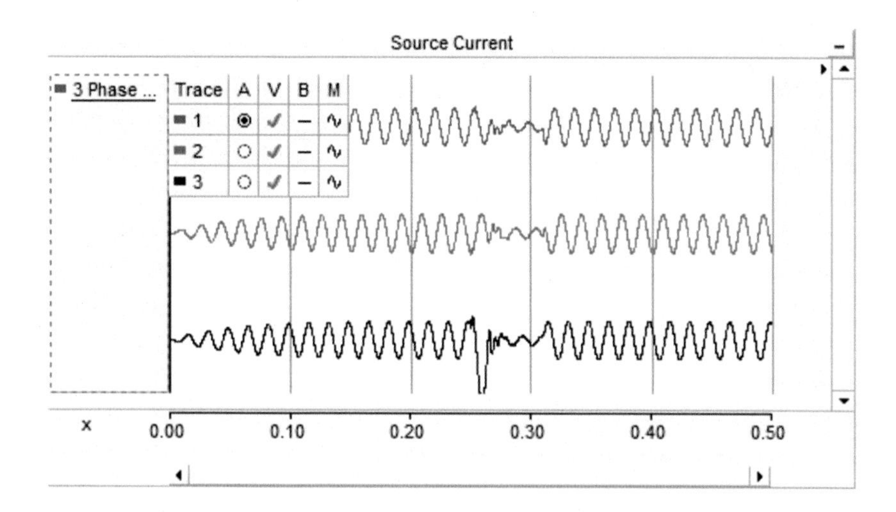

Invert Colors: Choosing this fetches a black color at rear of the graph.

Show Grid: Choosing this fetches grid line for the $X-Y$-coordinates in the main grid.

Show Cross-Hair: Choosing this fetches cross-hair condition.

Auto Curve Colors: Choosing this fetches for using self-loading color of curve at graph.

Show X-Intercept: Choosing this fetches X-intercept line and is ever at zero instances, and not able to be varied at overlay graphs.

Show Bands: Choosing this fetches a several rear window color in between aggregate of trends vs. time single chart, for simple observable distinction between them.

8.3.4 Curves

A curve is a special runtime object best defined as a graphical representation of a data point series, where each point is associated with a plot phase for the simulation. Curves are generated by linking to an output channel node to which a collection of data signals scalar or array has been inputted. Curves can also be multi-dimensional, that is, a single curve can contain several sub- curves or traces, where each trace corresponds to a single value of the series.

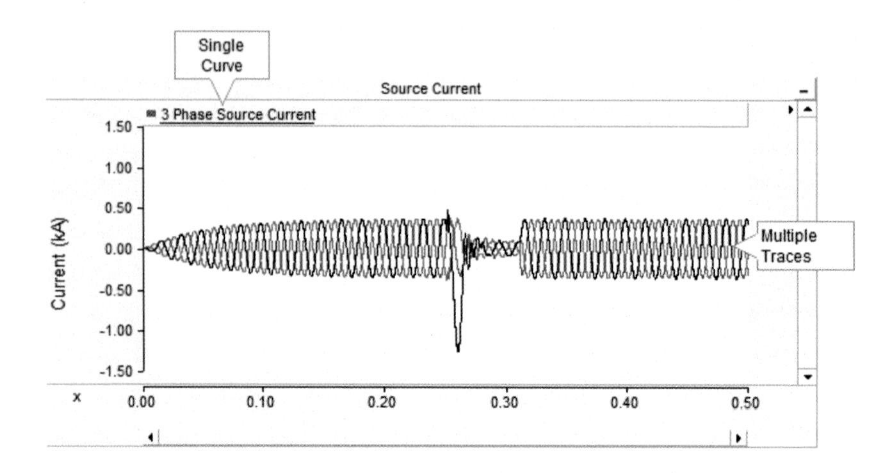

"ADD" Fresh Curve to Graph

"ADD" function to with graph is given by a two various methods:

1. *Drag and Drop Method*: Holding down the Ctrl key, left-click and hold at the output channel element in which we would like to pull up the curve. Drag the pointer graph, then release the mouse button. Observe the drag and drop for further information on it.
2. *Graphs/Meters/Controls way*: Right-click in the output channel constituent in which we pull up the curve. Choose Graphs/Meters/Controls | Add as Curve. Choose the desirable graph by left-clicking, and after that, right-click and then choose Paste Curve.

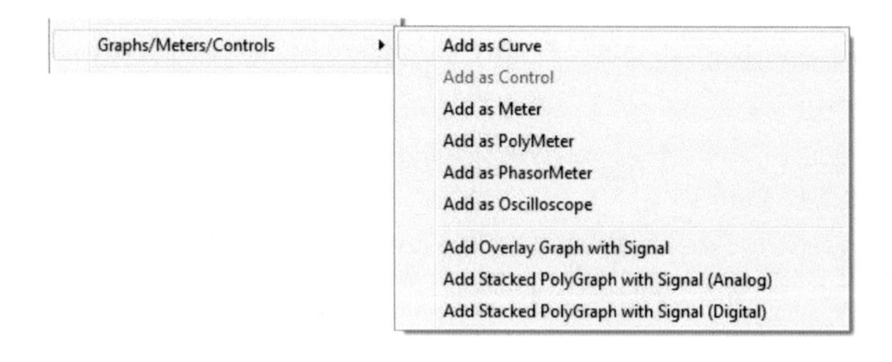

8.3.4.1 Curve legends

After the curve has been included with the graph, curve heading appears in the caption:

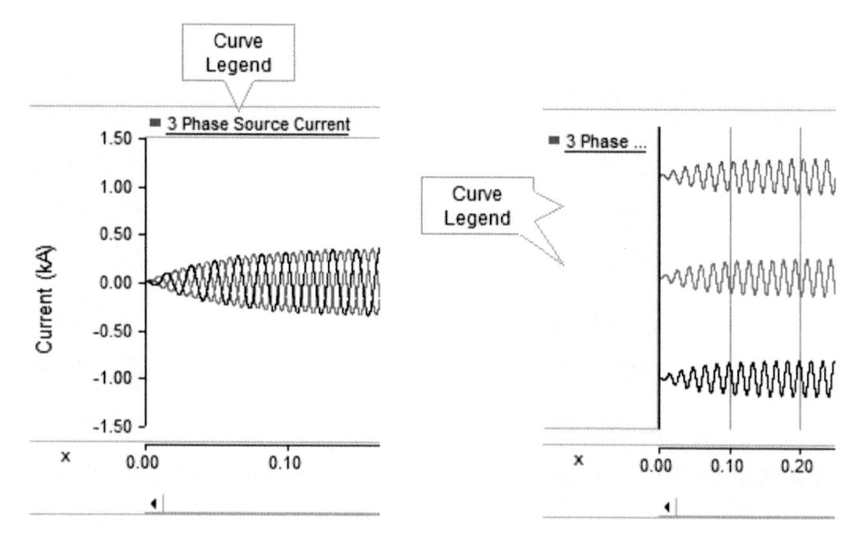

8.3.4.2 Curve order

Once multiple curves have been added to a graph, you may change the order in which they appear. Ordering curves can be accomplished in one of two ways:

1. *Drag and Drop*: *Left-click and hold* over the curve in the curve legend. Drag the mouse pointer to a new position in the curve legend and release the mouse button. See Drag and Drop for more details on this.
2. *Right-Click Menu*: *Right-click* over the corresponding curve legend and select one of the following from the pop-up menu: *Move to the Start* or *Move to the End*.

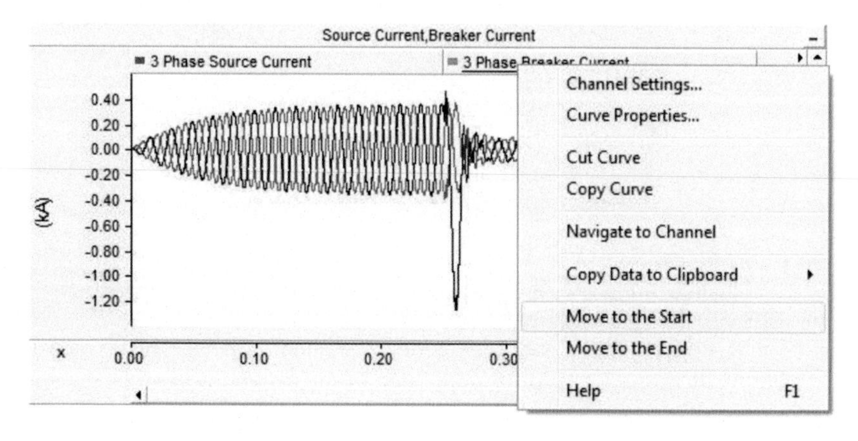

8.3.4.3 Cut/Copy/Paste an existing curve

Cut or Copy of Graph: Can be done by right-clicking on the graph which is to be cut (removed) or to be copied and choose it accordingly, after cut or copy, we can paste in the required graph.

Copy Data in the Clipboard: If the simulation result yields curve data, copy the entire or a component part of the entire data in the clipboard by selecting Copy Data in the clipboard; after that, select any one of the pop-up menus as follows:

"All": copy the entire curve information addressable
"Visible Area": copy all curve information viewable at graph.
"Between Markers": copy all curve information that exits in between markers.

As *Comma Separated Variables (*.csv)* formatting for simple movement in the simple available data investigation software.

8.3.4.4 Adjusting curve properties

Polygraph is selected by double-clicking or by right-clicking over graph and choose Edit options. These will display the Poly Graph attributes window:

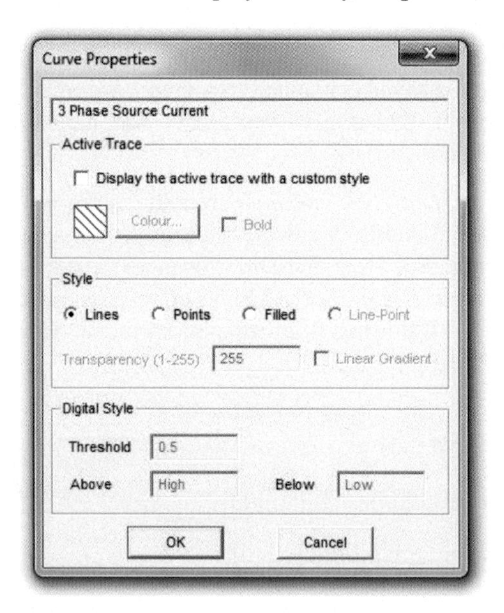

8.3.5 Active trace

The listings below explain the factor in this segment:

Exhibit atavistic trace with a customized manner: Choose this choice if we want modification of color or dimension of the active trace.

Color: choose a representation of color for the trace, then choose the OK button in the color dialog box. This choice is active only when the display of active trace for customized fashion is chosen.

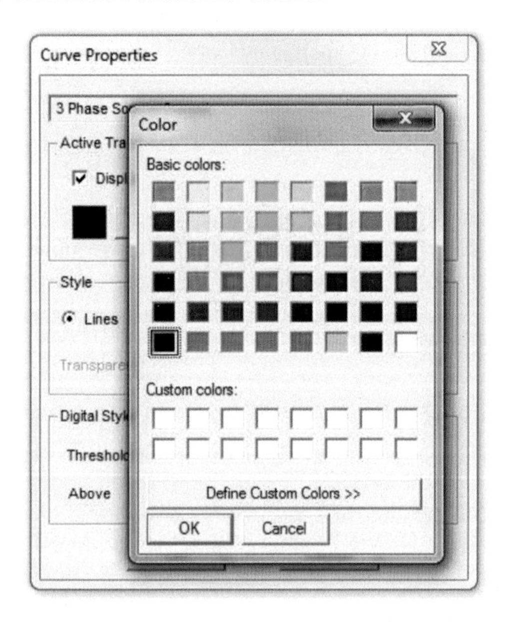

Bold: Choose this choice when the trace is applied. This choice is active only when the display of active trace for customized fashion is chosen.

8.3.5.1 Style

The following listings explain the factor in this segment:

Lines: Exhibits the curve as standardized line.

Points: Exhibits the curve as serial point with respect to set plot stride.

Fill: Fill up area with curve (lines in between curve and 0.0) with color.

Transparency (1–255): This option gives the transparency of the occupied area below a curve. This is changeable if the filled area is activated.

Linear Gradient: choose this choice for a lineal gradient outcome on the occupied area below curve. This is changeable if the filled area is dynamic.

8.3.5.2 Digital style

The choices are only reasoned if the curve is a constituent of a polygraph. It is the criterion for attributes that the curve traces while in digital mode. The below attributes explain the invariables:

Threshold: Measure of which alters the exhibit state of the curve.

Above/Below: State of the curve while the attributed quantity is up/down of the set outset, respectively.

8.3.5.3 Changing channel settings

The channel attributes for the source results for a given curve could be approached straightaway from curve caption pop-up menu, choose (right-click) channel.

The origin result channel factor for a specific curve can be got straightaway in pop-up menu of curve legends, by choosing the Channel Setting (right-click the curve).

8.3.5.4 Synchronizing of output channel limitation with those of the graph

When more than a single curve are given as input to overlay graph the respective resultant channel attribute MIN/MAX corresponding output channel component *Min/Max* Limitation could be made to be in phase to graph on *Y*-coordinate; to do it, select (right-click) on overlay graph and choose sync channel margins to graph.

8.3.5.5 Curves origin from multiple occurrence modules

Module constituents can have more multiple instance, that is, if output channel is over the canvas of the module, it will bring forth a specific curve for all representation on of module, in spite of only one output channel component. When the curve related with the output channel is positioned with a graph that persists external to module, which is single curve occurrence is displayed.

All curve occurrences are named as a call. While the curve prevails on a graph that is originated from output channel which contained from a sizable figure of instances component, then it can be accessed and control the yield of an ample figure of specific calls. For acquiring it, left-click on the curve caption. A visible listing will pop up showing all root curves that are available.

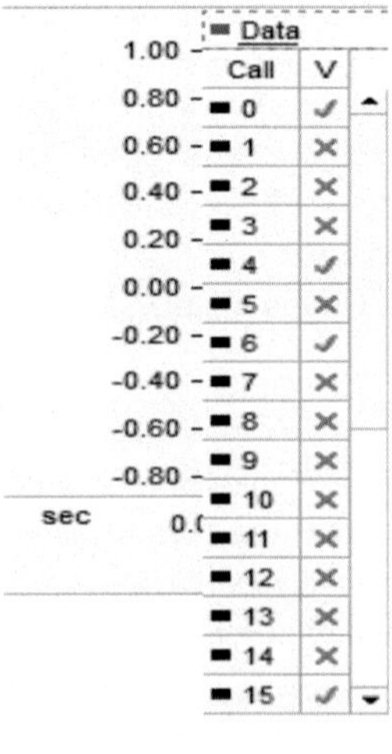

In the higher up image, it is evident the curve information is acquired from 16 assorted module attitudes, which are derived from same distinctness. In the aforementioned specific illustration, call 0.4, 6, and 15 are shown. We can prefer to show every call in one graph, or we can put the data curve in a number of graphs, and choose a specific call to exhibit in all graphs.

For further content on number of instance modules, get data in Multiple Instance Modules.

8.3.6 Traces

A set of signaling curves can be drawn online as an individual data plotted online as a single data, in which each set of component or "subcurve" is mentioned as trace. For each one, the trace can be enabled/disabled individually (i.e., exhibited or not exhibited).

8.3.6.1 Trace drop-down menu

Trace attributes and criterion may be approached with a specialized drop-down menu by left-clicking over the curve heading in the curve caption.

8.3.6.2 Modify trace attribute

Before modifying any trace attributes, it is important initially to put forward the trace drop-down menu as represented by the preceding section. This menu corresponds to four distinct file, each one permitted for a simple approach to definite trace attributes. These are represented as:

Trace	A	V	B	M
■ 1	⦿	✓	—	∿
■ 2	○	✓	—	∿
■ 3	○	✓	—	∿

Trace: Inform the trace number and related color environment. The number states to array index numbers of a single curve.

A: Represents active. Select the radio button in this column to select the active trace. The active trace will be the default focus when switching to the cross-hair mode. Also, only the properties of the active trace may be adjusted: See Adjusting Curve Properties.

V: Represents view. Left-click the respective check boxes in file to conceal/viewing respective traces. It could be concealed/viewed each trace by left-clicking on the V itself.

B: Represents bold. Left-click the respective check boxes in file bolding/unbolding respective traces, also bold/unbold all traces by left-clicking over B itself.

M: Represents mode. This utility is only effectual when the curve is shown in a polygraph. Left-click the one-on-one boxes in this file to do the trace mode of digital into analog. While in digital mode, the trace is shown in a specialized two-state formatting, and its state rely on either it is higher up or beneath a predetermined outset value.

8.3.7 Poly meters

A polymeter is an exceptional runtime constituent utilized generally for watching an individual, multiple-trace curve. The polymeter with real-time varying data show the ratio of each trace in bar type format named as a gauge. This gives solution overall appearance as spectrum analyzer output visualization. The powerfulness of the device is its quality to compact a large magnitude of data in little screening area, which is especially assistive when showing harmonic spectrums like data result from fast Fourier transform (FFT) element.

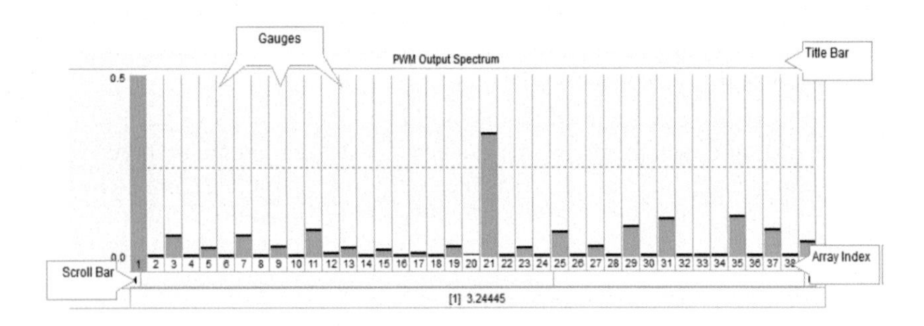

The relational gauge dimension is constant; hence when the polymeter is not broad to exhibit each data, and an inclined scroll bar is given. A set of index display is enclosed direct under the gauges for simple determination.

8.3.7.1 ADD polymeter

Adding is done by right-clicking over output channel element inside the schematic canvas, then choose Graphs/Meters/Controls | Add as Poly Meter. It will come out as an attachment with the mouse pointer, and then move it to the desired place, and left-clicking on canvas will add it.

8.3.7.2 Move and varying the size of a poly meter

To relocate a polymeter, left-click over the title bar and hold it, then locate the meter

For varying the size, left-click over title bar and hold it to select polymeter. Grips will be appearing just about the outside border as per diagram beneath.

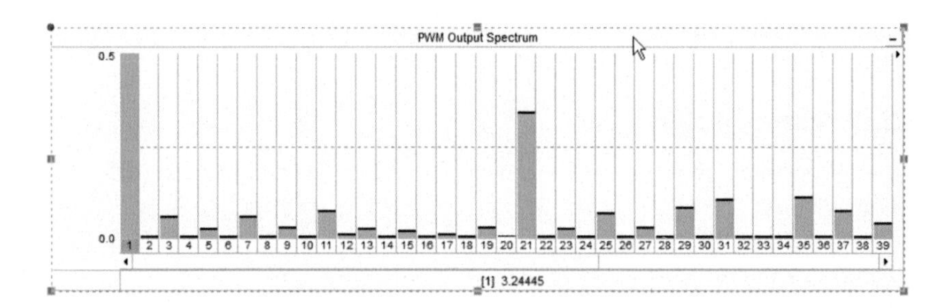

8.3.7.3 Cut/Copy polymeter

Right-click in polymeter title bar and choose Cut or Copy accordingly.

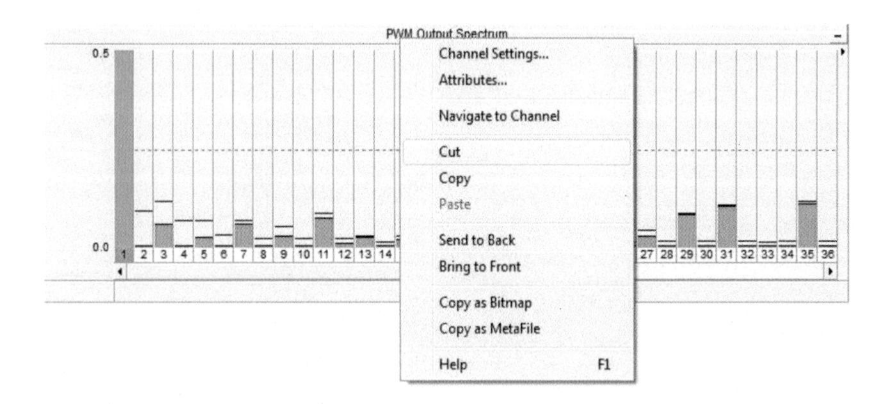

Paste Polymeter: Cut or copy a polymeter as represented preceding section. Right-click in an empty field in Schematic position and choose Paste, it can be done many number of times.

Copy Polymeter as Meta-File/Bitmap: The full polymeter exhibit could be made copy over Windows clipboard in meta-file (*.wmf) or bitmap (*.bmp) formatting. Choose polymeter title and choose Copy as Bit maps Meta-Files. Choosing report text files, paste the picture.

Maneuver to Channel: it could be navigated straight to the connected resultant O/P Channel section by choosing this choice by selecting Navigate to Channel. PSCAD itself will give output channel by highlighting it.

Modify Channel Setting: The *Y*-coordinate attributes of polymeter are formed to respective output channel attribute dialog corresponding Output

Channel which can be approached directly from the polymeter by choosing (right-click) the title bar and choosing Channel attributes.

Showing of Particular Data: The value of individualistic array elements may be shown in status bar in lower side of polymeter. To have a visual image of the value of that component, select (left-click) particular index figure in array scale.

8.3.8 Phasor meters

An extraordinary runtime object can be applied to show up to six different phasor values. The phasor meter shows phasor in a polar graph, in which the value and phase of every phasor act dynamical while simulating. The FFT spectrum of the output can be viewed without errors.

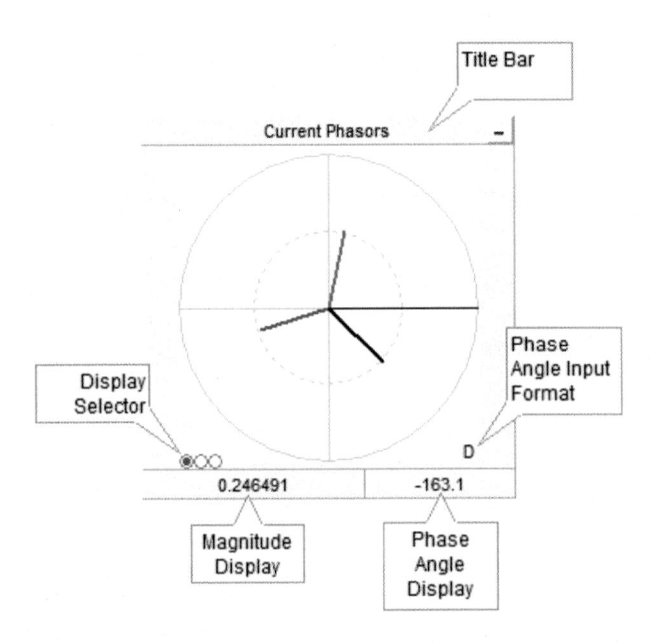

Phasor meters are exceptional objects that cannot be summed straight from the tool bar. Every phasor meter is directly connected to a one-on-one curve from a singular output channel factor. When phasor meter having curve with minimum of two traces (order of magnitude and phase angular value) is the negligible responsibility for equipment to work in good order. The subsequent subdivision explains the way to set up data signaling for demonstration in the phasor meter.

8.3.8.1 Adding a phasor meter

To adjoin a phasor meter, on output channel module, right-click over canvas and choose Graphs/Meters/Controls | Add as phasor meter. The phasor meter will come into view enclosed with pointer of mouse. Shift the pointer desired place the new meter to exist in and then left-click to rest on canvas.

8.3.8.2 Move and varying the size of a phasor meter

To relocate a phasor meter, left-click over title bar and hold it, then locate the meter to where it and free the mouse.

For varying the size, left-click over title bar and hold it to select meter. Grips will be appearing just about the outside border as per diagram beneath.

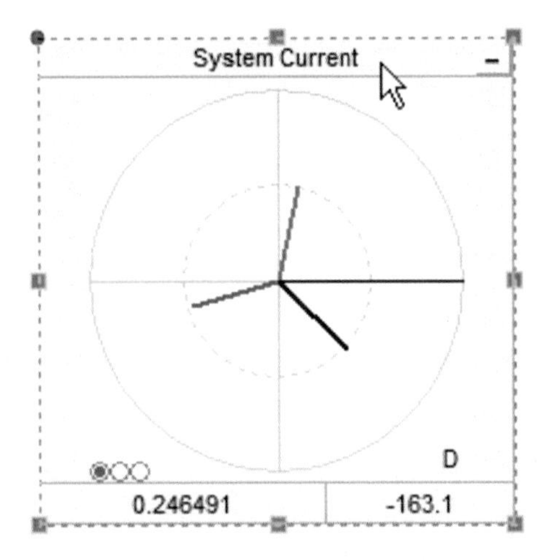

Paste Phasor Meter: Cut or copy a polymeter as represented preceding section. Right-click in an empty field in Schematic position and choose Paste, and it can be done many number of times.

Copy Phasor Meter as Meta-File/Bitmap: The full polymeter exhibit could be copied over Windows clipboard in meta-file (*.wmf) or bitmap (*.bmp) formatting. Choose phasor meter title and choose Copy as Bit maps or Meta-Files. Choosing report text files and then paste the picture.

Maneuver to Channel: it could be navigated straight to the connected resultant O/P Channel section by choosing this choice by selecting Navigate to Channel. PSCAD itself will give the output channel by highlighting it.

Modify Channel Setting: The *Y*-coordinate attributes of phasor meter are formed to respective output channel attribute dialog corresponding output channel which can be approached directly from the phasor meter by choosing (right-click) the title bar and choosing Channel attributes.

Showing of Particular Data: The value of individualistic array elements may be shown in status bar in lower side of phasor meter. To have visual image of the value of that component, select (left-click) particular index figure in array scale.

8.3.8.3 "Adjust" phase angle inputs to format

The layout that the inward phase angular information is in must be precise and can be simply done by toggling the D/R exhibit in the lower-right place of the graph, where D = Degrees and R = Radians.

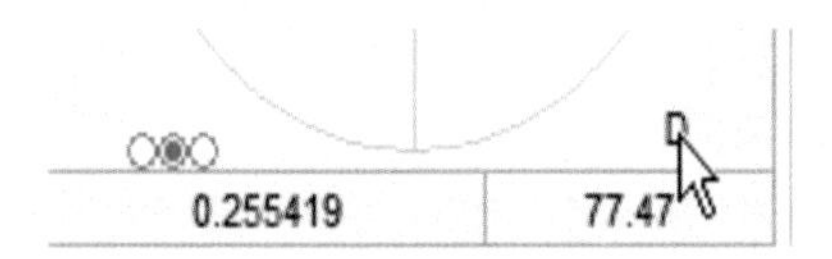

8.3.9 Oscilloscopes

It is a unique runtime object that is applied for imitating the trigger impacts of oscilloscope in a time-varying, recurring signals as AC V/I, at a specified frequency,

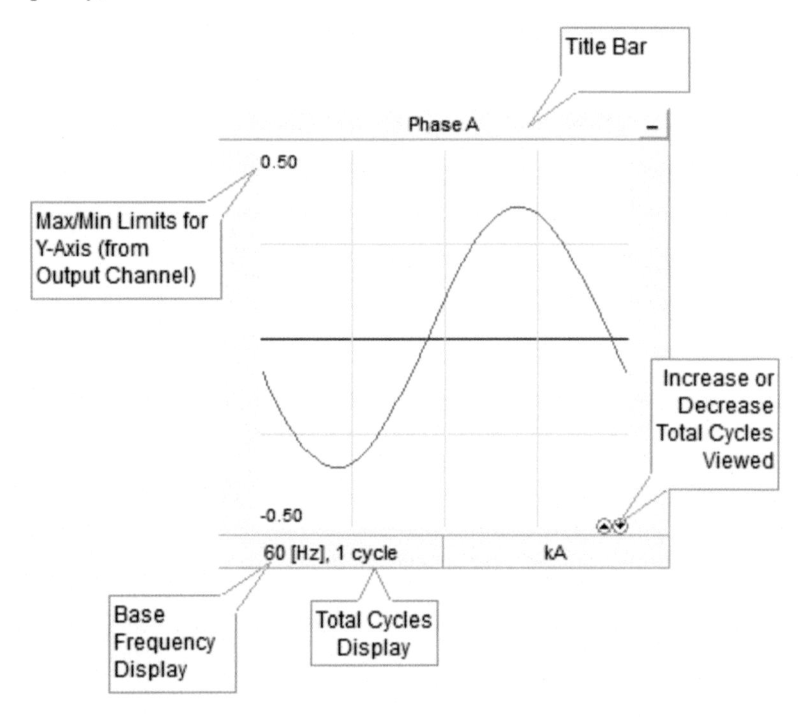

Oscilloscopes cannot be drawn from the tool bar since it is unique. Every object is in a straight line correlated with the curve commencing from single output channel constituent. It supports array signals—that is, the curves, which contains a number of traces. The next segment explains the preparing of data impulses for exhibiting in oscilloscope.

8.3.9.1 Addition of oscilloscope

To adjoin an oscilloscope, on an output channel module in the canvas right-click and choose Graphs/Meters/Controls I Add as Oscilloscope. The oscilloscope will come into view enclosed with the mouse pointer. Shift the pointer to the desired place and then left-click to add on the canvas.

8.3.9.2 Moving and varying the size of oscilloscope

To relocate an oscilloscope, left-click over the title bar and hold it, then locate the meter and free the mouse.

For varying the size, left-click over the title bar and hold it to select the meter. Grip will be appearing just about the outside border as per the following diagram:

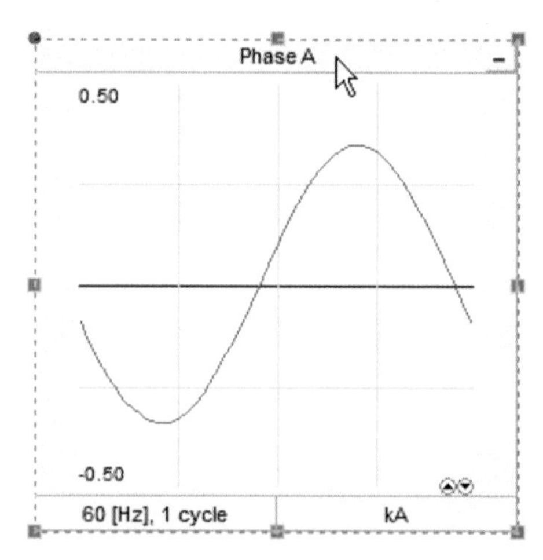

8.3.9.3 Cut/Copy oscilloscope

Right-click in *oscilloscope* title bar and choose Cut or Copy accordingly.

Paste Oscilloscope: Cut or copy as represented by the preceding section. Right-click in an empty field in schematic position and choose Paste, and it can be done many times.

Copy Oscilloscope as Meta-File/Bitmap: The full Oscilloscope exhibit could be made copy over Windows clipboard in meta-file (*.wmf) or bitmap (*.bmp) formatting. Choose Oscilloscope title and choose Copy as Bitmaps or Meta-Files. Choosing report text files and then paste picture.

Maneuver to Channel: It could be navigated straight to the connected resultant O/P Channel section by choosing this choice by selecting Navigate to Channel. PSCAD itself will give output channel by highlighting it.

Modify Channel Setting: The *Y*-coordinate attributes of Phasor meter are formed to respective output channel attribute dialog corresponding output channel which can be approached directly from the Phasor meter by choosing (right-click) the title bar and choosing Channel attributes.

8.3.9.4 Increment/decrement the overall shown period

The overall shown cycles is varied at status bar below the oscilloscope: choose respective display selector (left-click), where buttons correspond increment and decrement accordingly. In similar manner the work can be done (right-click) on the equipment title bar and choosing any of the increasing single cycle or decreasing single cycle.

8.3.10 *XY* Plots

XY Plots consist of a graph frame and also single, special graph window with the purpose of drawing a curve against one other. An *XY* plot can adapt various curves on every *X*- and *Y*-coordinates, and consists of changing zoom and polar grid property.

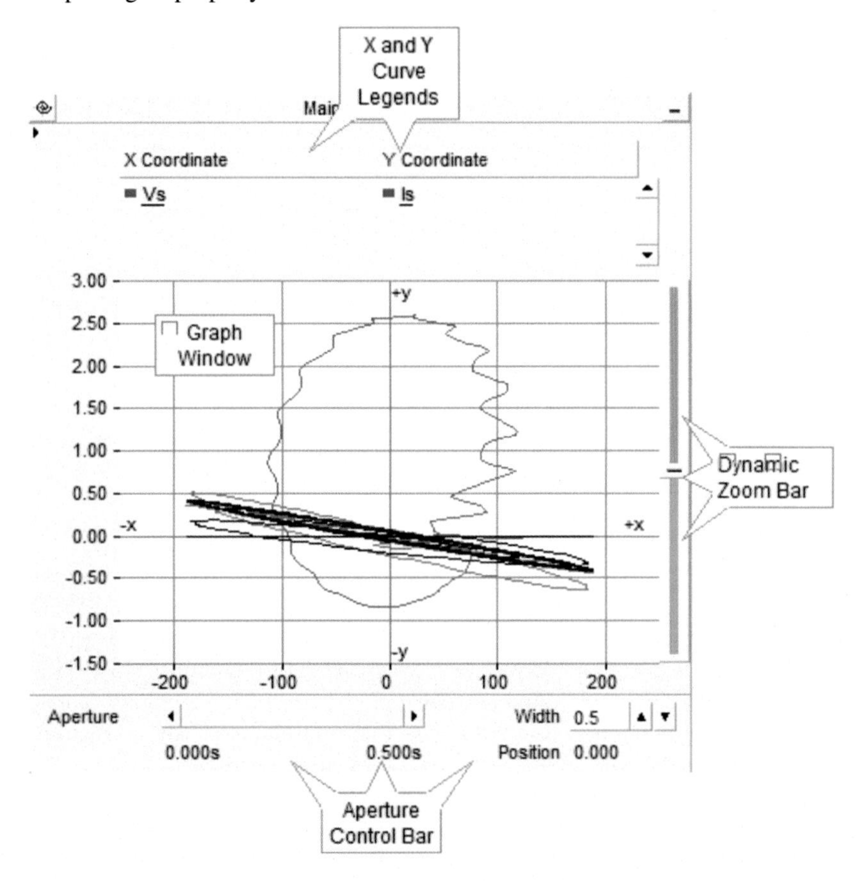

Even though it is applied to draw one input against other signals, each of these depend on the similar time domain, and hence it is possible to move within the information in the time scale: It also contains a time period scale aperture control bar, placed in lowermost of plot framework.

8.3.10.1 *Addition of* XY *plot*

To adjoin a *XY plot*, on an output channel module, in the canvas right-click and choose and add as *XY plot*, or chose *XY* plot button in constituent of ribbon control.

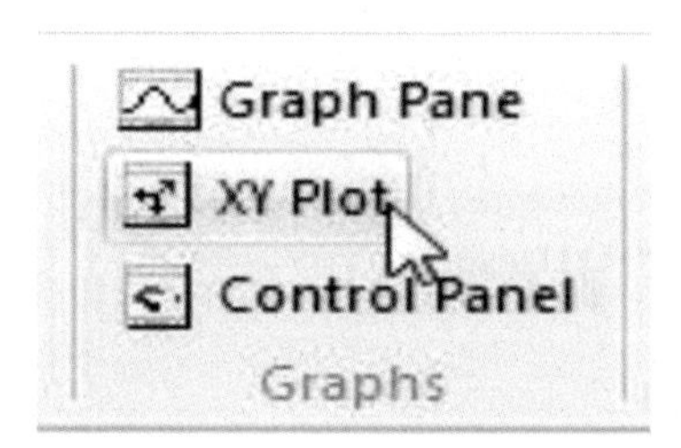

8.3.10.2 *Move and varying the size of* XY *plot*

To relocate an *XY plot*, left-click over the title bar and hold it, then locate the meter to where it should go and free the mouse.

For varying the size, left-click over title bar and hold it to select meter. Grip will be appearing just about the outside border as per diagram beneath.

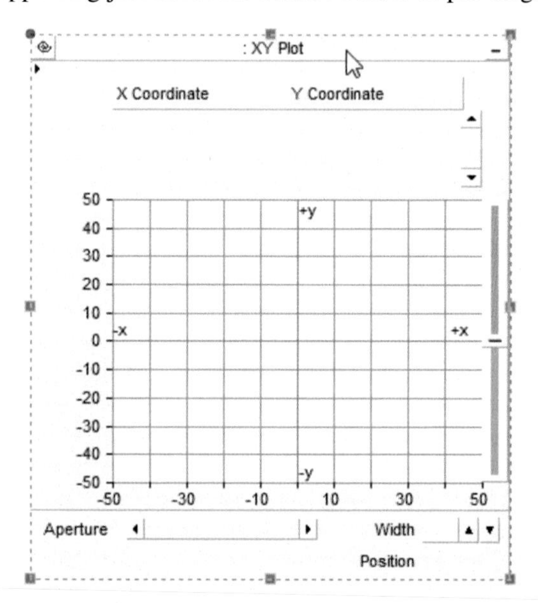

8.3.10.3 *Cut/Copy* XY *plot*

Right-click in *XY plot* title bar and choose Cut or Copy accordingly.

Paste XY Plot Frame: Cut or copy a as represented preceding section. Right-click in an empty field in Schematic position and choose Paste, it can be done many number of times

Copy Data to Clipboard: When simulation is done the *XY* plot comprise curve data, we have choice of copy whole or a part of data on clipboard.

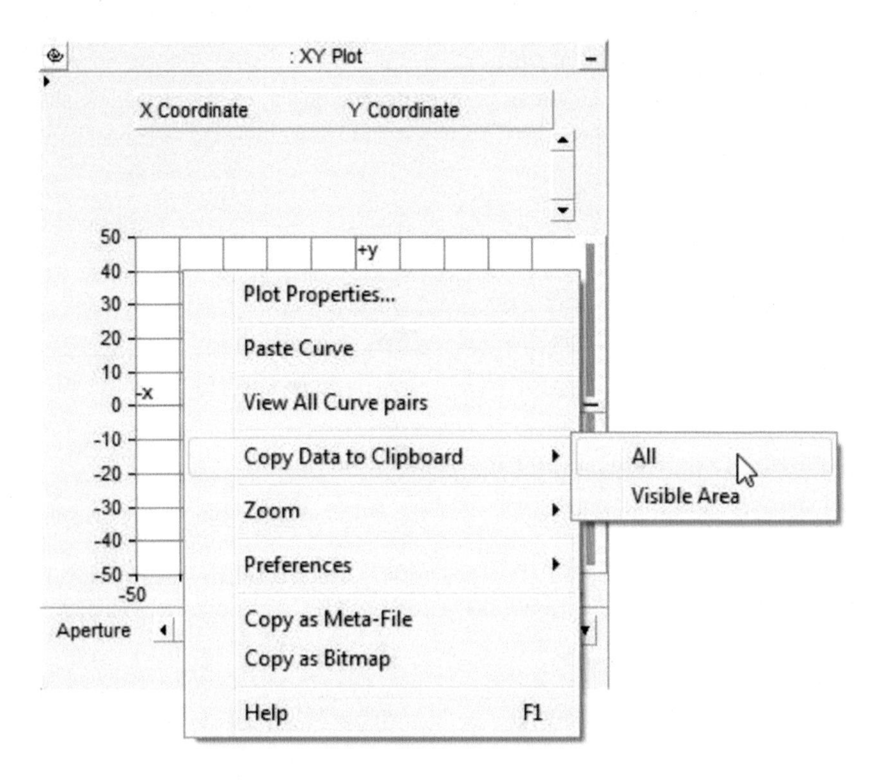

8.3.10.4 *Adjusting* XY *plot frame properties*

To approach the Plot Frame Attribute, choose (left-click) the plot title bar, or choose Plot Frame Attribute (right-click). This brings Plot Frame Attribute dialog window.

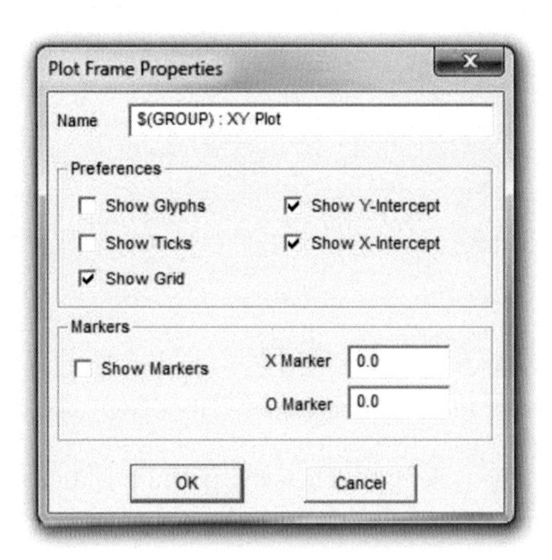

Name: Create a heading to *XY* plot, the option text will be shown a bit cryptic: such syntax is applied as a naming rule for classification system of objects in working area. For more content can be got from Grouping of Runtime Objects.

Preferences:

Show

Markers: Exhibits the *X* and *O* markers of every graph.

Glyphs: Exhibits glyph symbols on every curve of frame.

Ticks: Exhibits ticks along the *Y*-intercepts in every graph.

Grid: Exhibits the grid in every graph.

Y-Intercept: Exhibits the *Y*-intercept at every graph.

X-Intercept: Exhibits the *X*-intercepts at every graph.

8.3.10.5 Adjusting plot properties

Left double-click over the plot area (white part), or right-click over the plot area and select Plot Properties. This should bring up the Plot Properties dialog window.

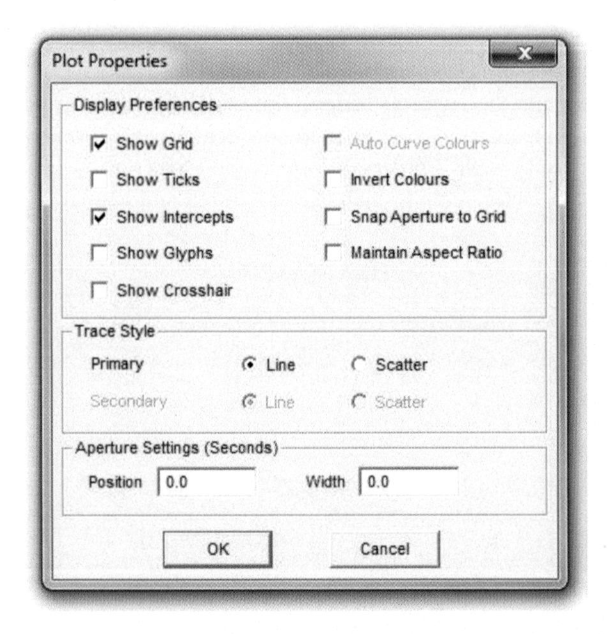

There are various parameters that may be edited though this window, each of which are described later.

Preferences:

Show Grid: Fetches grid line for the *X*−*Y*-coordinates main grid.

Show Ticks: Fetches main grid tick mark in *Y*-axis cutting mark.

Show X-Intercept: Fetches *X*-intercept line and is ever at zero instances, and not able to be varied at overlay graphs.

Show Glyphs: Fetches glyph sign in every curve of graph.

Show Cross-Hair: Fetches cross-hair condition.
Invert Colors: Fetches a black color at rear of the graph.
Auto Curve Colors: Fetches for using self-loading color of curve at graph.
Invert Colors: Fetches a graph black scene.
Snap Aperture to Grid: Fetches changing aperture alteration, which catch with the major grid during zoom.
Maintain Aspect Ratio: Choosing this option keep the aspect ratio curve plot (*X*- and *Y*-coordinates) every time the plot frame is altered. When it is activated, curve will elongate or shrink with respect to original size.
Trace Style:

Primary: Choose whether for drawing traces as line or Scatter viewing. Scatter will merely adjoin only one dot for every *X-Y*-coordinate.
Aperture Settings (seconds):

Position: choose starting position in second in aperture window.
Width: choose dimension in second of the aperture window.
Style
The listings, as under explains the factor in this segment:

Lines: Exhibits the curve as standardized line.
Points: Exhibits the curve as serial point with respect to set plot stride.
Fill: Fill up area with curve (lines in between curve and 0.0) with color.
Transparency (1–255): this option gives the transparency of the occupied area below a curve. This is changeable if filled and activated.

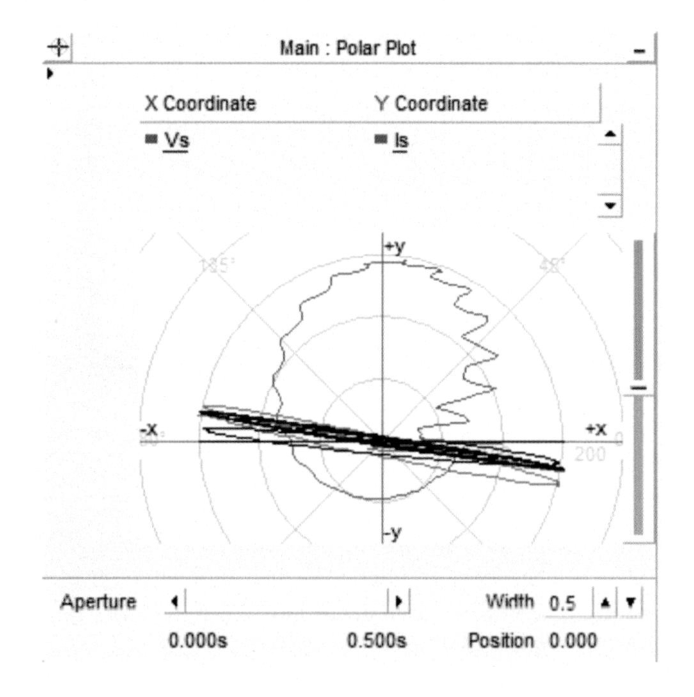

8.4 Power lines and cables

Tower line and UG cable segments are dealt in two sections: the description of the transmission division includes the admittance/impedance information, earth impedance figures, and geometrical location of each tower and lines; all interconnecting equipment with the remaining power system, including their machinery (in terminals). If the OHT line is modeled, without the interconnection method, the interfacing equipment not essential.

8.4.1 Overhead lines erection

PASCAD has remote ends technique for OHT erection, which consists of transmission line design module having two OHT Interconnecting machinery, which has the send/receive terminals systems. The principle of the interconnecting components is linking the OHT line with total electrical system networking.

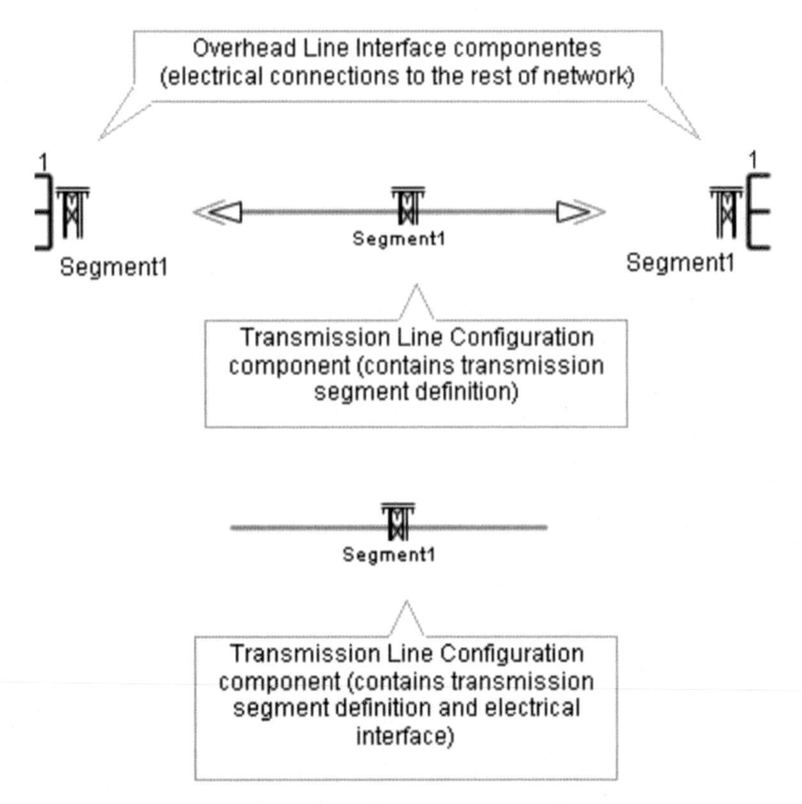

8.4.1.1 *Constructing overhead line*

The OHT design module and its two OHT interconnecting equipment linked by input data values make sure subsequent actions are employed:

1. Section Name: make sure that this module OHT model and its connecting equipment name input data are same (Case sensitive)
2. Quantity of Conductor: make sure that Ensure that the quantity of Conductor in either side of line model and its elements and connecting elements are the same.

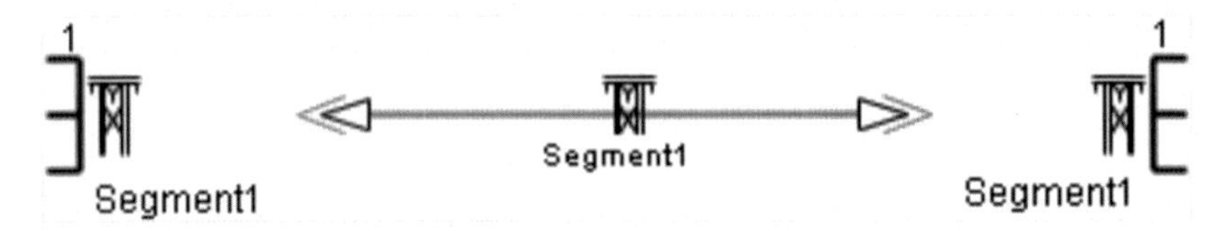

8.4.1.2 Constructing OHT line (direct method)

Revise the data of the design section and make sure that line End method | Direct Connection is chosen. This technique come automatically while the module is formed initially.

8.4.1.3 Edit the OHT line data

The data on OHT can attune straightforwardly inside the line design module by (right-click) section and choose Edit parameter.

8.4.2 Construction of underground cable system

The Direct Connection technique is not applied in cable system design. It has only remote ends technique, in which cable design section having two cable connecting components, in place of sending/ receiving terminals of cable. The principle of the cable interconnecting equipment is to give the termination within the whole electrical system. A UG cable system is depicted in the following drawing:

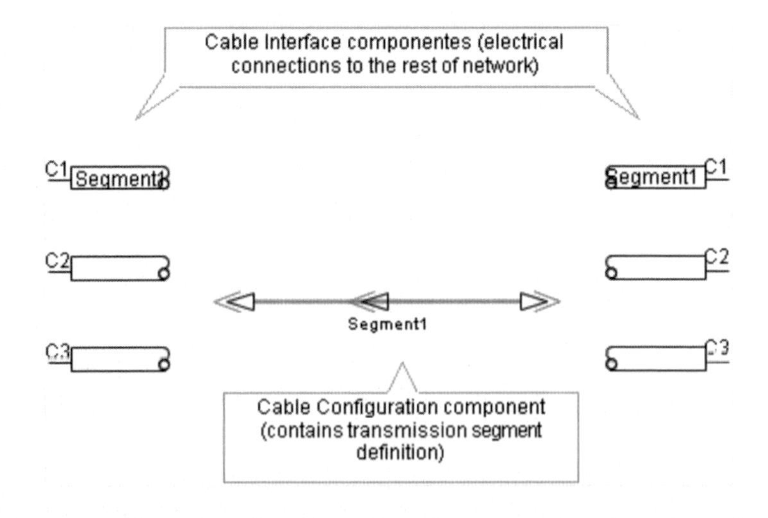

8.4.2.1 *Constructing a underground cable system*

Choose single cable arrangement constituent and two cable crossing-point machinery on the page. The common way to include the straightforward method is to choose equipment tab in the ribbon control bar.

The cable design element and two cable termination equipment are interconnected with input data. It is confirmed that actions below are adopted:

1. Section Name: make sure that this module UG cable model and its connecting equipment name input data are same (case sensitive).
2. Quantity of UG cable: Make sure that the quantity of cable in either side of the model. Every cable may have various conduction layers; it is a random number, and its parameters are disabled in the design. It is necessary to detail the figure of conducting layer for every cable in the cable interface equipment. This information must be equal with the cable segment cross-section which has been specified.

8.4.2.2 *Altering cable data*

The cable data can be modified straightaway with cable design module by selecting part and choosing Edit Parameters (right-click).

8.4.3 Addition of a tower component

A Tower can be included with editor by two methods

1. The simple ribbon control bar by selecting any accessible towers and position it on the schematic canvas.
2. Context menu: On the schematic tab, in blank area of the window move pointer right-click and choose "Add Tower Cross-Section." A submenu will come into view, which includes a record every predesigned collection of projects, which contains a record every transmission line tower apparatus accessible in those specific files. Choose a tower which is added in auto.

The tower machinery can be included by copying and pasting straight from the master documents. Choose the library in diagram outlook, choose tower equipment, and then choose (right-click) Copy and then paste on empty area.

Numerous towers could be included in single pattern. Make sure that conductors are given numbers in order number inside the tower apparatus, also the *X*-coordinate of the new towers in that segment are accustomed. The conductors included with supplementary towers should reflect in the respective OH line terminal equipment.

8.4.3.1 *"Edit" tower attributes*

Tower attribute modified by related tower data dialog window. Choose tower component (without choosing it) and choose Edit Parameters to the right to use this dialog.

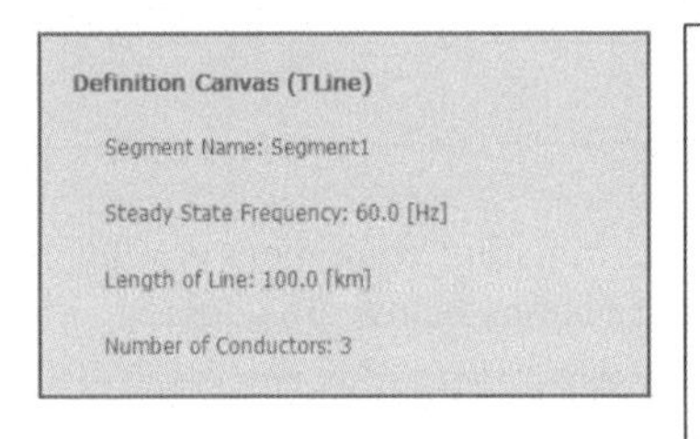

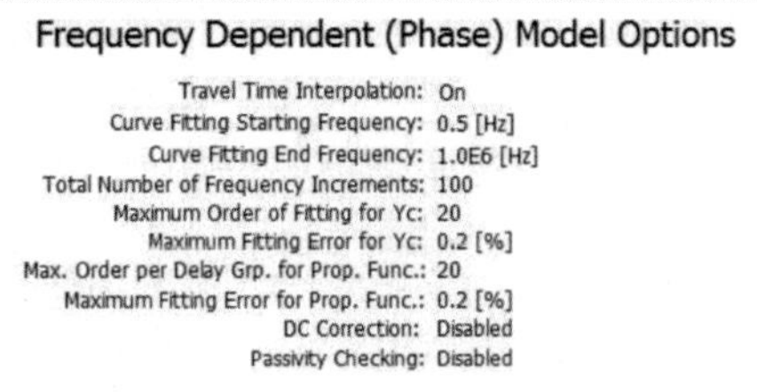

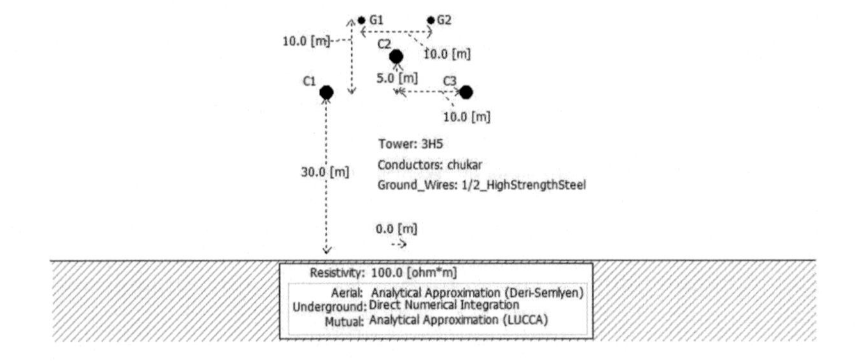

8.4.4 Addition of a Cable Cross-Section constituent

A Cable Cross-Section constituent can be included with editor by two methods.

1. The simple ribbon control bar by selecting any accessible Cable Cross-Section constituent and position it on the schematic canvas.
2. Context menu: On the schematic tab, in blank area of the window move pointer right-click and choose "Add Tower Cross-Section." A submenu will come into view which includes a record every predesigned collection of projects which contains a record every Cable Cross-Section constituent which is accessible in those specific files. Choose a tower which is added in auto.

The Cable Cross-Section constituent can be included by copying and pasting straight from the master documents. Choose the library in diagram outlook, choose a Cable Cross-Section constituent, then choose (right-click) copy and then paste on empty area

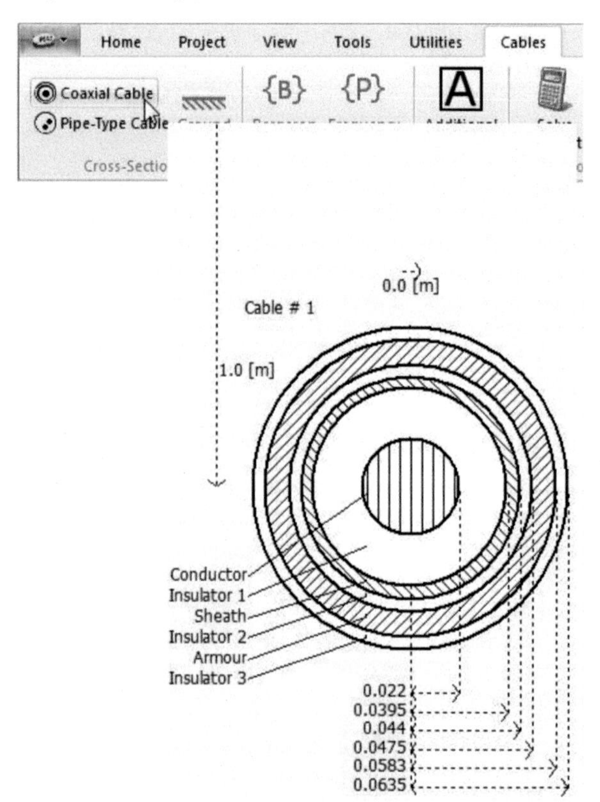

Numerous cables could be included in a single pattern. Make sure that cables are given numbers in order number inside the cable apparatus; also the *X*-coordinate of the new cables in that segment are accustomed. The conductors included with supplementary cables should reflect in the respective cable terminal equipment.

8.4.4.1 "Edit" cross-section parameter

Cross-section parameters attributes are modified by related to its data dialog window. Choose its component (without choosing it) choose Edit Parameters to the right to use this dialog.

8.4.4.2 The Bergeron Model

This form is meant to indicate the inductive and capacitive elements of the circuits. It is precise for particular frequency. Also, this model is best applicable in the requirements for the load flow studies in that particular frequency and for relay coordination

While applying Bergeron forms, it is not compulsorily to apply for tower constituent in order to show the OH lines of the network. When the 3-phis simulated the parameters of impedance and admittance may applied, and after that input the data, straight by giving the Y, Z constituent.

8.4.4.3 *The frequency-dependent mode*

This (mode) form indicates data that depends on the system frequency of every data and not on a particular frequency as indicated in Bergeron mode. This applies technique to find solution for line constant parameters and takes a constant transforms. Hence it is precise for power lines of ideally transposed conductors or for two horizontally constructed lines systems or lone conductor system.

8.4.4.4 *The frequency-depending (phase) mode*

This model deals the frequency dependency of all data parameters like reproduction of above but, the Frequency-Dependent (Phase) mode produces constant conversion problem by shortest formulation exists at phase domain. Hence it is more exact for every transmission models, together with unbalanced transmission line coordinates.

8.4.5 Addition of line model

Line model mechanism could be included in the edition option in two different methods.

1. The simple ribbon control bar by choosing the existing model and keep in schematic canvas, more details can be got by addition modules to a project.
2. Application of context menu, in which the pointer is moved in the blank area and then the model is selected (right-click) a submenu pops up with a listing of every model components available in the master library. Choosing a model will be added by itself.

It can be copied and pasted model elements straightaway from the library project, by opening schematic view then opening a line model component, (right-click) and choose copy then pasting on (right-click) in desired area.

8.4.5.1 *Edit line parameters model*

It can be altered with respective model data dialog box (right-click) in the line model component (without selecting it) then choose Edit Parameters to enable this dialog.

8.5 Debugging and refining the project work

Building or Compiling Message

The building or compilation progression is easily done with the following ways.

1. Build the Source and Data File for the project: In this PSCAD collects every module explanations in the project and analyzes it. The outcome of this is the formation of resources FORTRAN (*.f) and Data (*.dta) files. If any troubles subsist in modules, a fault or caution message pops up. PSCAD compile all sections, which are highlighted for it, detecting the errors and solving in every section; only next step will be preceded after resolving all errors
2. Making of Map File: After every sections has been constructed the corresponding local node and its related system of the modules have been built, their respective local nodes and subsystems must be linked collectively with every module which is carried creating the project Map (*. map) file. If any prohibited with interlink within sections, or as relevant faults and advice messages will emerge here.
3. Create Make File: The Make (*.mak) file is created as instruction file for FORTRAN compiler the issues in the process will be shown here.
4. Finding solution for transmission part: The concluding procedure previous to the simulation executive file is generated to resolve every transmission lines and cables in work. For every transmission part, PSCAD generates OHT line inputs (*.tli) or a UG cable inputs (*.cli) file. PSCAD after that finds solution for the segment of Line Constants Program (LCP). In case if the solution of the given problem could not be found due to logic failure and the LCP solution fails, a fault or caution communication will pop up.

8.5.1 General output window messages

There are numerous fault and caution messages which are produced and shown at Output Window. A message source can originate directly from the PSCAD or EMTDC software. It can also present by its subsystems, also all of communication will not happen if a project is carried out with care from the first place.

The underlined message will occur while executing the project.

8.5.2 Warning

This message comes up if PSCAD finds an opened electrical circuit element - typically caused by circuit node which doesn't being associated to any of circuit. If an open circuit of the network is to be carried out the best way to deal with this warning is to be connected with the very high value resistive component in the range of Mega ohms to ground at node. This makes sure that the arithmetical strength and it influence on the simulation final output will be negligible insignificant.

"A key has not been properly defined" or "does not exist in the component data."

This pops up when the program detects REAL signal is given as the input if the integer is to be anticipated as the input is given; PSCAD will automatically change the REAL signal data to the adjacent numeral.

If PSCAD finds a data of similar name is produced from additional source. This fault is most commonly caused when module is interacts with definite inner variables are duplicated, so it duplicates the inner resultant variable.

Fault message given if PASCAD finds that dimensions <dim_1> is given in input and request for the output of a dimensional value of <dim_2>. This mistake usually happens with the power electronic devices, where it anticipate a 2-dimensional input gate signal while used for the processing inputs,

When PSCAD finds array of signal category disparity. Error warning will be given during simulation. For instance, if a signal array is a clear-type integer, and the client tries to tap off a single part with the Data Signal Array Tap constituent assigned with type real (or vice-versa); also, if the array of type real is input into a module, where the exterior input is given as an integer array.

Error linked with application of the Breakout module: Grounding cannot be linked to the Breakout terminals. The Breakout was originally intended to map numerous links in scalar side to a single array. As the ground nodes could not be mapped, the compiler gives out this warning. The recommended work is intended for a Current Meter as a series element linking between Breakout terminal and ground. The entitled "Valid Connections" in the Breakout component online gives more information.

The node with 3-φ with part of this module are not the real electrical node, but models that takes the number of node where it is linked. A unique situation, which could not be indicated as an "unbalanced" is called, refers to electrical nodes but not real impedance. The fundamental law to keep in mind is that three lines on the $3-\varphi$ plane should contain minimum of single series impedance.

To view Build and Data Files: Every time a simulation is designed, numerous files are formed and saved in the linked project transitory folder. A few of the files, like FORTRAN, Data/Map, are applied to debug. These can be noted straight in PSCAD.

Component Ordering: PSCAD by itself has an elegant algorithm, which by itself keep in succession of all modules concerned in EMTDC mechanisms. This is processed by itself to make sure that the variable data values are processed with the exact order, and the step process times are fewer: The algorithm repeatedly scans the whole simulation and gives the order ranks to the entire project, and then assigns sequence numbers to all accessible module. In a broad sense, input constant data are taken to the apex of order, and results are pushed down.

This algorithm is left to be free with its own parameters at all times; in spite of it, a large number of moments where it is required to put off and physically arrange the modules while removing the flaws. This happens if it is liked to manually organize the response point. A response for control may brought in by introducing a Feedback Loop choose module in the signal flow path.

8.5.3 Exhibiting the string Numbers

Before physically bringing any component, the case should be compiled, and make sure that sequence of numbers set, is made possible in canvas setting dialog: in order to pop-up canvas options dialog. Then select in the blank part of schematic canvas (right-click) and then choose canvas setting.

Colors: The code for specific module can be placed in two locations inside the system dynamics (i.e., dynamics subroutine, DSDYN or DSOUT) As a consequence, the sequence numbers are colored coding such that the it can graphically find out the place of code which are detailed as under.

Color Legends: Aqua: The part code lies inside DSDYN for the present element.

Olive: This code lies inside DSOUT for the present elements.

To set series Number Physically: To manually make series numbers, initially make sure that its option is displaying, and the Sequence Ordering choice is set to Manual Assignment in canvas options dialog. In the component choose the sequence (right-click) from pop-up.

8.5.4 Showing signal place

PSCAD uses icons available on link and wire terminal points to give permission for graphical finding of where time step lag are existing. Also, the icon colors are utilized to correspond to signal type.

Control Signal Caption:

Green: Inform that signal is REAL
Blue: Inform that signal is INTEGER
Magenta: Inform that signal is LOGICAL

Electrical Signal Legend:

Green: Inform that its operational node
Brown: Inform that it is a ground node
Gray: Inform that it is deleted node
Red: Inform that it is separated node

8.5.4.1 Virtual control wires

This can be applied to present a seeable "virtual connection" for two or many Data Label modules with similar names inside a particular module also it appears as a dotted lines, straight in between respective Data Description.

8.5.4.2 Virtual filters wire

It can be filtered that the presentation of virtual wires are supported with similar signal. These applies in places where there is disorderly quantity of data connections which makes it challenging to categorize and to discover the signals which are required the most. Separate out the virtual wires exhibit by addition of the signal name which are required to be seen, comma isolated, into Virtual Wires Filter choice in canvas options.

Color Legend:

Green: Inform that signal is REAL
Blue: Inform that signal is INTEGER
Magenta: Inform that signal is LOGICAL

8.5.5 Control signal tract

It is applied to see the control signal flow way (i.e., originated source till the sink) pursuing compiling the project. The index will be shown in wire section, which are portion of a control signal route. It is viewed as arrowheads direct on the wire.

Green: Inform that signal is REAL
Blue: Inform that signal is INTEGER
Magenta: Inform that signal is LOGICAL

8.5.6 Creating Library (*.lib) and Object (*.obj) Files

Object (.obj) Files*

Commonly, If a source file is connected with a project, the FORTRAN compiler utilized to roll up the work will by itself make a compiled object (*.obj) record for ever connected source file. It is kept in the project transient folder. It may be selected to render customer with compiled object files, in place of source code. This is good if a single or multiple source files are implicated. Bigger simulation work projects on the other hand, can have number of source file relations, and give an object file for every source file will swiftly develop into complicated. To solve this trouble is to combine every source routines to a sole file. This work tiresome, and will produce troubles in current progress of the source code.

Static Library (.lib) Files*

A further easy and competent way out for combining various merging source files together is to merge every individual object files as one

compiled library (*.lib) file. PSCAD has options for a trouble-free way for producing a "compiled library" file for all (*.pslx) file, if that connects to source files are available inside the library project. In library works, this task is done by the application of File Reference components.
Generating a Library (.lib) File*

The primary consideration prior to generate a library file is having FORTRAN compiler, which will form compiled files, and can or cannot be well-matched for applying along the supplementary compilers. Alternatively, it is imperative to recognize which kind of FORTRAN compiler the customer is using, in order that files given are compatible. The majority of PSCAD application, customers will generate the same file for every serving FORTRAN compiler. These files are located in the suitable directories as explained in supplementary library (*.lib) and object (*.obj) files.

Step 1:

Generate a fresh library project as explained in Generating a New Project (or correct an on-hand library), then connect every source file to be incorporated in the compiled library (*.lib) file by means of File location components.

Step 2:

In library forename in the Workspace (right-click) choose Create Compiled Library (*.lib).

When the generate Compiled Library (*.lib) function is applied, PSCAD will generate a momentary folder for the library project situated in the identical address list as the project (*.pslx) file. Within that it is positioned the compiled library file (*.lib), and also an self-object file (*.obj) for each connected source file.

8.5.6.1 Bringing in dynamic link library (*.dll) files

It is feasible to comprise dynamic link library (*.dll) files when carrying out projects, even if associations to the files ought to be given by means of import library (*.lib) file. Alternatively, the import library file is the one, which should be straightforwardly connected to in the precise method as is explained for connecting stationary library files in supplementary to library (*.lib) and Object (*.obj) Files.

Example. I.

Step 1: Generate a new work by means of either the *Menu* or *Toolbar*. A new project should emerge in the *Workspace settings* permitted *no name [psc]*. In this Workspace setting (right-click) select Save As... and provide a project name. Form a folder called *c:...... /Pscad Training/Tutorial_01*. Save the work as *Project01.psc*

Step 2. Start the main page for fresh case. Develop a problem to solve the inrush current, case study of, during charging a transformer. The constituent data is as depicted below. The rating of Transformer is 66/12.47 kV (Fig. 8.1).

Step 3. Draw the currents (Ia) and voltages (E_66) in the High Voltage terminals of the transformer. *N.B:* Ia and Ea has the three waveforms for the $3-\varphi$ systems (Fig. 8.2).

Step 4. The Low Voltage side of transformer in no load condition. The breaker make time is at 0.5 second and charges transformer at 66 kV HT bush.

Inrushing is associated to core saturating current. Checking that saturation is incorporated in the project model applied for this work, and its value in inrush current magnitude is based on the "point on wave" breaker "ON" conditions. Employ a manual switch to control the breaker. Reminder: the position on wave dependent on o inrush crest (Fig. 8.3).

Step 5. Adjust the work to comprise a 12.47 kV/0.5 mega volt amp (MVA) (Wound rotor type) induction machine. This project will be applied to learn the procedure of switch ON the Induction motor. Thc module data is as depicted.

Step 6. Input the data of the module.

Step 7. Draw the current on both ends of either of transformer (ia and ib).

Step 8. The input torque to the Induction motor is equivalent to 80% of the square of the speed. Obtain this with the use of control modules. Use it to apply the in the equation.

$$T_m = 0.8 \cdot w^2$$

Step 9. The breaker is switched on at 0.2 second to start the motor.

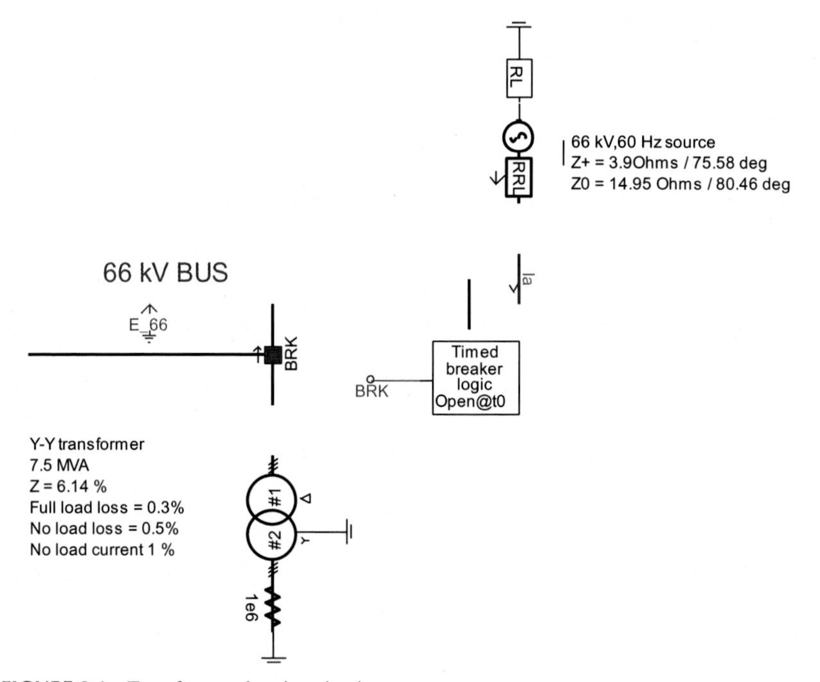

FIGURE 8.1 Transformer charging circuit.

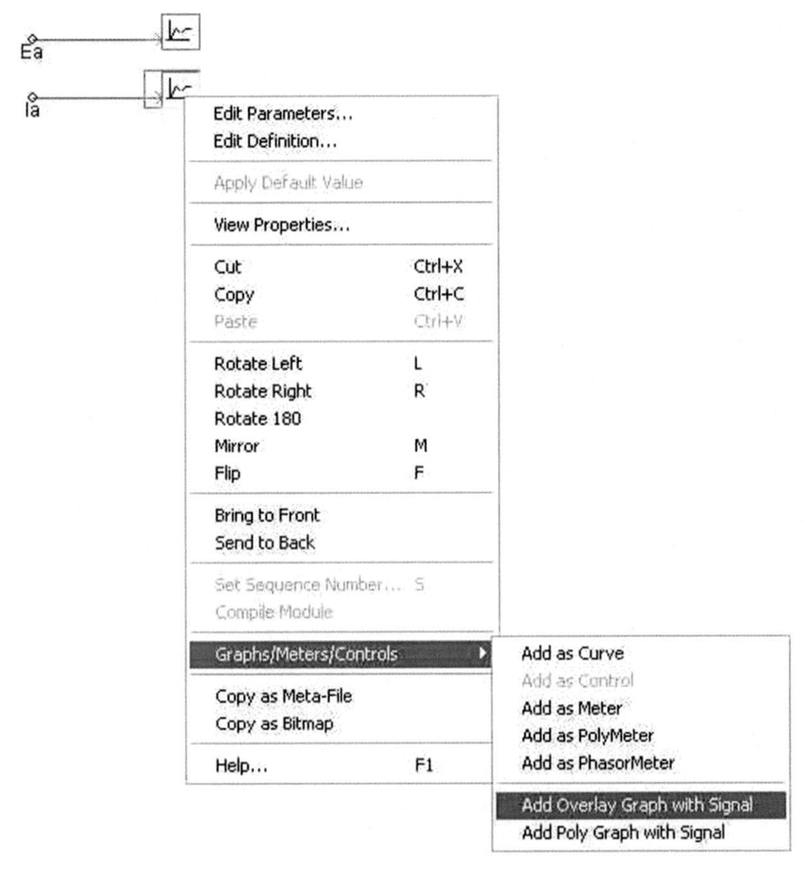

FIGURE 8.2 Procedure to create a graph with a given signal.

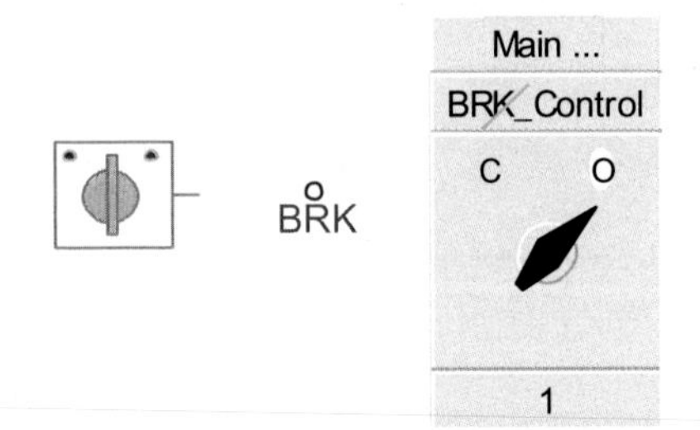

FIGURE 8.3 Two-state switch attached to a control panel.

Step 10. Plot the machine speed, the torque by both mechanically/Electrically created.

Example II

Step 1: Generate a new work by means of either the *Menu* or *Toolbar*. A new project should emerge in the *Workspace settings* permitted *no name [psc].* In this Workspace setting (right-click) select Save As... and provide a project name

Step 2. Start the main page for fresh case. Develop a problem to solve a simplified two-area power system as depicted in the picture as under. A 55 km OHT line links Station A with a 100 MW wind Energy system. Every other links to Station "A" are shown by a corresponding 230 kV source. The same level of source impedance is calculated from a steady state fault case at 60 Hz. The OHT is shown by its series reactance. The transformer is depicted by its impedance, taken from the 230 kV side (Fig. 8.4),

Step 3. The wind Energy system is shown by the network equivalent. The positive sequence impedance of the source at 33 kV is 1 Ω at 89 degrees.

Step 4. The voltage of the equivalent impedance of wind farm is 35 kV. The phase angle is 7 degrees. resolve thc power flow in between lines.

Step 5. design the real and reactive power flowing in the terminals of line at either ends. The signal can be got from the voltage source equivalent as internal outputs.

Step 6. Apply correct scales inside the Output Channels' to alter PU values to Active/Reactive powers validate the results.

Step 7. Alter the source at the point of termination A to organize its data externally. Include a power panel to identify these quantity (Fig. 8.5).

Step 8. Alter circuit to comprise breakers, its control, measuring instruments and the PSCAD "fault component." The project will be like the diagram as in Fig. 8.6. Plot, E1, I1, and the rms quantity of E1.

Step 9. Simulation is done for the A-G fault and its initiation time is 0.4 second. The fault period is 0.5 second. N.B. dc offset is I1.

Step 10. Add an FFT block in the work as depicted at Fig. 8.7. change I1 to its sequence modules. Make sure that the outputs of FFT for various results faulty variety. Insert a "poly meter" to watch the frequency range.

Example III.

Step 1. Build a folder named c:....../Pscad/.......

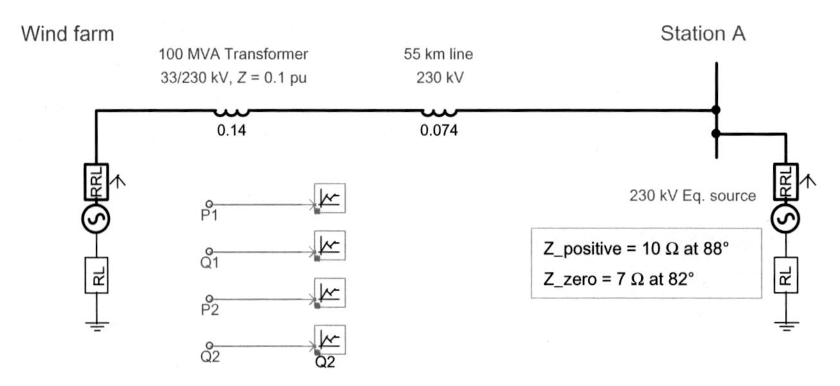

FIGURE 8.4 Two-area system.

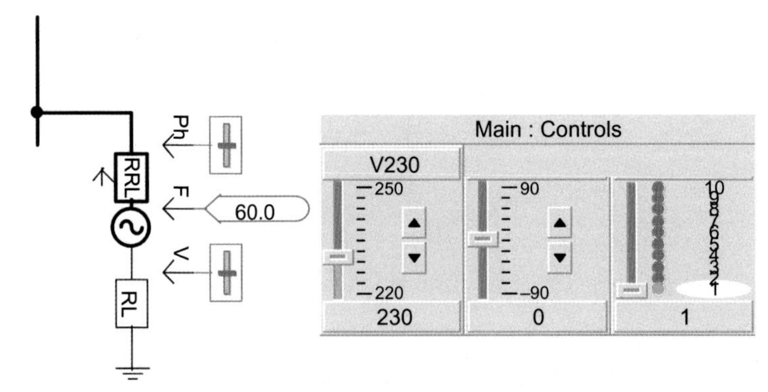

FIGURE 8.5 External control of the source parameters.

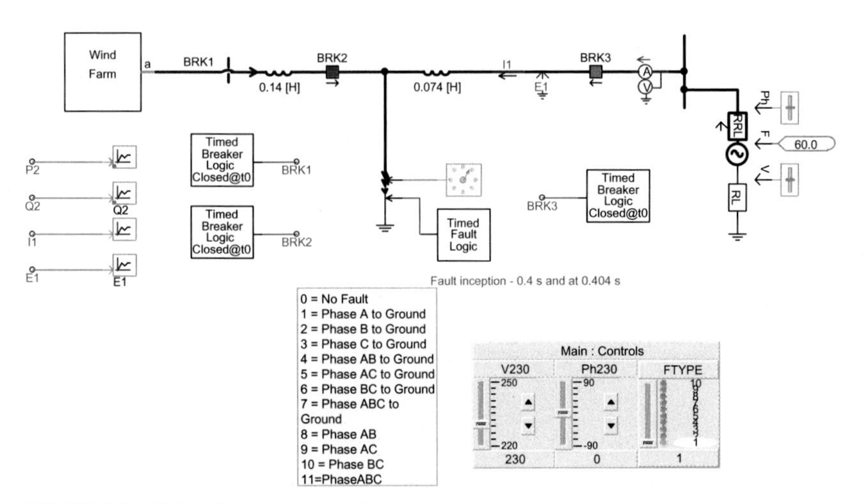

FIGURE 8.6 Meters, breakers and faults.

The Power supply system company aims for additional 300 MVars of capacitor bank for producing VAr component in a station A to maintain the 230 kV bus voltage. A brief learning is necessary to plan apparatus of this system.

Mathematical analysis and simulation are necessary to resolve the standards/ratings of the connected limiting reactors (inrush and outrush). Adjust the model case to comprise another folio as depicted in Fig. 8.8.

The electrical circuitry within the additional leaf indicates a 230 kV capacitors array having four step per line (observe related drawings). Every stage is having capacity of 25 Mvar/ph and are having solid Earthling system. The inrushing/outrushing reactor capacity are to be ascertained such that switch

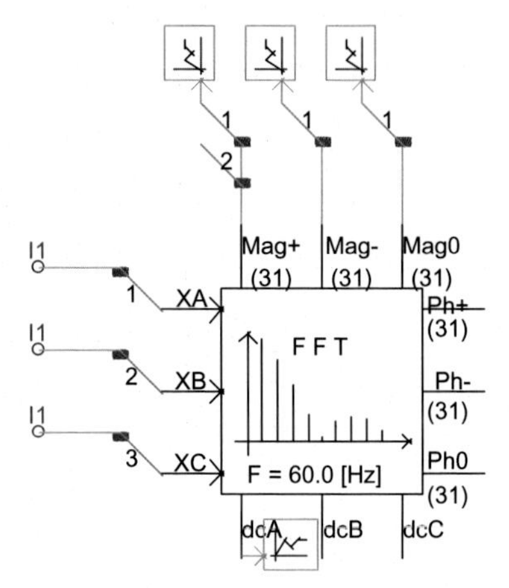

FIGURE 8.7 FFT Block. *FFT*, Fast Fourier transform.

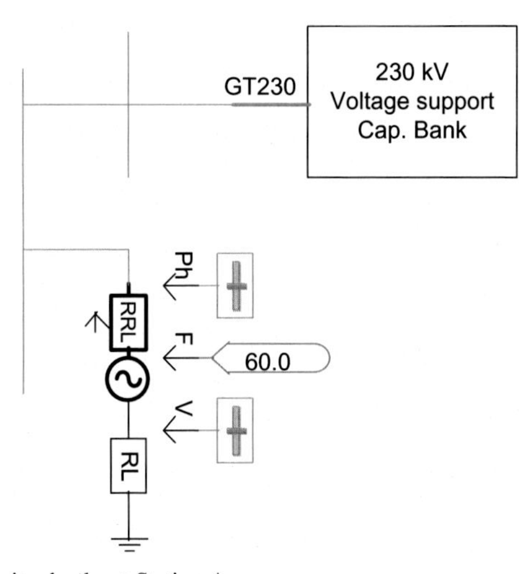

FIGURE 8.8 Capacitor banks at Station A.

in disturbances does not go beyond the capacity of breaker and fallows inside IEEE regulations.

Step 2. Apply hand-operated breaker activity to bring the R1, R2, and R4 in service, and quantify currents at breakers.

Step 3. Include a breaker time element for controlling R3, quantify the I in R3.

Step 4. Add meters to quantify the V/I of scheme of the outrushing reactor.

Step 5. Trial the working condition of R1 closed, R2 and R4 open, with R3 closed at 0.2 second.

Step 6. Discover the max measure and frequency of oscillation of I in R3.

Step 7. Discover the max measure and frequency of oscillation of Iin the outrushing reactor.

Step 8. The maximum inrushing I rely on POW operation, and it is observed to an assurance of breaker fulfills the TRV and di/dt potentiality.

Step 9. Apply the Run element to standardize the R3 ON duration, and collect the data of the I in Breaker R3 and in main feeder.

Step 10. Find the impedance range applying the "Harmonic Impedance" factor which is vital in the designing of capacitor banks. The inclusion of the capacitor will give out system resonances, which is unacceptable.

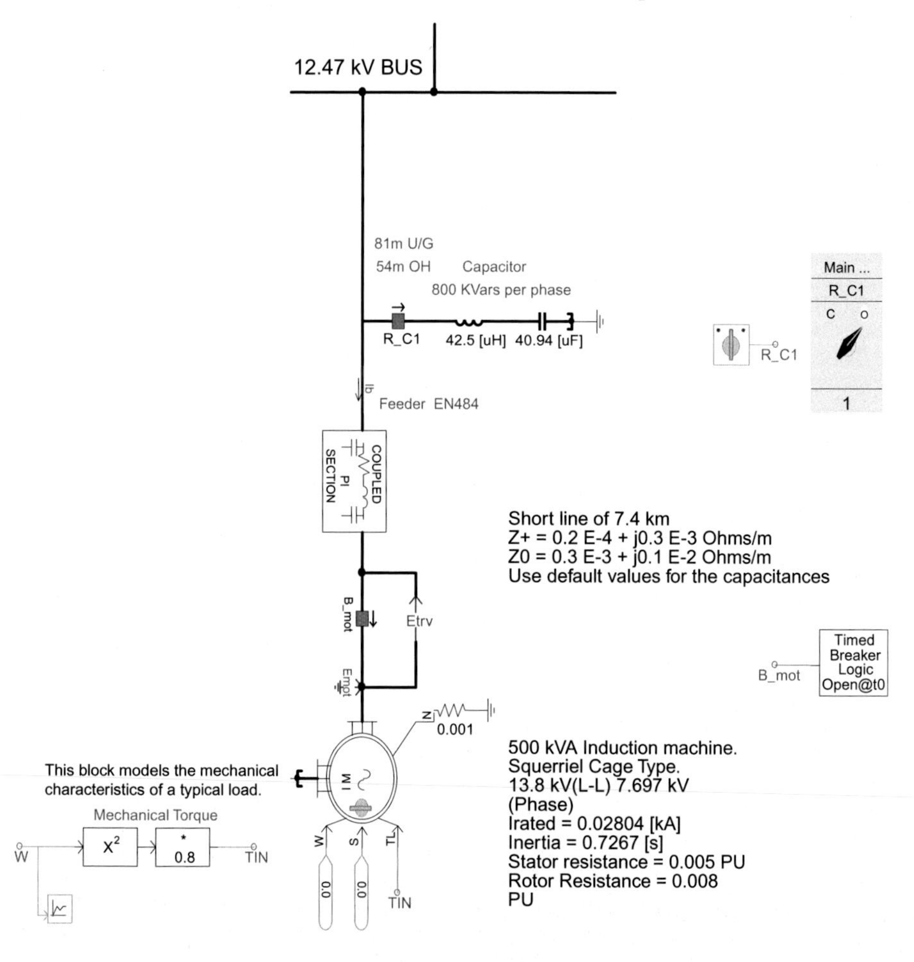

Step 11. Alter the circuit as per Fig. 8.9 to admit surge arrestors.

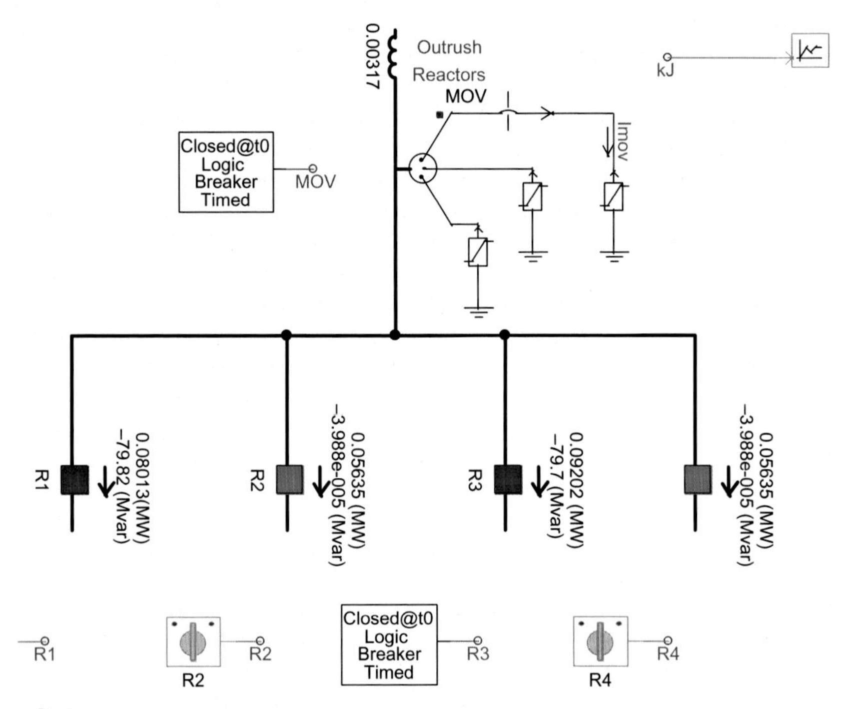

FIGURE 8.9 Surge arresters.

The surge arrestors must safeguard capacitors from switch time voltage surges. Restrike of its breaker will make higher voltage transients which are normally the main consideration for the choice of metal oxide varistors (MOVs). Deal the data written record for the MOV model.

Step 12. At the start Breaker R3 switched ON and is switched off at 0.204 second but restrikes at 0.2124 second. Detect the power aggregation of MOV, phase A.

Example IV

Step 1. Build a folder named c:. . .. . ./Pscad/. . .. . ..

Switch off breaker R3, and keep all others ON and do the "fault" constituent inoperationality.

Almost all transient scrutiny needs the precise model of transformers and OHT lines. Transformer inrushing I needs the precise model of non-linear iron core. Switching transient learning needs the model of OHT lines for adding the influence of frequency-influenced line parametric quantity and traveling wave occurrence.

Step 2. Apply elaborate equivalent model to correspond the 33/230 kV transformer and a 55 km OHT line. The transformer has a Star/Star design

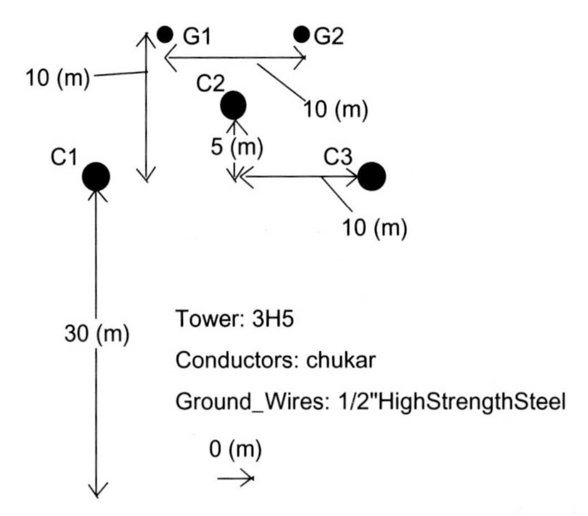

FIGURE 8.10 230 kV Transmission tower.

has three separate 1-ph units and has 1% I in idle unloaded condition and Cu loss of 0.003 and 0.002 pu correspondingly.

The line design is depicted as detailed below. Apply frequency-depending phase design to symbolize the line (Figs. 8.10 and 8.11).

Step 3. Switch OFF the breakers 2 and 3. The transformer is charged with no load on no while switching ON switch gear 1 and then find the inrush I @ 0.15 second.

Step 4. Include a 1 Ω resistance in serial way in 33 kV winding and discover the outcome signal.

Step 5. Take the "single pole operation" condition of the switch gear. Switch ON each phase in during the time of voltage of the individual phase is at a peak. Find outcome.

Line Charging Examination:

Step 6. Switch ON the switch gear 1 and Switch OFF 3. Consider the multiple run constituent to monitor the working of switch gear 2 (OFF condition while starting). The ON intendance B1 is found from the aggregate run (Fig. 8.12).

Step 7. The Switch gear ON Timing (B1) should be changed for each run. The breaker is opened 0.15 second after its closing operation. Set the number of tests to switch for 10 consecutive times instances on a 60 Hz wave shape. Note the Max V E1 at the receiving terminal.

Lines on the same right of way:

A 130 km OHT system links power producing plant C and a substation. This line is laid symmetrical to the 55 km line with Station A and a Wind generators for 20 km from substation A. The terminal voltages of generators are step up to the transmission system voltage of 11/230 kV, with $Y{-}Y$ bank.

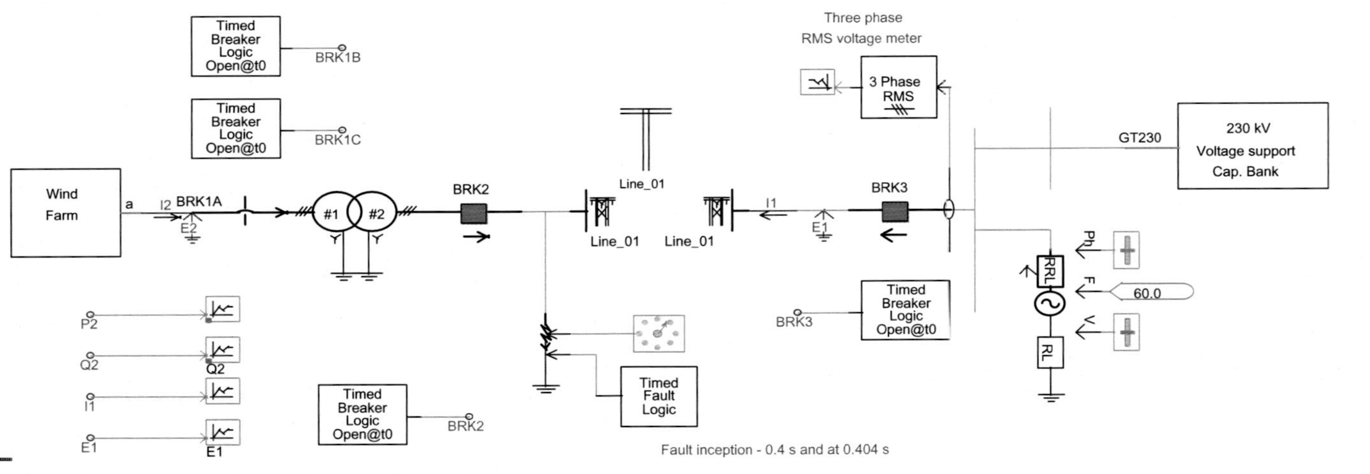

FIGURE 8.11 Two-area system model for a transient study.

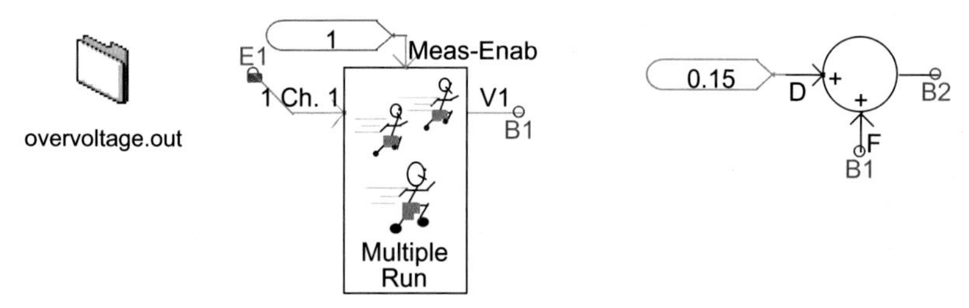

FIGURE 8.12 Multiple run component for breaker control.

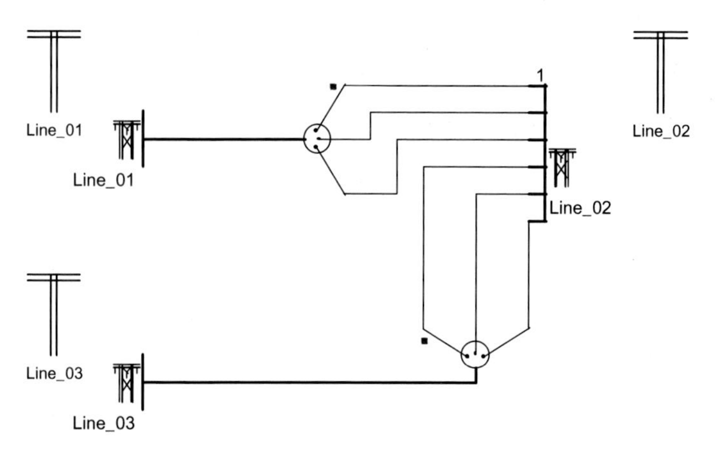

FIGURE 8.13 Line arrangement inside the subpage.

Step 8. Broaden the framework to consider the 130 km OHT system and the generating station as depicted in the single line diagram 4. The OHT lines are organized in directional leaf as depicted as in 5 (Fig. 8.13).

Single-line diagram (SLD).4. Three area system

Step 9. The Vin the same source impedance of the voltage source correspond the four alternators at power generating source is C is 12 kV at 21 degrees.

Step 10. Apply Mathcad working sheet to validate outcome.

Step 11. Alter the design of the 11/230 kV transformer to correspond a D-*Y* connection. Vary the 11 kV generating voltage angle to indicate this alteration.

Example V

Step 1. Build a folder named c:. . . . . ./Pscad/. . . . . .

The induction generators are applied in wind farms with terminal voltage of 33 kV. The aggregate MVA of the installation is 100 MVA. Substitute the

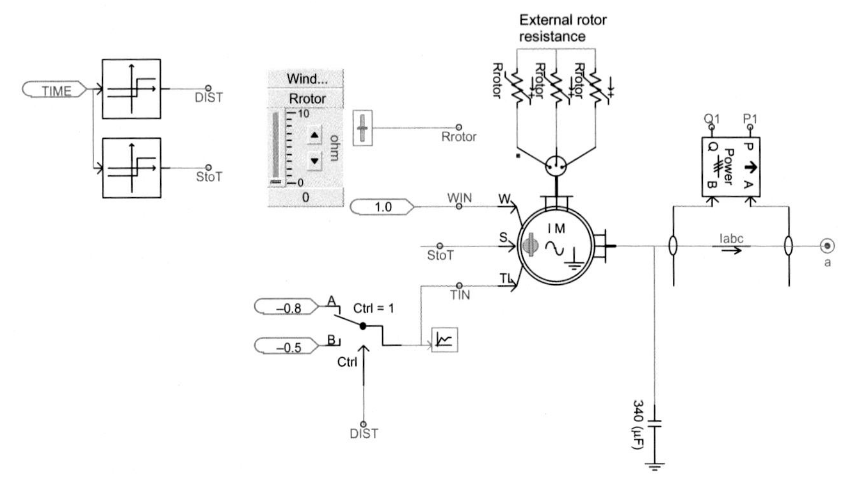

FIGURE 8.14 Induction generator.

equal source with an elaborate kind of an induction generator. Presume all units of at the wind system are functions with same condition. The system is indicated in SLD as depicted in diagram 1 (Fig. 8.14).

Step 2. Put ON the switch gear 1 at 2 seconds, and also all other in ON condition

Step 3. Work out the rate of the shunt capacitance needed to keep the actual MVA flowing. Apply Mathcad computation. Reduce the time durations to 25 μs.

Step 5. Address in which way the wind energy system is linked with system, applying *BRKA* suitably, at 1 second.

Step 6. Observe line currents in the system part while the wind generator is attached to the system. Alter the startup acceleration of the turbine to 0.6 pu and iterate the modeling.

A Soft Starter depicted in Fig. 8.15 is utilized to control the startup I while linking the wind generator with the power lines, The back to back thyristors are utilized to limit the voltage given with generator during its acceleration. The firing angle characters are presented in the file "softstart. txt" and its controls in Fig. 8.16.

Step 7. Notice the startup I with/without soft start.

Example VI

Step 1. Build a folder named c:. . . . . ./Pscad/. . . . . .

Step 2. Interpret the elementary idea of the double fed link.

Step 3. Determine the character of various control in modules in the simulation.

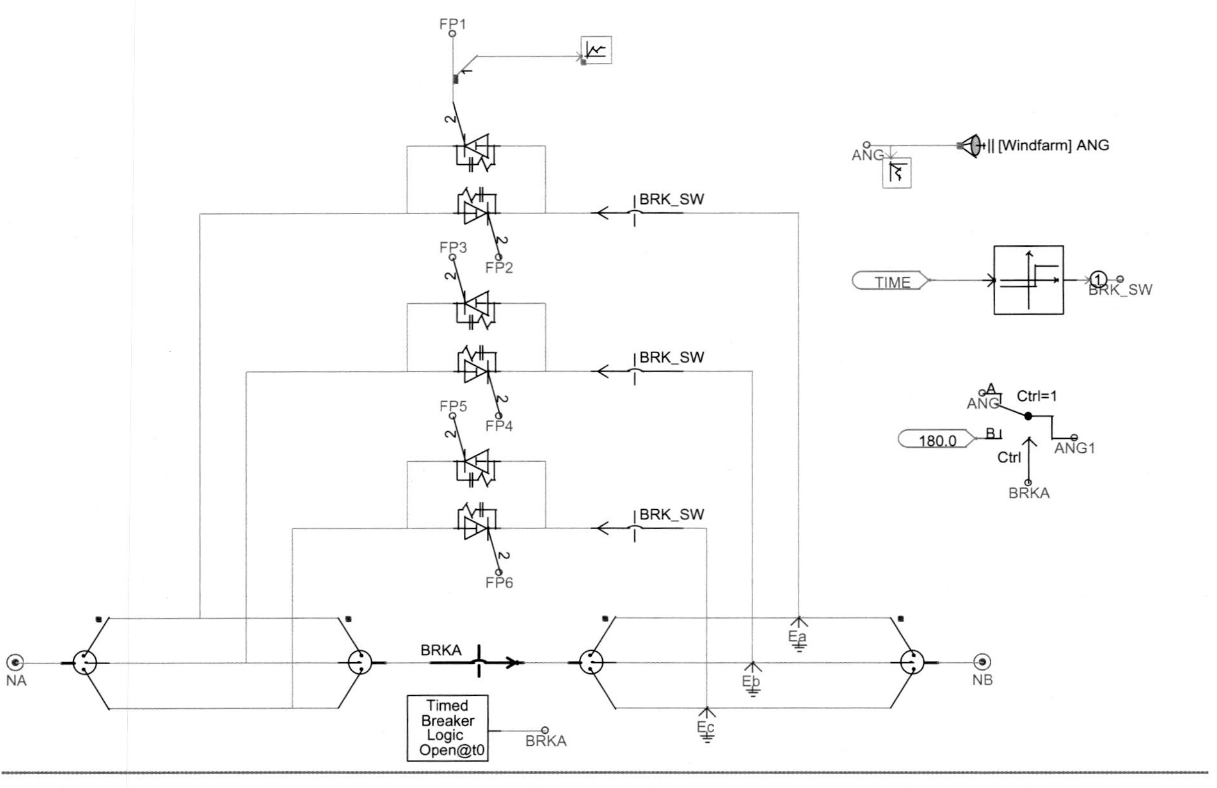

FIGURE 8.15 Soft Starter.

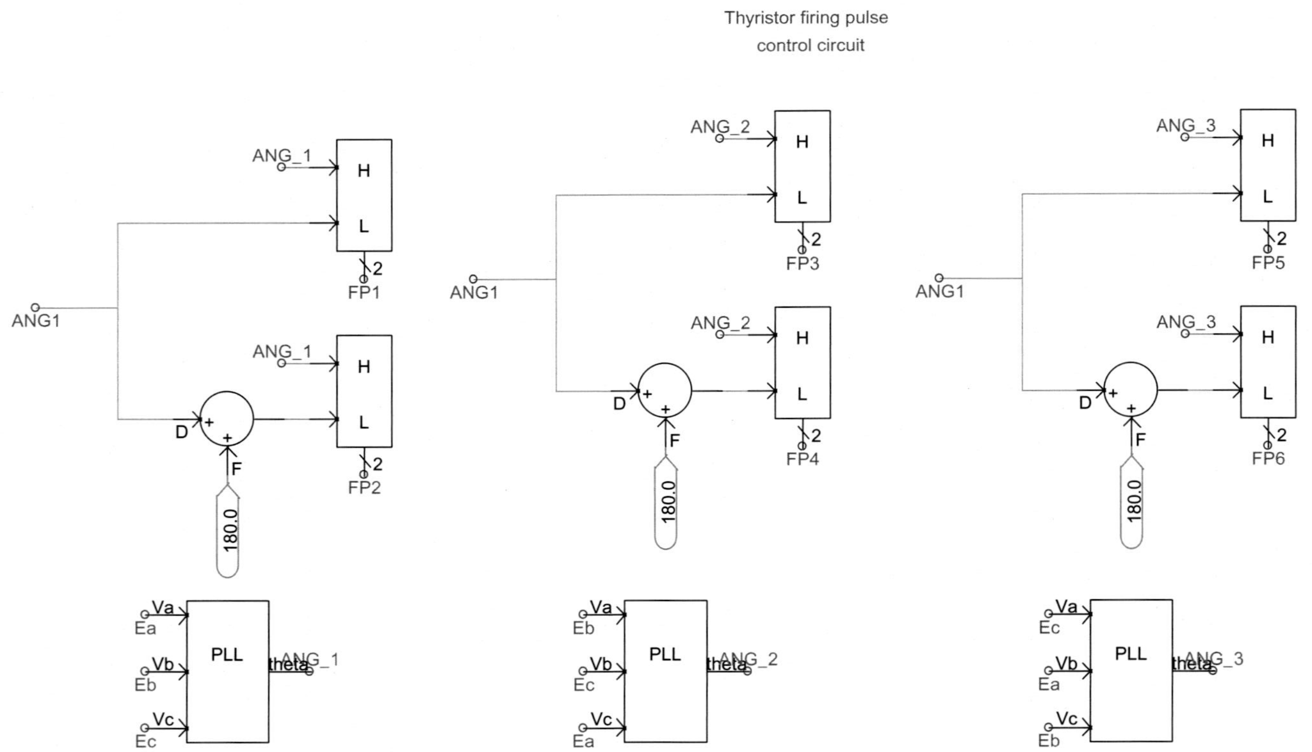

Thyristor firing pulse control circuit

FIGURE 8.16 Firing controls.

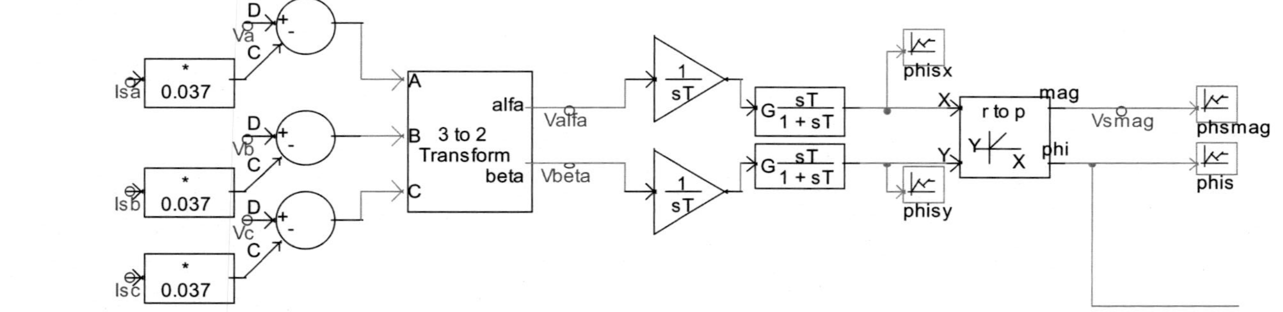

FIGURE 8.17 Stator flux vector.

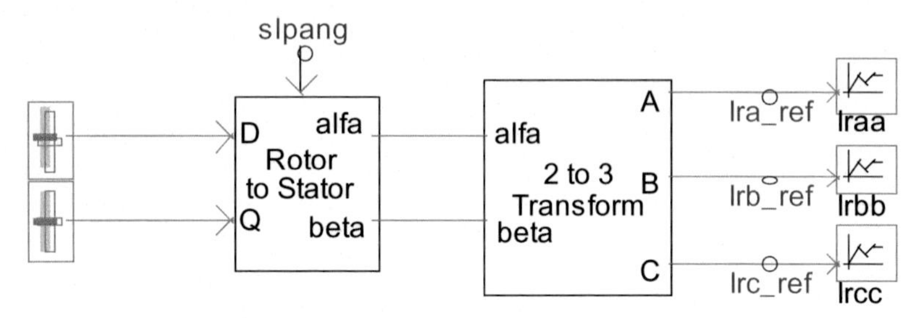

FIGURE 8.18 Rotor reference currents.

Step 5. Test the working of the two examples (Figs. 8.17 and 8.18).

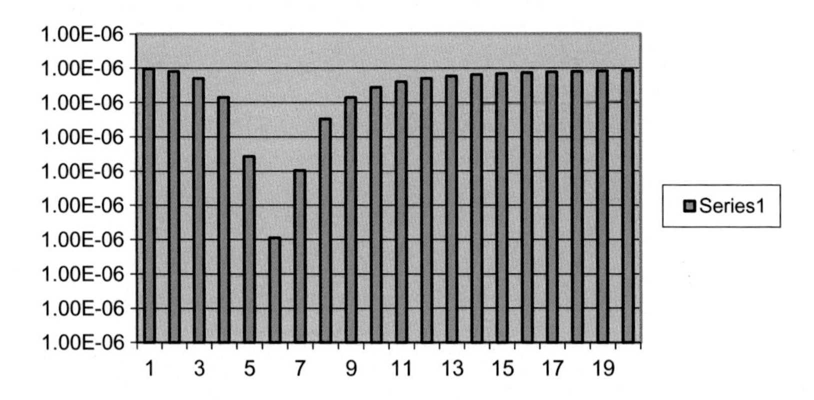

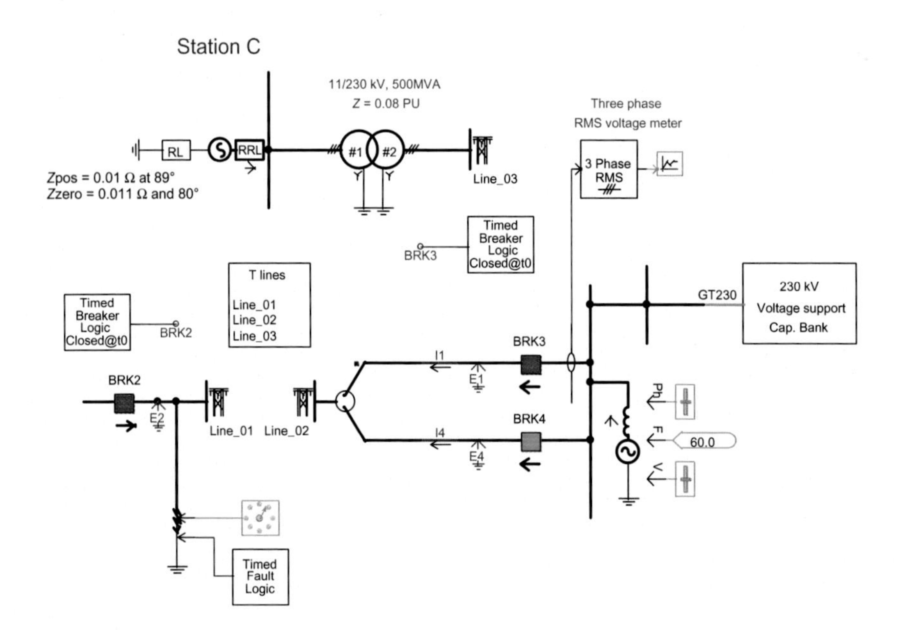

8.6 Summary

The readers can get knowledge of the quick, precise, and user-friendly power system simulation technique for the design, investigation, maximizing, and confirmation of power system/power electronic control strategy. This section provides good techniques and predesigned examples for absolute and precise examination of power systems. Encapsulating endless usage, PSCAD is becoming the design tool option for business, industrialized, and new innovations organization universally. Hence, the student will gain higher knowledge while carrying out these actual problems.

8.7 Review questions

1. Concisely state the PSCAD software.
2. How to validate THD by applying PSCAD software.
3. Enumerate the plotting tools utilized in PSCAD.
4. Provide few representation for overlay graphs and poly graphs.
5. State some of the employment of poly meters and phasor meters in PSCAD.
6. Give brief notes on oscilloscopes and *XY* Plots.
7. Enumerate the ways to construct OHT lines applying PSCAD software?
8. How to lay the UG cable system with PSCAD software?
9. How to implement tower section with the help of ribbon control bar.
10. Compose every procedure to include Cable Cross-Section Element.
11. State the four-step compilation procedure in PSCAD.
12. Enumerate the methods for making library files with PSCAD software?
13. Compose brief notes to create object (*.obj) files.

Chapter 9

PVSYST

Chapter Outline

9.1 Introduction

Finding a sustainable power source is the ultimate goal of humanity. Because of the high utilization and the decreasing accessibility of petroleum fuel resources sustainable power (solar, windmill, waterpower, and so on) is a topic of interest over the last couple of years. Solar power is a rising sustainable power source utilized everywhere throughout the globe and on a large scale. The intensity of the sun captured by the earth is more prominent compared to the present utilization rate on the globe of every power source. Hence, solar power can give resolution to all the current and upcoming issues identified with power. Solar energy is the transformation of sun rays into power, either in a direct way using photovoltaic (PV) or in a roundabout way utilizing concentrated energy or concentrated solar PV. PVs transform light into the electric flow utilizing the PV. A housetop PV power plant, or PV framework, has the power created from the solar panels placed on the housetop of a private or business building or construction. A PV framework comprises a PV exhibit, battery, and components for a power conditioner. The PV framework changes the solar power into DC energy. On the off chance that air conditioner loads are utilized, the framework expects inverter to change over the DC mode into AC mode.

There are two sorts in PV framework, for example, matrix-associated and independent. Grid-associated PV frameworks feed power straight to the electrical system, corresponding to the customary power source. Grid-associated frameworks create clean power close to the point of utilization without the distribution or transmission losses or the requirement for the batteries. Its level of performance relies upon the climatic conditions,

Software Tools for the Simulation of Electrical Systems. DOI: https://doi.org/10.1016/B978-0-12-819416-4.00009-0

direction, and tendency of the PV exhibit, and the performance of the inverter. Though an independent framework includes no connection with the utility network, the created electricity is associated with the loads. On the off chance that the PV exhibit does not supply the load, a memory gadget is required. Generally, this is actually a battery, since the battery pack stores power when the electricity provided by the PV module surpasses energy demand and discharges it again while the supply of PV is not enough. This independent PV power creation will be used in the homes for the purpose of electricity. A wide assortment of apparatuses subsist for the examination and dimensioning of the grid-associated and independent PV frameworks. Framework designers and fitters use more basic instruments for measuring the PV framework. For the most part, researchers and architects normally utilize included simulation devices for leveraging. Programming tools identified with PV frameworks can be categorized into the preattainability investigation, measuring, and recreation.

9.2 Solar photovoltaic system

Since the demand for sunlight-based electric frameworks has grown, dynamic manufacturers are including solar PV as an alternative for their clients. This outline of solar PV frameworks will give the developer a fundamental comprehension of:

- analyzing a construction location for the potential of solar;
- usual grid-associated PV framework setups and segments;
- considerations in choosing segments;
- considerations in structure and establishment of a PV framework;
- typical expenses and the labor needed to set up a PV framework;
- electric along with building code prerequisites;
- where to discover more data; and
- emphasis will be set on data and it will be valuable in including a grid-associated PV framework in an offer for private or business buildings.

There are two sorts in PV framework, for example, matrix-associated and independent. Grid-associated PV frameworks feed power straight to the electrical system, corresponding to the customary power source. Grid-associated frameworks create clean power close to the point of utilization, without the distribution or transmission losses or the requirement for the batteries. Its level of performance relies upon the climatic conditions, direction and tendency of the PV exhibit, and the performance of the inverter. Though an independent framework includes no connection with the utility network, the created electricity is associated with the loads. On the off chance that the PV exhibit does not supply the load; a memory gadget is required. Generally, this is actually a battery, since the battery pack stores power when the electricity provided by the PV module surpasses energy demand and discharges it again while the supply of PV is not enough. This independent PV power creation will be utilized in the homes for the purpose of electricity. A wide assortment of apparatuses subsist for the examination and dimensioning of the Grid-associated and independent PV frameworks. Framework designers and fitters use more basic instruments for measuring the PV framework. For the most part, researchers and architects normally utilize included simulation devices for leveraging. Programming tools identified with PV frameworks can be categorized into the preattainability investigation, measuring, and recreation.

- Evaluating a site for solar PV potential: The site ought to have enough sun for matrix-associated PV frameworks to work well. Whereas in shady conditions, the facts confirm that PVs generate just 5%–30% of their greatest yield. But since solar power PVs become a little less productive when sweltering, our cooler seasons and the long and hot summer days assist to make up for the shady days.
- Analyzing a construction site: Not each construction site will be appropriate for the solar panel establishment. The initial phase in the plan of a PV framework in deciding whether the site you are thinking about has great solar power potential. A few inquiries you should pose are:
  - Is the establishment site liberated from hiding by close by trees, huge buildings or different deterrents?
  - Can the PV framework be situated for acceptable execution?
  - Does the rooftop or property contain enough zone to fit the solar panels?
  - If the panel will be mounted on the roof, what sort of rooftop is it plus what is the condition of the rooftop?
- Mounting area: Solar panels are generally mounted on rooftops. On the off chance that the rooftop region is not accessible, PV modules may be mounted on poles, ground, wall, or set up as a major aspect of the shaded structures.
- Shading: PV panels are unfavorably influenced by concealing. A well-planned PV framework needs distinct and unhampered access to the beams of the sun from around 9:00 a.m. to 4:00 p.m., consistently. Indeed, even a bit of shadow, for example, the shadow of a tree, can essentially lessen the energy yield of a solar panel. Shade from the structure itself—because of vents, upper room fans, lookout windows, peaks, or shades—should likewise be maintained a strategic distance from. Remember that territory might be unshaded in the day, however, shaded at some other day. Additionally, a site and that is unshaded in the late spring might be shaded in the winter season because of long shadows during the winter.
- Direction: In northern scopes, customary wisdom of PV modules are directed toward the south. Whereas, in southern scopes, PV modules are obviously directed toward the North.

Be that as it may, the tilt or direction of a rooftop should not be immaculate in light of the fact that solar modules generate 95% of their complete energy while within 20 degrees of the direction of the sun. Rooftops that face toward the east or west might likewise be satisfactory. For instance, due west-bound housetop solar PV framework, tilted at about 20 degrees, will deliver around 88% of the force as a single pointing south at a similar area. Flat rooftops function well on the grounds that the PV modules may also be mounted on the frame and tilted up to the south side. Ideal direction can be affected by conventional climate changes.

- Tilt: Normally, the ideal tilt of a PV exhibit in the Pacific Northwest is equivalent to the geographic scope around −15 degrees to accomplish a yearly optimized yield of energy. An expanded tilt inclines toward source yield during the winter and a diminished tilt inclines toward yield in the mid-year. By the by, it is suggested that modules be set up at a similar pitch as a slanting rooftop, whatever the incline is, principally for aesthetic purposes, yet in light of the fact that the tilt is exceptionally lenient.
- Required location: Residential along with small business frameworks need as meager as 50 ft^2 for a setup to about 1000 ft^2. Generally, each 1000 W of PV need 100 ft^2 of authority territory for modules utilizing crystalline silicon (as of now the most widely recognized PV cell).

Every 1000 W of PV modules may produce around 1000 kWh every year. When utilizing less proficient modules, for example, nebulous silicon or some other light coat sorts, the zone should be roughly multiplied. On the off chance that your area restrains the physical measurement of your framework, you might need to set up a framework that utilizes more productive PV modules. Remember that entrance space surrounding the modules may add around 20% to the needed zone.

- Rooftop types: For rooftop-mounted frameworks, conventionally roof shingles are simplest to operate with and slate tile and tile rooftops are the most troublesome. By the by, it is conceivable to set up PV modules on every rooftop type. On the off chance that the rooftop will require supplanting in 5 or 10 years, it ought to be supplanted when the PV systems setup for avoiding the expense of expelling and reinstalling this PV framework

Building integrated PV (BIPV) modules that can be incorporated into the rooftop itself may be considered for new development or for a more established rooftop needing supplanting. While BIPV items right now have excellent value, costs are considered to decline.

9.2.1 Types of photovoltaic system

The types of PV framework can be comprehensively characterized by answers to the accompanying inquiries:

- Will that be associated with the transmission network of the utility?
- Will that generate exchanging flow (AC) or direct flow (DC) power, or even both?
- Will that have back up for the battery?
- Will that have a diesel, gas or propane producer set as back up?

Now we will concentrate on frameworks that are associated with the utility transmission network, differently alluded to as utility-associated, network-associated, matrix-interconnected, network-tied, or network intertied frameworks. These frameworks produce a similar nature of substituting flow (AC) power as is given by the utility. The power produced by a network-associated framework is utilized initially to control the AC electrical requirements of the residential or business. The surplus force that is produced is bolstered or "pushed" over the transmission network of the electric utility. Any of the construction's capacity necessities that are not satisfied by the PV framework are controlled by the transmission matrix. In this manner, the framework can be considered as a virtual battery power for the construction.

9.2.1.1 *Types of common system*

Various new PV frameworks being set up are network-associated private frameworks with no battery backups. Numerous matrix-associated AC frameworks is likewise being set up in business or open offices. The network-associated frameworks are of two sorts, in spite of the fact that others subsist. These are:

- Grid-associated AC framework without battery or producer back up.
- Grid-associated AC framework with battery backups.

Model setups of frameworks with or without batteries are appeared in Figs. 9.1 and 9.2. Notice that there are normal minor variations from the setups appeared, in spite of the fact that the fundamental capacities and general plans are comparable.

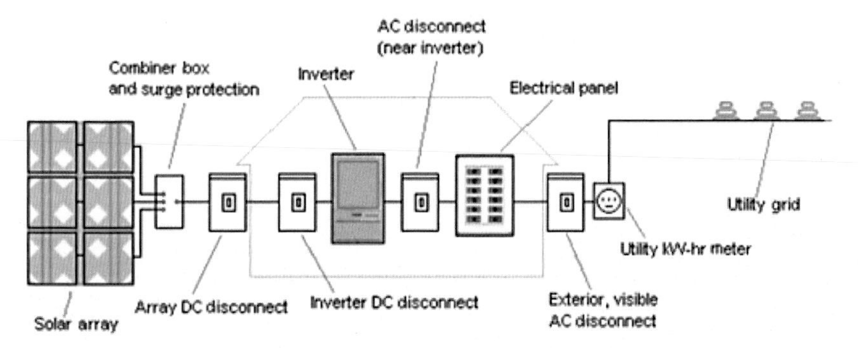

FIGURE 9.1 Network-associated AC photovoltaic framework with no battery backups.

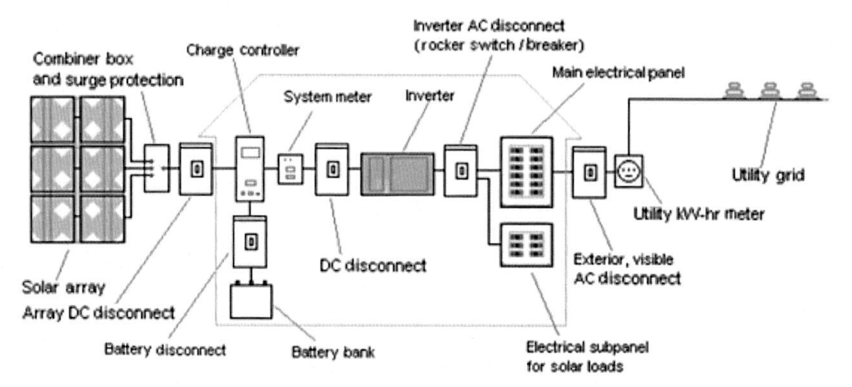

FIGURE 9.2 Network-associated AC photovoltaic framework with battery backups.

Are battery banks really required?

- The easiest, generally dependable, and the most affordable arrangement does not have battery backups. With no batteries, a network-associated PV framework will close when the utility energy blackout happens. Battery backup keep up capacity to a few or the entirety of the electric gear, for example, lighting, refrigerator, or fans, in any event, when the utility energy blackout happens. A network-associated framework may likewise have generating back up if the offices cannot endure power blackouts. Using battery backups, power blackouts might not get noticed. In any case, including batteries to a framework accompanies a few hindrances that should be weighed over the upside of power backups.

  These cons are:
- Batteries devour power while charging and releasing, diminishing the proficiency and yield of the PV framework by around 10% for lead-corrosive batteries.
- Batteries enhance the intricacy of the framework. Both initial cost and establishment costs are expanded.
- Various low-cost batteries need maintenance.
- Batteries ordinarily should be supplanted before different parts of the framework and at significant cost.

Framework elements

Essential elements of network-associated PV frameworks with or with no batteries are:

- PV modules (solar)
- Array attaching racks
- Grounding gear
- Combiner boxes
- Surge protective (regularly a component in the combiner boxes)

- Inverters
- Meters: framework along with kilowatt-hour
- *Disconnects:*
- Array DC
- Inverter DC
- Inverter AC and exterior AC

In the event that the framework incorporates batteries, it may likewise need:

- Battery power bank with wiring and housing construction
- Charge controllers
- Battery disconnects

9.2.2 Solar panel modulus

The core of a PV framework is the solar panel mode. Numerous PV cells are cabled together by the maker to deliver a solar panel module. When set up at a site, the solar modules are combined together in a sequence to make strings. These strings of modulus are associated with corresponding to create a cluster.

- Types of module: Rigid and flat formed modules are, as of now, the most usual and the greater part of these are made out of silicon. Since silicon cells contain a nuclear structure that is single-crystalline (monocrystalline), polycrystalline (multicrystalline) or shapeless (fine-coat silicon). Different cell materials utilized in solar-based modules are cadmium telluride or CdTe, generally articulated "CadTel" along with copper indium diselenide or CIS. A few modules are made utilizing mixes of these substances. A model is a thin-film of nebulous silicon kept onto the single-crystalline silicon's substrate.
- BIPV: PV innovation has been incorporated into material tiles, adaptable material shingles, material layers, glue laminates solely for metal standing-crease rooftops, windows, and different BIPV items. BIPV modules are commonly more costly than unbending level modules; however, they are foreseen to, in the long run, lessen, by and large, expenses of a PV framework in view of their double purpose.
- Rated output: Grid-associated private PV frameworks utilize modules with the rated output running from 100 to 300 W. The smallest modules like 10 W are utilized for different applications. The rated output is the most extreme force the module can create with 1000 W of daylight for each square meter at the module temperature of about 25 °C (77°F) in hot air. Real conditions will seldom coordinate the rated conditions thus the real power yield will quite often be less.
- System voltage PV: Modern frameworks with no batteries are commonly cabled to give 235–600 V. Within the battery-based frameworks, the

pattern is likewise toward utilization of higher cluster voltages, albeit many various controllers, despite everything need low voltages like 12, 24, and 48 V to coordinate the battery string's voltage.

- Utilizing manufacturer's item information to contrast modules: Because module expenses and efficiencies keep on changing as innovation and assembling strategies improve, it is hard to give general suggestions that will be valid into the near future with respect to, for instance, which sort of module is least expensive or the finest, by and large, decision. It is ideal to make contrasts dependent on current data gave by makers, joined with the particular prerequisites of the application.
- Two figures that are helpful in contrasting modules are the cost for every watt along with the rated yield per region (or proficiency). When glancing through a producer's index of solar panel modules, you will regularly discover the rated output, the general elements of the modules, and its cost. Discover the expense per watt by separating the module's cost by the rated yield in watts. Discover the watts per zone, by separating the rated yield by its zone.
- Module expense for every watt: Generally, thin-coat modules have the least expenses compared to crystalline silicons for modules of comparative power.
- Module proficiency (watts for each area): Modules with greater proficiency will contain a higher proportion of watts to region. The greater the productivity, the smaller the territory (e.g., fewer modules) are going to be needed to accomplish a similar force yield of an exhibit. Set up and racking expenses will be a lot less with progressively productive modules, yet this should be weighed over the greater expense of the modules. Indistinct silicon, thin-coat CdTe along with CIS modules contain rated efficiencies and that are a lot less compared to crystalline silicon, yet enhancements in productivity proceed.
- Warranty: This is critical to check guarantee times of all parts of the framework, which includes solar panel modules. Various modules are truly sturdy, enduring, and can face serious climate, including extraordinary warmth, cold, and hailstorm. Mirroring this life span, many silicon modules convey 20- to 25-year maker guarantees.
- Array-mount rack: Arrays are generally mounted on rooftops or over steel posts established in concrete. In specific applications, they might be mounted at the floor level or over walls. Solar panel modules can likewise be mounted to fill in as part of the entirety of a shade composition, for example, a yard cover. On rooftop-mounted frameworks, the PV exhibit is commonly mounted on attached racks, corresponding to the rooftop for stylishness and remained off a few inches against the rooftop level to permit wind stream that is going to keep them cool and useful.
- Adjustability: The leaning of slanted housetop clusters is generally not changed, because this is badly designed, by and large, and often perilous.

Notwithstanding, most mounting racks are customizable, permitting resetting of the angles of the PV modules regularly.

- Tracking: The mounted poles of the PV exhibits can fuse GPS devices that permits the cluster to consequently keep track the sun. Followed PV exhibits can build the framework's every day rated yield by 25%–40%. Notwithstanding the expanded energy yield, the following frameworks typically are not supported by the expanded expense and multifaceted nature of the framework.
- General set up notes: Accurate rooftop mounting may be time-consuming, contingent upon the sort of rooftop and the way mounting sections are set up and fixed. It is ideal to follow the suggestions of the material worker, racking framework providers and module producers. Module producers will give subtleties of help prerequisites to their modules. A decent racking provider will furnish code-agreeable designing details with their item. Disallowing purposes, in any case, it is conventional to have one help section for each 100 W/PV modules. Specific consideration must be provided to verifying the cluster directly to the auxiliary individuals from the rooftop and to climate fixing of rooftop infiltrations. All insights about joining the mounting sections to the rooftop and fixing surrounding them are best affirmed and brought about by the material worker with the goal that the rooftop guarantee will not be exploited.
- Asphalt rooftops: For asphalt rooftops, all mounts should be verified to the rooftop with steel slack jolts, joint inside the rafters. Mount sorts incorporate support poles along with L-sections. Support poles are favored in light of the fact that they are intended to give a decent seal over the boots. Support poles are mounted accurately after the rooftop decking is implemented and prior to the rooftop material is set up. Support poles and rooftop jacks might be set up by the material contractual worker or the group accountable for spreading out the exhibit mounting framework. The material contractual worker at that point flashes the poles as they set up the rooftop. It is exceptionally normal to set up mounts when the rooftop is set up, drill through the asphalt structure material to set up the jolts. The sealant is applied into the jolts without blazing. Now, the upper layer of material ought to be painstakingly lifted back to infuse sealant under the material. While this is considerably less work compared to when flashed, except if performed by the material worker, this technique may exploit the guarantee on the rooftop.
- Metal rooftops: There are a few kinds of standing-crease metal rooftop items, including vertical crease, level crease and delta crease items. Right now, specific clasps, alluded to as S-5 clasps, are accessible to append exhibits with no entrances to vertical and level crease rooftops and some other standing-crease rooftop profiles. The clasps make the setup of the solar exhibit a generally simple issue contrasted with some other rooftop kind. Conversely,

clasps for delta crease metal rooftops are not accessible. For those rooftops, it is important to deduct into the material, set up boots surrounding the mounting poles, and afterward seal the entrance. This being unwanted and time-consuming, it is ideal to determine ahead of time a vertical or flat crease metal rooftop or other rooftop kind matching with S-5 clasps.

- Other roof kinds: While it is conceivable to set up a PV exhibit on shake, tiles and slate rooftops, these rooftop types represent certain issues. Communicate with the racking framework provider for data on items and set up techniques for these rooftop types. Work straight with the material worker before requesting the racking framework. Likewise search for rooftop incorporated modules which can be utilized with tiles or slate rooftops.
- Roof ventilation and fans: We recommend setting up rooftop ventilations and plumbing vents, along with fans at the north of the rooftop to stay away from impedance with the solar cluster. This will likewise lessen the potential for unintentional concealing of the exhibit.
- Grounding tools: Grounding tools give a well-characterized, low-obstruction way from your framework to the ground to shield your framework from current floods from lightning or hardware breakdowns. Grounding additionally balances out voltages and gives a typical reference point. Also, the grounding bridle is generally situated on the rooftop.
- Equipment ground: Equipment ground gives security from shocks and currents brought about by a ground deficiency. A ground issue happens when the current flow conductor contacts with the casing or body of an apparatus or electrical boxes. All framework segments and any uncovered metal, including receptacle, equipment box, PV mounting tools, and appliances frame must be grounded.
- System ground: System ground needs taking a single conductor from a two-cable framework and associating it with the ground. Within a DC framework, this implies holding the negative-charged conductor to the ground at a single point in the framework. This must be practiced within the inverter, and not at the PV exhibit.
- Combiner boxes: Cables from different PV modules or even strings are hurried to the combiner boxes, regularly situated on the rooftop. These cables might be single conduit pigtails with connectors and that is precabled onto the PV module. The yield of the combiner boxes is one bigger two-cable conductor in course. The combiner box ordinarily incorporates a protection breaker or wire for every string and may incorporate a surge safety as well.
- Surge safety: Surge safety help to secure your framework from power floods that may happen if the PV framework or close by electrical cables are struck heavily by lightning. The energy surge is an expansion in voltage altogether over the voltage design.

9.2.3 Meters along with instrumentation

Basically, two sorts of meters are utilized in PV frameworks: utility photon meters and system meters.

- Utility photon meters: The utility photon meters estimates power conveyed to/from the matrix. In homes with solar systems, utilities commonly set up a bidirectional meter with a digitalized screen that monitors power in the two directions. A few utilities will permit you to utilize an ordinary meter that may turn backward. In this instance, utility meters turn forward once you draw power from the network and in reverse when your framework is sustaining or "pushing" power onto the matrix.
- System meter: The framework meter evaluates and shows framework execution and its status. Checked points may incorporate electricity generation by modules, power utilized, and batteries. It is conceivable to work a framework without system meters; however meters are unequivocally prescribed. Current charge controllers include framework checking capacities; thus a different framework meter may not be fundamental.

9.2.4 Inverter

The Inverter deals with four essential points of power conditioner:

- Transforming the DC power originating from the PV modules or storage packs to AC power
- Making sure that AC cycle's frequency is 60 cycles for every second
- Diminishing voltage vacillations
- Making sure that the state of the AC waves is fitting for the implementation, for example, a sine wave for network-associated frameworks.

9.2.4.1 Method for choosing a network-associated inverter: the accompanying elements ought to be considered for a network-associated inverter

- The UL1741 list of the inverter for using in a matrix intelligent software
- The voltage of the approaching DC currents from the solar cluster or storage pack.
- The DC window of the PV exhibit
- Elements showing the nature of the inverter, for example, high effectiveness and great frequency, and voltage guidelines
- Extra inverter highlights, for example, meters, pointer lights, and vital security disconnects
- Producer guarantee, which is ordinarily 5–10 years
- MPPT (maximum power point tracking) capacity, which amplifies power yield

Various network-associated inverters may be set up outside, while various off-matrix inverters are not weather-resistant. There are basically two kinds of framework intuitive inverters: those intended for utilization with batteries along with those intended for a framework with no batteries.

- Quality health: Inverters for network-associated frameworks generate superior to utility-quality.
- Energy: For matrix-association, the inverter should have the terms "Utility-Interactive" issued straightforwardly on the list name.
- Voltage inputs: The DC voltage of the inverter input window should compare with the ostensible voltage of the sun-powered exhibit, generally 235–600 V for frameworks with no batteries plus 12/24/48 V for battery-based frameworks.
- AC power yield—Grid-associated frameworks are estimated by the force yield of the PV cluster, instead of the load prerequisites of the construction. This is on the grounds that any power necessities over what a matrix-associated PV framework can give are naturally drawn from the network.
- Surge storage: The beginning surge of a gear, for example, in engines, is not a preference in measuring network-associated inverters. While beginning, an engine may attract as much as multiple times its rated watts. For network-associated frameworks, this beginning surge is naturally drawn from the matrix.
- Frequency along with voltage directive: Superior quality inverters are going to create close to steady yield voltage along with frequency.
- Effectiveness: New inverters normally utilized in private and business frameworks have top efficiencies of about 92%–94%, rated by their makers. Genuine field conditions, for the most part, bring about in general efficiencies of around 88%–92%. Battery-based inverters frameworks have somewhat lower effectiveness.
- Integral security disconnect: The AC disconnects in various inverter versions cannot meet prerequisites of the electric utilities (see segment "Disconnect"). Accordingly, different outside AC disconnects might be needed regardless of whether one is incorporated for the inverter. Every inverter that is UL recorded for network association incorporates both DC (PV inputs) and AC (inverter yield) disconnects. In more reliable inverters, the inverters area may be expelled independently from the DC disconnect and AC disconnect, encouraging fix.
- MPPT: current nonbattery inverters incorporate most extreme power points tracking. Therefore MPPT consequently modifies framework voltage with the end goal that the PV cluster works at its most extreme power points. For battery-based frameworks, this element has as of late been consolidated into superior charge controllers.
- Chargers for inverters: For battery-based frameworks, inverters are accessible with production line-coordinated charge controllers, alluded to inverter-chargers. Make certain to choose inverter-chargers which is rated for framework association, in any case. In case of a network power blackout, utilization of an inverter-charger that is not set up for network

association would bring about overcharge and harming the batteries, called "cooking the battery."

- Automatic offloading: For battery-based frameworks, the inverter may consequently shed any pointless loads in case of a utility force blackout. Sunlight-based loads, for example, the loads that will be kept fueled up throughout the blackout, are associated with a different electrical subboard. A battery-based framework should be intended to control these basic loads.
- Warranties: Inverters regularly convey guarantees of 5 years, despite the fact that the business is pushing toward the 10-year guarantee. The transformer and strong state segments of the inverter are vulnerable to overheating and harm from power surges, lessening its life span.
- Disconnect: Regulated and manual security disconnects shield the cabling and segments from power spikes and other gear glitches. They additionally guarantee the framework can be securely closed down and framework parts can be expelled for support and fix. For network-associated frameworks, security disconnects guarantee that the producing gear is segregated from the network, which is significant for the security of utility work force. When all is said in done, disconnects is required for every resource of energy or power bank gadget in the framework. For every one of the capacities recorded underneath, it is not constantly important to give a different disconnect. For instance, if the inverter is found outside, a solitary DC disconnect may serve the capacity of both the cluster DC disconnects and the inverter DC disconnects. Prior to precluding a different disconnect, in any case, contemplate if this is ever bringing about a perilous condition while performing upkeep on any segment. Likewise, think about the ease of the disconnects area. An uneasy found disconnect may prompt the inclination to leave the force on during support, bringing about a safety risk.
- Cluster DC disconnects: The cluster DC disconnects, additionally known as the PV disconnects, is utilized to securely intrude on the progression of power from the PV exhibit for support or investigating. The cluster DC disconnects may likewise have coordinated circuit breakers or wires to ensure against power spikes.
- Inverter DC disconnects: Including the inverter AC disconnects, the inverter DC disconnects are utilized to securely detach the inverter from the remainder of the framework. Much of the time, the inverter DC separate will likewise fill in as the cluster DC disconnects.
- Inverter AC disconnects: The inverter AC disconnects disengage the PV framework from the structure's electrical cabling and the network. Much of the time, the AC disconnects are set up inside the structure's primary electrical board. In any case, if the inverter is not situated close to the electrical board, an extra AC disconnect ought to be set up close to the inverters.
- Exterior AC disconnects: Utilities ordinarily need an outside AC disconnect that are lockable, have obvious sharp edges and are mounted by the utility meters so they are available to utility staff. The AC disconnect

situated in the electrical board or vital to the inverter may not fulfill these prerequisites. One option that is as adequate to certain utilities as an available AC disconnect is simply the evacuation of the meter, yet this is not the standard. Preceding obtaining gear, counsel the electric utility to decide their prerequisites for interconnection.

- Batteries DC disconnects: In the battery-based framework, the batteries DC disconnects are utilized to securely detach the storage bank from the remainder of the framework.

9.2.4.2 *Storage pack*

Batteries save DC electrical power for use later. This power stockpiling includes some major disadvantages, nonetheless, because batteries decrease the proficiency and yield of the PV framework, regularly by around 10% for lead-corrosive batteries. Hence, batteries additionally increment the multifaceted nature and expense of the framework.

Kinds of batteries normally utilized in PV frameworks are lead-corrosive battery and alkaline battery.

- Lead-corrosive battery: Lead-corrosive batteries are generally basic in PV frameworks in normal and fixed lead-corrosive batteries are most normally utilized in network-associated frameworks. Fixed batteries are spill proof and do not need occasional upkeep. Overflowed lead-corrosive batteries are generally the most economical yet need refined water at any rate month to month to renew water lost throughout the ordinary charging procedure. There are two sorts of fixed lead-corrosive batteries: fixed permeable glasses mat (AGM) and the second is gel cells. AGM lead-corrosive batteries have known to be the business standard, since they work without maintenance and especially appropriate for network-tied frameworks where batteries are normally stored at a full condition of charges. Gel-cell battery, intended for freeze-reluctance, is commonly a poor decision in light of the fact that the overcharging will for all time harm the batteries.
- Alkaline battery: Due to their generally significant expense, antacid batteries are just prescribed where incredibly chilly temperatures (like −50°F or even less) are envisioned or for some business or modern applications needing their favorable circumstances over lead-corrosive batteries. These points of interest incorporate resistance of freezing and high temperatures, lower keep up prerequisites, and the capacity to be completely released or over-charged with no damage.

9.2.4.3 *Sizing storage banks*

For matrix-associated frameworks, batteries are generally estimated for moderately brief timeframe periods with 8 hours being conventional. Size may differ, be that as it may, contingent upon the specific needs of an office and the length of electricity blackouts anticipated. For correlation, storage banks for off-network frameworks are normally estimated for 1–3 shady days.

Solar modules interaction: The sunlight-based exhibit should have a greater voltage compared to the storage bank so as to completely charge the battery. For frameworks with battery backups, give specific consideration to the evaluated voltage of the modules, additionally known as the maximum power points, within the electrical determinations. It is significant that the voltage is sufficiently high compared with the voltage of a completely energized battery. For instance, the rated voltage somewhere in the range of 16.5–17.5 V is normal for a 12 V framework utilizing the liquid lead-corrosive battery. Greater voltages might be needed for long cabling distance among the module and the charge controllers and storage banks.

9.2.4.4 Charge controllers

The charge controllers, at times alluded to as PV controllers or battery chargers, are just essential in frameworks with battery backups. The essential capacity of the charge controllers is to forestall overcharge of the battery. Most likewise incorporate the low voltage disconnects that forestall over-releasing batteries. Moreover, charge controller keeps charge from depleting back to sun-based modules around evening time. Some advanced charge controller comprises maximum power points tracking, which advances the PV cluster's yield, expanding the power it generates.

9.2.4.5 Kinds of charge controller

There are basically two kinds of controllers: shunt charge and series charge. A shunt charge controller sidesteps current surrounding completely energized batteries and by a force transistor or obstruction radiator where abundance power is changed over into heat. Shunt charge controllers are easy and reasonable; however, they are intended for little frameworks. Series charge controllers stop the progression of current through opening the circuitry among the battery and the PV cluster. Series charge controllers might be the single-stage or pulsed kind. Single-stage controllers are little and modest and have a more prominent load taking of limit compared to shunt-type charge controllers. Pulse kind controllers and a kind of shunt charge controller alluded to as a multistage controllers (e.g., three-staged controllers) have schedules that enhance battery charge rates to broaden battery life. Various charge controllers are currently three-staged controllers. Selection of charge controllers is based on the following points:

- Selecting: Charge controller is chosen dependent on:
- PV exhibit voltage: The DC voltage inputs of the controller should match the ostensible voltage of the sunlight-based cluster.
- PV exhibit currents: The controller should be estimated to deal with the most extreme current created by the PV cluster.
- Interaction with inverters: Since most charge controllers have also been set up in off-network frameworks, their standard settings might not be proper for a matrix-associated framework. The charge controllers must be

installed to such an extent that it does not meddle with the best possible activity of the inverter. Specifically, the controller should be installed with the end goal that charging the battery from the PV cluster outweighs charging from the network. For more data, contact the maker.

- Interaction with battery: The charge controllers must be chosen to convey the charging currents proper for the kind of batteries utilized in the framework. For instance, on the 12 V framework, surged lead-corrosive batteries contain a voltage about 14.6–15.0 V once completely charged, while fixed lead-corrosive batteries are completely charged at 14.1 V. Allude to the battery producer for the charging prerequisites of specific batteries.

9.3 Introduction to PVSYST

Solar is a climate-induced power source, and it fluctuates fundamentally in time and the territory. Arranging and executing the solar photovoltaic framework is a fairly demanding multistage process incorporating assessment of the solar prospect of a site, evaluation of sunlight-based source, in general plausibility, plan, reproduction, enhancement of framework's output, and log-term execution. These days different apparatuses and databases are accessible for a PV framework configuration, measuring, demonstrating, reproduction, and execution evaluation. In any case, the present problem is that just one apparatus without anyone else's input cannot implement far-reaching investigation of a PV framework because of extraordinary intricacy of the procedure. In this manner, it is normal practice to join the input information and results from a few demonstrating, estimating, and structuring devices alongside estimations from a site to obtain the most dependable outcomes. It relies upon an area, programming accessibility, and professional experience which device or mix of instruments and datasets to be utilized in a PV venture. It ought to be noticed that current programming and reproduction models are continually being enhanced and overhauled.

PVSYST is a committed PC programming for PV frameworks. The product was created by the University of Geneva. It incorporates prepracticality, estimating and recreation support for all PV frameworks. After characterized the area and loads, the client chooses the various parts from an item database and the product naturally figures the size of the framework. Right now, maximization and cost examination of the solar power plants at private, business housetop just as on utility scales in India is to be talked about. Plan, reproduction, optimization and expense investigation is going on reenactment office like PVSYST.

Further, PVSYST and C programs are utilized for the measuring of the solar-powered PV plants. It is intended to help the designer in the measuring of PV set ups Jaydeep V. Ramoliya et al. (2015) exhibited the simulating process of a network-associated solar PV framework utilizing of the program PVSYST and their presentation was assessed. Jones K. Chacko (2015) explored the main considerations which influence the exhibition of the solar PV modules three distinct plans of solar PV module are taken on an independent framework and compared diverse board course of action that will limit the floor region and expand power production through following the sun. By the written literature review, obviously a similar investigation of private and utility scales PV framework required for effective utilization of PV frameworks. Cost correlation will likewise give a superior knowledge to the effective utilization of a PV framework.

PVSYST is a product package that permits the users to utilize full-feature investigation and examination of a PV venture. PVSYST incorporates reenactment of a PV framework with an assessment of its preachievability, estimation and monetary examination, regardless of whether it is a network-associated, independent, siphoning or DC matrix framework. Meteorological information is given by Meteonorm 7.1 to around 1200 topographical locales. Meteonorm 7.1 comprises month to month estimated and hourly orchestrated information. Month to month irradiance value is midpoints of irradiance estimations during the time of 1960–91. For the most part, meteorological stations within the PVSYST are referred to the genuine ones, in any case, the information is added between the three closest stations. To acquire hourly quantities, PVSYST employs engineered production to a month to month estimated information. Month to month meter information from Meteonorm 7.1 incorporates worldwide and diffused light, temperatures and wind speed. Hourly information can be likewise developed by utilizing another information source within PVSYST straightforwardly, anyway it is asserted that Meteonorm provides progressively sensible and solid outcomes because of its enhanced version for temperatures and wind speed quantities. The same as Meteonorm 7.1, different quantity, interjected or blended meteorological information from these sources are Satellite, US TMY2/3 NASA-SSE, SolarGIS, and many others are accessible for reproduction in PVSYST. It is additionally conceivable to import client characterized information comprising a collection of parameters recorded in the table underneath.

PVSYST has a packaged information foundation of PV framework parts including as of now accessible and conventional module, inverters and analyzers. Physically characterized parts can be utilized in recreation also. Design of the framework is finished by the product naturally when the client characterizes a venture territory or required set up limits and picks a module along with an inverter. In light of these sources of info PVSYST suggests a framework design, and in this way preliminary recreation can be operated. The ideal measuring is finished by satisfactory over-loaded loss throughout the year, for example, the apportion of an exhibit ostensible capacity to ostensible AC intensity of an inverter. The ideal measuring normally suggests an over-size of electric power apportion by a factor of 1.2. Therefore PVSYST permits the client to characterize and control different elements and losses, for example, wiring loss, confound between modules, loss because of temperature, soil and numerous others as per the mounting framework, site condition, and inaccessibility. Shading loss, as among the most basic parameter influencing framework execution, can be characterized with 3D manager.

Far shading may be fixed by PVSYST naturally dependent on skyline concealing from land information, imported from some other database or a site image or drawn physically by the client.

Close to concealing investigation execute by PVSYST is continually being enhanced because of its unstable and untrustworthy execution. The client can characterize close obstructions by freehand or even by utilizing items structure 3D device, run and spare a concealing scene to be utilized in the reenactment. Close to concealing development is somewhat perplexing and requesting, in this way a few phenomena are not precisely determined and dependent on the presumptions (e.g., division for electrical impact).

When all needed and wanted parameters are fixed, the reenactment figures power distribution consistently. In this manner, the assessment of the framework benefit and quality should be possible dependent on absolute power creation (MWh/y), execution proportion (%) and particular power (kWh/kWp) as a relationship between the generation figure and illumination accessible at the site with provided direction. The potential enhancement of the framework execution can be founded on figures from the point by point loss outline that comprises primary energies and addition or loss in the reenactment procedure. Different simulation variations can be achieved and looked at inside the task. Monetary assessment of the framework can be utilized by setting venture, financing and advance parameters. As such, the client will characterize the expense of the parts, (e.g., PV modules, inverter, cabling, mounting framework), charges, endowments, and credit term and loan fee. Carbon balance, as an exhibition normal for the framework, can be assessed inside a fiscal investigation device. The carbon balance gauges the CO_2 emanations spared on account of the PV framework activity. The computation depends on life cycle emissions as CO_2 emanations (tons) related to power sum or part all through the complete life cycle concluding generation, creation, activity, support, removal.

9.3.1 Case study utilizing PVSYST application

The way toward planning a PV system within PVSYST incorporates the accompanying essential steps:

1. Project: Characterize the area and meteorological information.
2. Direction: Characterize module azimuth along with tilting.
3. Systems: Pick the framework modules, inverter, and electrical structure.
4. Modules Design: Make the electrical strings associations as indicated by the 3D scenes.
5. Detailed Loss: Make sure the jumble losses are set at 0% for the Solar PV framework.
6. Simulation: See a synopsis of the framework's power yield.

FIGURE 9.3 PVSYST interface window.

Project's designation
File name
Project's name
Site File
Meteo File
Ready for simulation
Meteo database
Project settings

FIGURE 9.4 Project description window.

9.3.1.1 *Stage 1: Project*

Select *Project Design* then *Grid Connected* within the interface PVSYST display

Press on the *New Project* option and afterward press on the *Choose Site* option to choose the right site along with meteorological files (Fig. 9.3) (the meteorological document will regularly be connected with the picked site).

So as to add another site into PVSYST, press on *Databases* in the primary PVSYST display and afterward *Geographical Site* (Fig. 9.4).

Pick *New* and find your site over the *Interactive Map* menu, or write the topographical area in the inquiry box. Press *Import* and afterward press on the *Import* option with a sun symbol. Press *OK* and afterward *Save*. Once saved the hourly qualities, press *Yes* and afterward *Close* and *Exit*. You would now be able to pick the new meteorological sites for your task (Fig. 9.5).

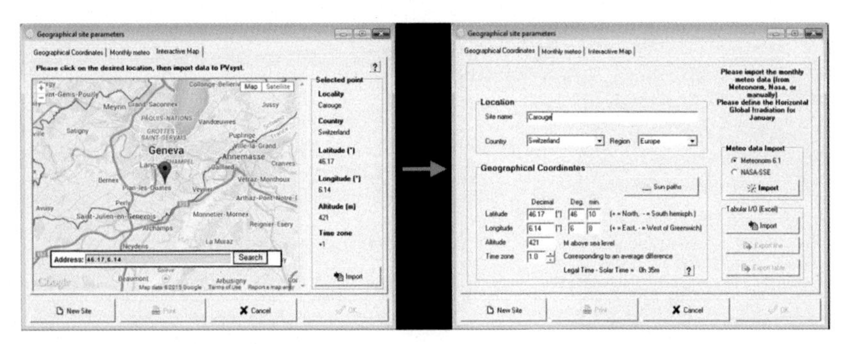

FIGURE 9.5 Interactive map menu.

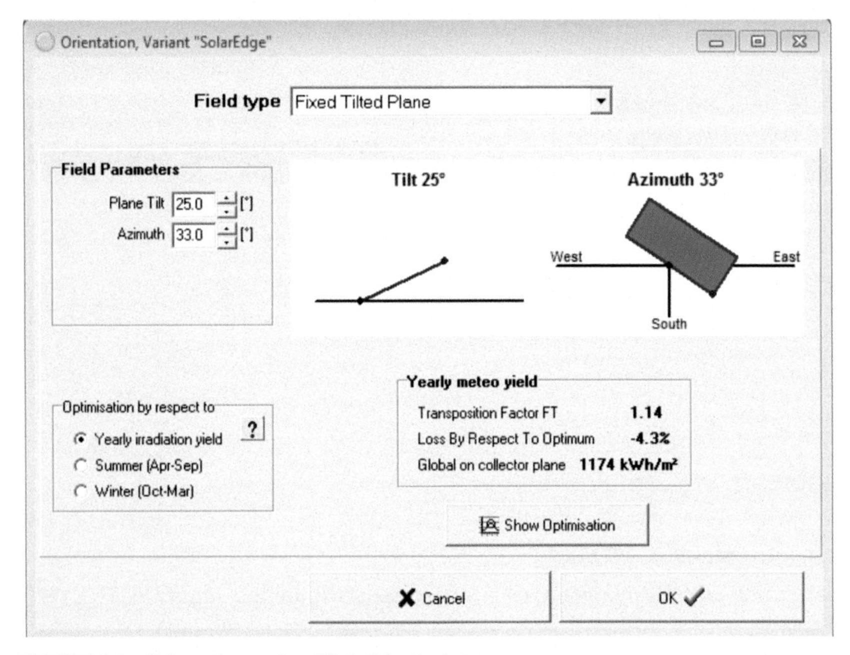

FIGURE 9.6 Orientation, variant "SolarEdge" window

9.3.1.2 Stage 2: Orientation

When the topographical area and climate record have been picked, you will characterize the cluster's azimuth and tilting. PVSYST provides various alternatives to fit different sorts of ventures, including basic fixed tilted planes, numerous directions (up to eight directions), occasional tilt modifications, "boundless sheds" for enormous frameworks, sun panels (module mounted to exteriors of structures) just as different sorts of following clusters, single and twofold pivot (Fig. 9.6).

9.3.1.3 Stage 3: Systems

The principle framework parameters, which include modules and inverter models, framework limit, string length and so forth, are characterized in the System display.

Essential design

In the wake of picking PV modules, verify the *Use Optimizer* and choose a fitting force streamlining agent starting from the drop menu. You at that point choose the inverter appropriate for your venture. In the event that you intend to associate over one module for each enhancer, utilize the *Optimizer input* appeared underneath to determine what number of modules to interface and the sort of association (arrangement/equal):

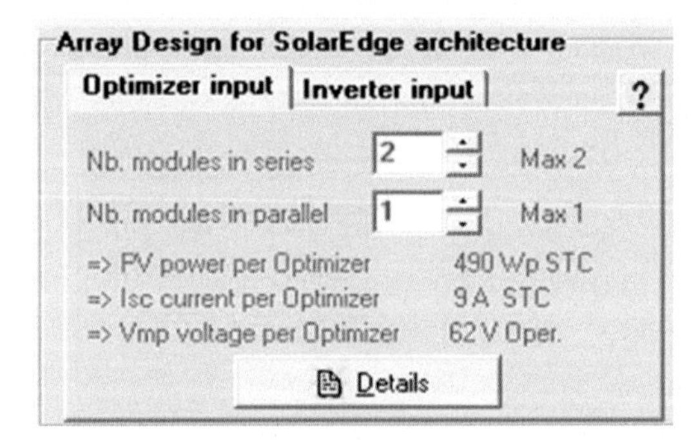

The *Inverter Input* within the inverter display records the quantity of power enhancers for every string and not the number of all the modules. Along these lines, for instance, while interfacing 2 modules for each force streamliner the string lengths would be 20, which means 40 modules:

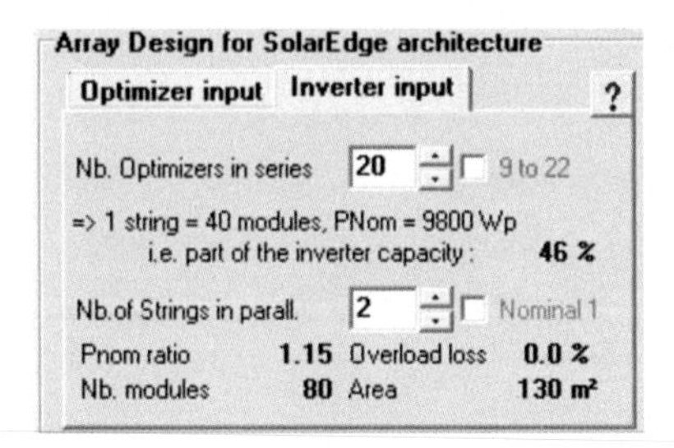

Lopsided string design

A typical constraint of a customary framework is that every string associating with an MPPT should be of a similar length. The chosen SolarEdge framework has no confinement because MPPT is executed at the module-level utilizing the SolarEdge power streamlining. Various strings can be associated with a single inverter, paying little mind to lengths, modules azimuth, tilts or module types. So as to suit this capacity PVSYST has included a SolarEdge-explicit String Configuration display. This display permits the distribution of the string to the inverters, as per the task needs and framework configuration limits. The accompanying three models disclose how to make an electrical plan with the SolarEdge framework.

Model 1 Imagine the SolarEdge SE6000H inverters with about 23 modules isolated into two strings: first with eleven modules, and the second with twelve modules.

In this plan, utilize two PVSYST Subclusters: one for every string. Every Subcluster will contain a single "concerned inverters," in light of the fact that both Subexhibits concern (associate with) the 1 SE6000H. One Subcluster will have one string of 12 P370 electric analyzers, and the second Subexhibit will contain a string of eleven P370 power enhancers. The Subexhibits will resemble this:

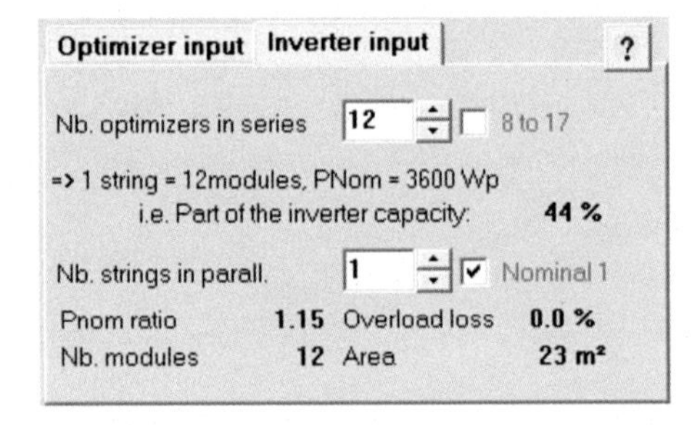

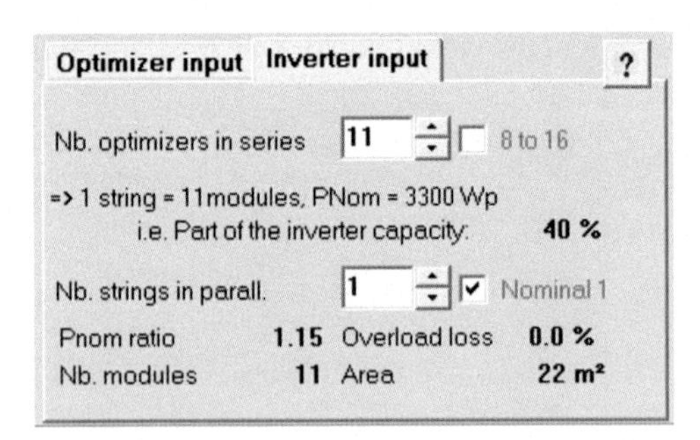

Press on the Strings Configuration to make sure that the framework is designed appropriately:

InvertersInputsTab

Existing sub-arrays

N	Opt ser/parall	Nb. inv. inputs		Adjusts sub-arrays
Sub-array #1	12	1	1 SE6000H	OK
Sub-array #2	11	1	1 SE6000H	OK

Design parameters

Reinitializes Inverter List

Max. number of strings 4

Nominal PNom ratio 1.55

Show sub-arrays

Show nb. optimizers in series

Inverters input specification

		String#1	String#2	String#3	String#4	PNom PV	PNomRatio
Inverter #1	SE6000H	Sub-array #1	Sub-array #2			6.28 kW	1.05

The "Inverter Inputs Specification" region displays the strings are recorded accurately: a string from Subexhibit #1 and the second string from Subcluster #2.

Model 2 Imagine a SolarEdge framework with 1 SE27.6K associated with a 104x300 W module. We will utilize the P700 electric streamlining associated with two modules in sequence. This will bring about a plan of three strings as two strings with seventeen electric streamlining (34 module) and a string with eighteen force enhancers (36 module). The Subexhibits will resemble this:

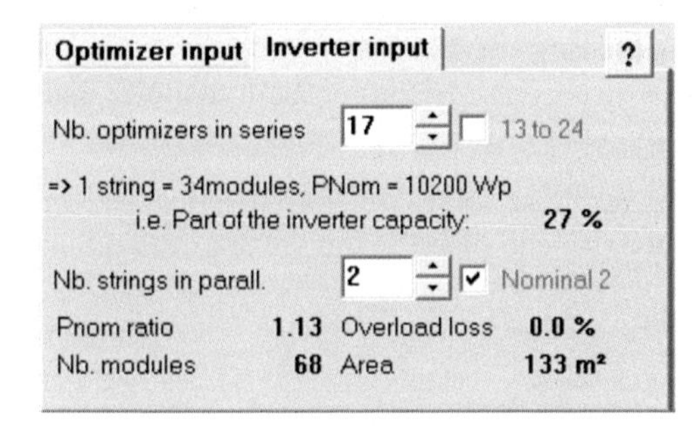

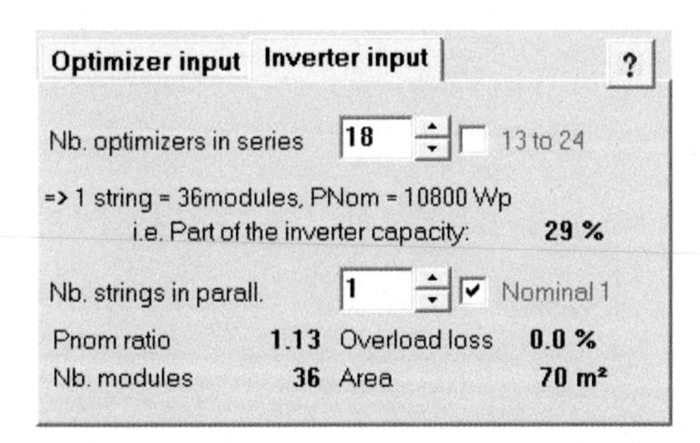

As in the past, every Subcluster will contain one "concerned inverter" since both Subexhibits associate with similar inverters.

The Strings Configuration display will resemble this:

InvertersInputsTab

Existing sub-arrays

N Opt ser/parall Nb. inv. inputs

Adjusts sub-arrays

Sub-array #1 17 2 1 SE27.6K OK

Sub-array #2 18 1 1 SE27.6K OK

Design parameters

Reinitializes Inverter List

Max. number of strings 12

Nominal PNom ratio 2.35

Show sub-arrays

Show nb. optimizers in series

Inverters input specification

		String#1	String#2	String#3	String#4	PNom PV	PNomRatio
Inverter #1	SE27.6K	Sub-array #1	Sub-array #1	Sub-array #2		28.41 kW	1.03

Two strings arc from Subcluster #1 and the second string is from Subexhibit #2.

Model 3 Imagine a 176 kW business framework with four SE33.3K inverters associated with 586 300 W modules. We will utilize the P700 power enhancer associated with two modules in sequence. The picked structure for this venture is:

Inverters 1–2: three strings of 24 P700 or 3x48 module
Inverters 3: three strings of 25 P700 or 3x50 modules
Inverters 4: two strings of 24 P700 or 2x48 modules and one string of 25 P700 or 1x50 modules

In spite of the fact that there are four inverters, so there are just two string lengths, hence, we can plan the framework in two Subclusters, as:

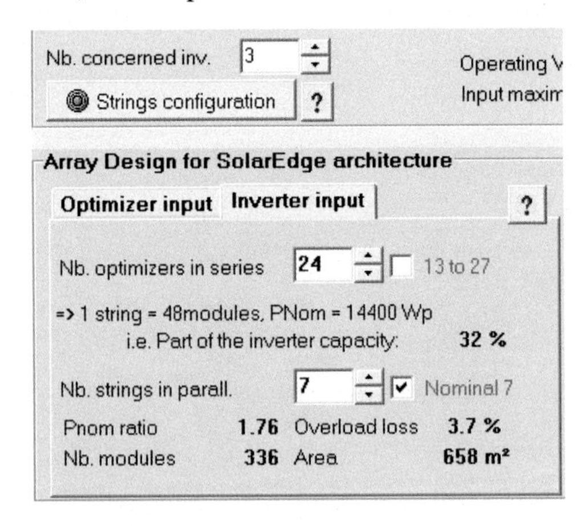

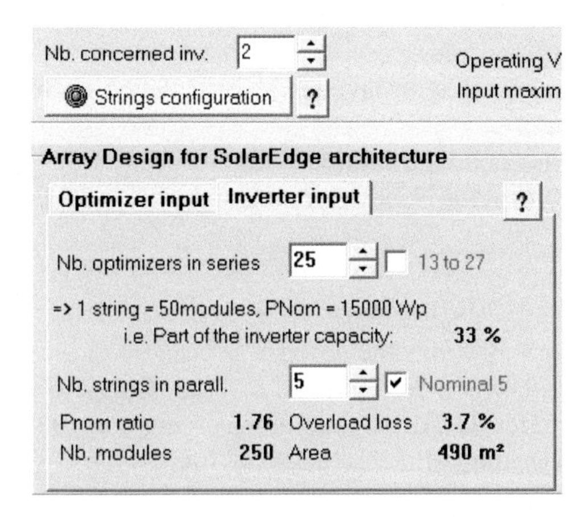

Notice the quantity of "concerned inverters" in every Subcluster. The main Subexhibit has three "concerned inverters": and it is on the grounds that the 24-analyzer strings interface with three inverters: inverter 1, 2, and 4. The subsequent Subexhibit has two "concerned inverters": inverter 3 and 4.

Within the Strings Configuration display, ensure the *Maximum quantity of strings* is fixed to three (the regular quantity of strings for inverters within this design) press on *Reinitialize Inverter Listing*. You would now be able to change the string association as indicated by the ideal electrical structure. At the point when completed, press on *Adjust Subexhibits* to employ the plan. On the off chance that all Subexhibits are set apart by a greenish "OK", the structure is substantial.

The outcome should resemble this:

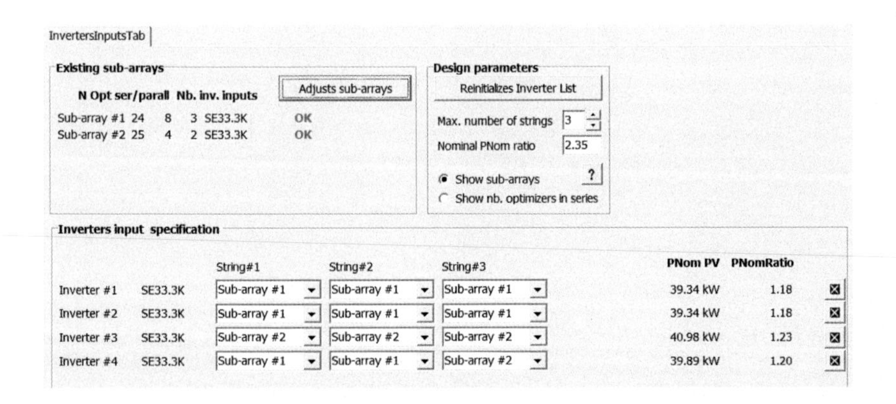

The strings associated with inverter 1 and 2 are included in Subcluster no.1 or 24xP700 modules
The strings associated with inverter 3 are included in Subcluster no.2 or 25xP700 modules
The strings associated with inverter 4 are included in Subcluster no.1 (two strings, 24xP700) and in Subexhibit no.2 (one string, 25xP700 module).

9.3.1.4 Stage 4: Module Design

The Inverter framework limits the power loss because of incomplete concealing when contrasted with a conventional inverter. On the off chance that the framework being referred to has concealing of any sort (trees, smokestacks, between push concealing) the 3D physical framework design ought to be intended to represent the power loss because of shade. If necessary, you can allude to the PVSYST assist documents for help with developing a 3D concealing display.

In contrast to a conventional framework, in which the creation of whole strings might be influenced by fractional concealing of as meager as a couple of cells of one module, SolarEdge limits the impacts of partial concealing to the concealed modules just, on account of its module-leveled MPPT. The module-leveled enhancement in PVSYST is considered when utilizing the Module Layout technique for ascertaining concealing electrical loss.

To initiate the Module Layout function, utilize the accompanying stages:

When the 3D scene is built, press on the *Module Layout* option in the primary interface:

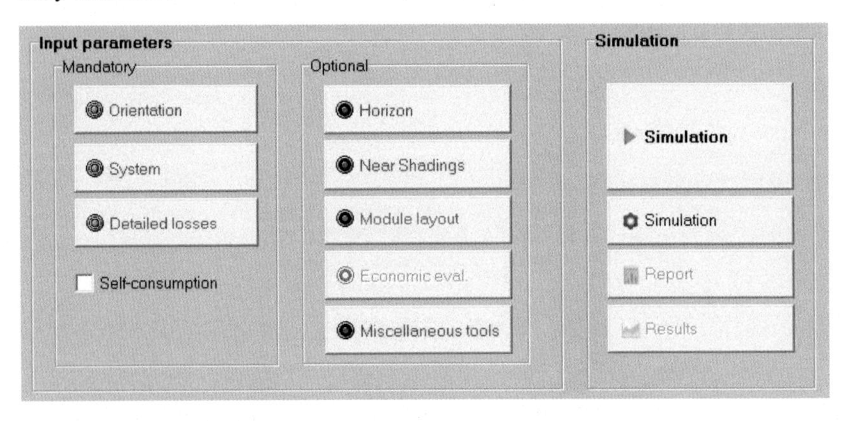

The Module Layout display represents a schematic design of the module from the 3D scenes:

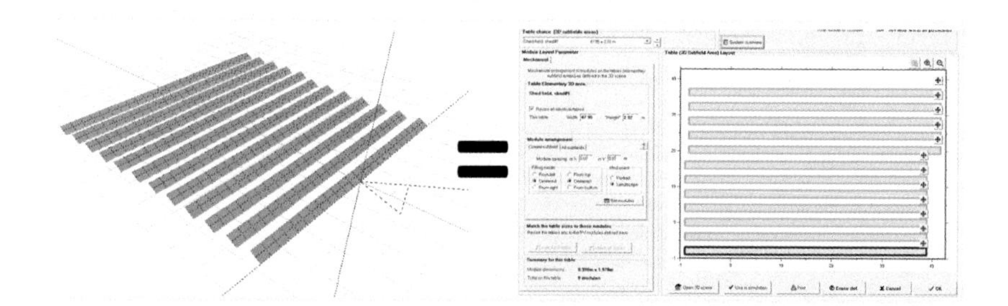

Within the Module Arrangement territory, choose the *All Subfields* menu, guarantee the right racking technique is picked (representation or landscape) and press the *Set all Modules* option. On the off chance that the 3D scenes coordinated the electrical structure (a similar quantity of modules is planned in both), another tab called *Electrical* will consequently open, demonstrating the accompanying string portion interface:

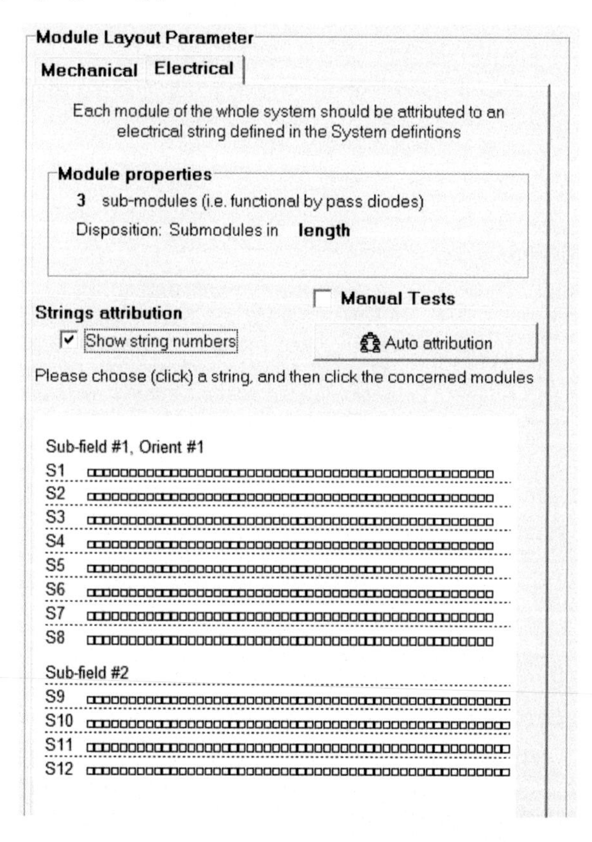

Press on *Auto Attribution*, then click *Distribute All*:

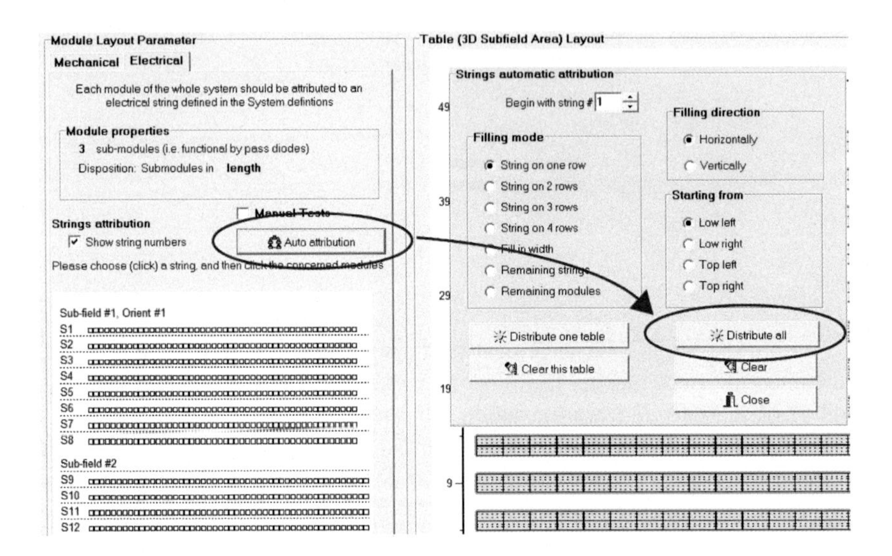

When the modules are allocated to their particular strings (and hued), press on *Close*. Press on *Use in Simulation* then, tap *OK* in order to exit.

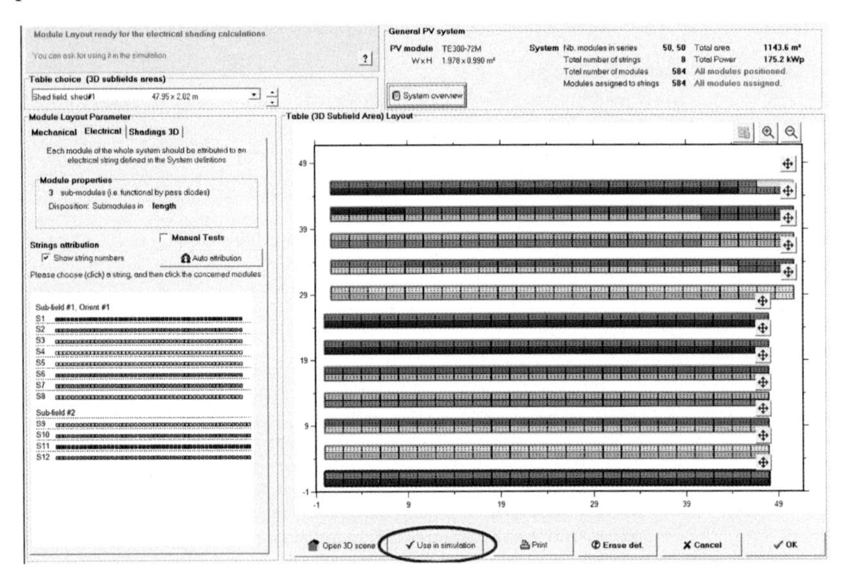

9.3.1.5 *Stage 5: Detailed loss—mismatched*

The inverter wipes out every loss that outcome from jumble, contrasted with a 2% for the first-year loss in a customary framework. Confound loss is brought about by the assembling resistance of the module, temperature contrasts during activity, lopsided soiling and various natural factors, for example, cloudy climate conditions, snow, leaves and so forth. While choosing an inverter, PVSYST naturally sets the mismatched loss to zero. For conventional frameworks, you should make this parameter up to 2% by tapping the *Detailed Losses* option at the base of the System display and choosing the *Module quality to LID to the Mismatch* menu.

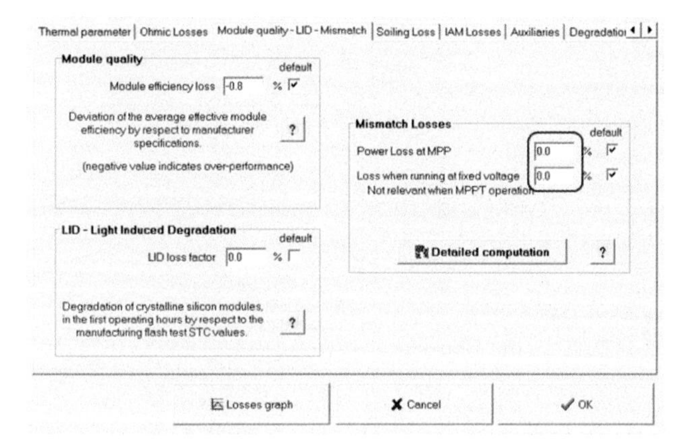

9.3.1.6 Stage 6: Simulation

When the framework parameters have been characterized, press on Simulation. When the simulation completes, a report will be accessible for review and printing. Moreover, the Simulation Settings display permits you to modify the time period of the recreation (from 1 day up to 1 year), just as trading hourly quantities of different parameters which include power, inverter proficiency, PV cluster's electrical conduct and much more. Contingent upon the multifaceted nature and size of the framework, the recreation procedure may take from a couple of moments to over 60 minutes.

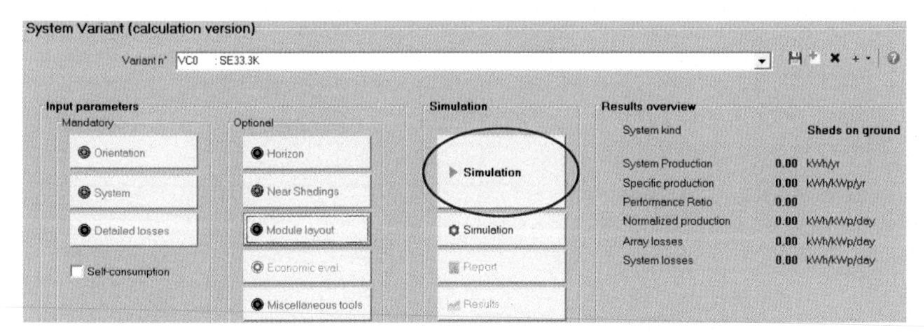

9.3.1.7 Attachment A: Mismatched growth with time

PVSYST provides a device that simulates the development of mismatch losses and modules degradation with time.

- The mismatched losses of a conventional framework will develop with time because of the lopsided rate of degradation among the modules within a string.

- So as to simulate the framework for a particular year of activity, initiate the degradation device from the basic interface display: press on the *Detailed Losses* option and select the *Degradation* menu. Any year may also be particularized by the users. In this model, we utilized year 20. Here it is:

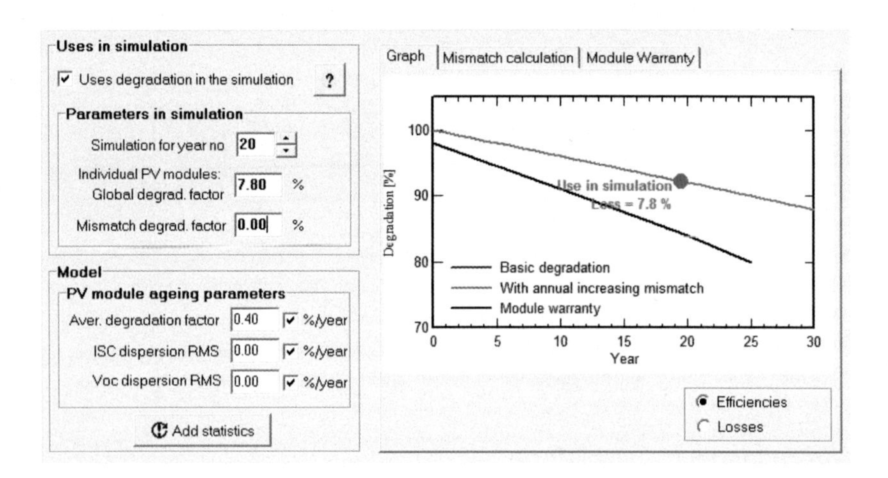

SolarEdge Systems

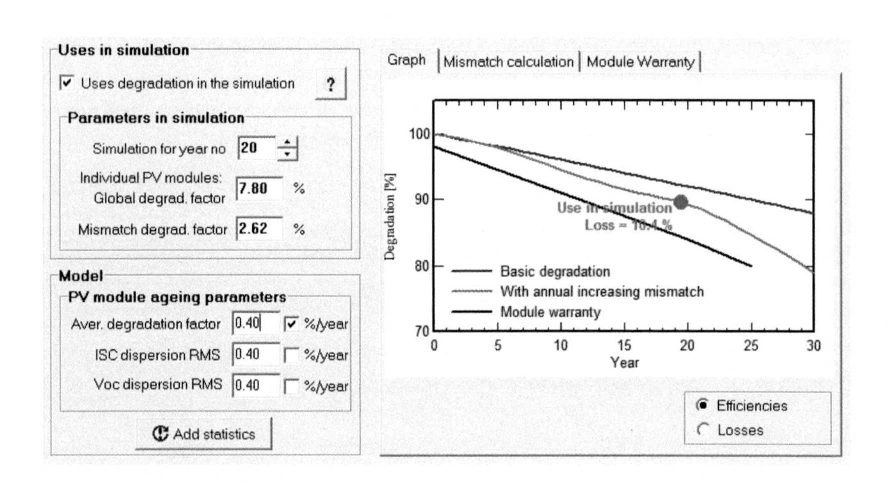

Traditional Systems

9.3.1.8 Attachment B: PVSYST losses parameter

PVSYST estimates various losses parameter throughout the simulation, as displayed in the loss figure underneath. This figure shows at the end of every

PVSYST report. With this, there is an outline of the loss's parameter, estimated successively.

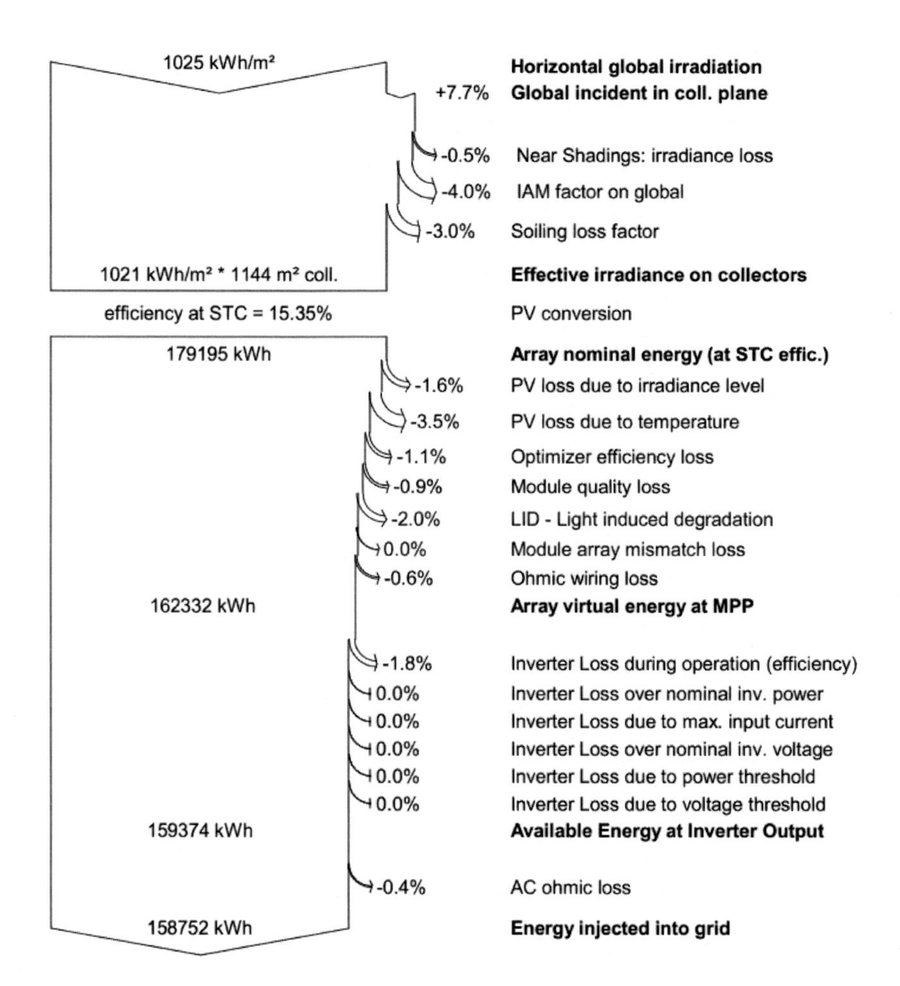

- *Horizontal worldwide illumination*: a mix of the worldwide diffused irradiance and the worldwide bar irradiance determined hourly more than a year on a surface level
- *Global occurrence in assortment planes*: utilizing a transposition version (Perez or even Hay), PVSYST increments or diminishes the even worldwide light contingent upon the azimuth and tilting. This is the real illumination arriving at the modules

- *Near Shadings: irradiance losses*: the losses because of direct concealing (influencing the shaft segment of the irradiance) and diffused concealing (influencing the diffused segment of the irradiance. For instance: close by articles, for example, the following column of modules within a grounding mount framework lessen the diffuse light, regardless of whether they do not cause direct concealing)
- *IAM element on worldwide*: IAM known as incidence angle modifier identifies with the reduction of irradiance arriving at the PV cell because of the sun beams' refraction when permeating the PV modules antireflective covering and glass. The greater the episode edge (regarding the sun's placement) the greater the losses
- *Soiling losses element*: any normal losses because of soiling (dirt, snowing, lcavcs and so forth.)
- *Effective irradiance on authorities*: the rest of the light after the recent comprehensive loss, increased by the PV region (the module territory as characterized in the PAN record)
- *PV conversion*: the modules proficiency at standard test condition (STC)
- *Cluster ostensible power (at STC proficiency)*: the PV transformation productivity increased by the viable irradiance on authorities
- *PV losses because of irradiance levels*: ascertains the diminished module effectiveness in lower light condition
- *PV losses because of temperature*: decreased module execution because of change in temperature. The temperature coefficients of the module just as the warm losses factor influence this loss. The warm losses factor is certainly not an experimentally decided quantity; instead it is fixed by the client as per experience and past estimations. An estimation of 20 W/ m^2 K is worthy for most frameworks. On the off chance that the modules are installed into a rooftop structure, the quantity can be fixed to 15. At the point once the modules are unsupported in a cool and breezy area, the quantity can be fixed to 29.
- *Optimizer productivity losses*: the proficiency related with the activity of the power streamlining.
- *Shading: electric losses as per strings*: notwithstanding the Near Shaded irradiance losses, this speaks to the power lost because of the electrical impact of concealing. For instance, in a customary framework with strings associated in equal, one concealed module can make an entire string be avoided because of a voltage befuddle (in a SolarEdge framework, the losses are bound to the concealed module as it were)
- *Module quality losses*: This parameter utilizes the assembling resistance esteems in the PAN document (e.g., $\pm 2\%$ resistance). The formula utilized is: The lower resilience in addition to a fourth of the

distinction among low and high resistance. So for instance, for the modules with a $\pm 2\%$ resilience, the quality losses are $[-2\% + (0.25 \times 4\%)] = -1\%$.

- *Module exhibit mismatched losses*: The power losses because of mismatch among modules within a string. The 2% of mismatch for customary string inverters comes from assembling resistance, soiling, temperature contrasts among modules, tilts or azimuth contrasts inside the string, and so on. For SolarEdge frameworks these losses will consistently be 0%
- *Ohmic cabling losses*: The voltage drops because of cabling obstruction is determined as one incentive for the entire framework, set by the client. The default estimation of 1.5% losses from STC is prescribed. This means a sensible genuine loss of about 0.6% over the DC for most frameworks
- *Inverter losses during activity (proficiency)*: The DC to AC change of the inverter's productivity, weighted for difference in electric levels throughout the year
- *Inverter losses over ostensible inverter power*: The force cutting in overloading condition (where the cluster creates more DC energy than the optimized AC yield of the inverter)
- *Inverter losses because of power edge*: The loss of power when the exhibit works beneath the inverter's base power edge (characterized in the OND record)
- *Inverter losses over ostensible voltage*: The power losses when the exhibit is delivering voltage underneath the maximum power point (MPP) range of the inverter
- *Inverter losses because of voltage limit*: The power losses when the cluster is delivering voltage over the MPP range of the inverter
- *AC ohmic losses*: Just like the DC wiring losses, a prescribed estimation of 1% AC losses regarding the STC quantity will deliver about 0.5% genuine power losses
- *External transformer losses*: Unless genuine parameters are accessible, the standard 0.1% iron losses along with 1% resistive losses are suggested

Example 1.
Step 1

- Open *PV_SYST* software
- *Project Design Standalone System*

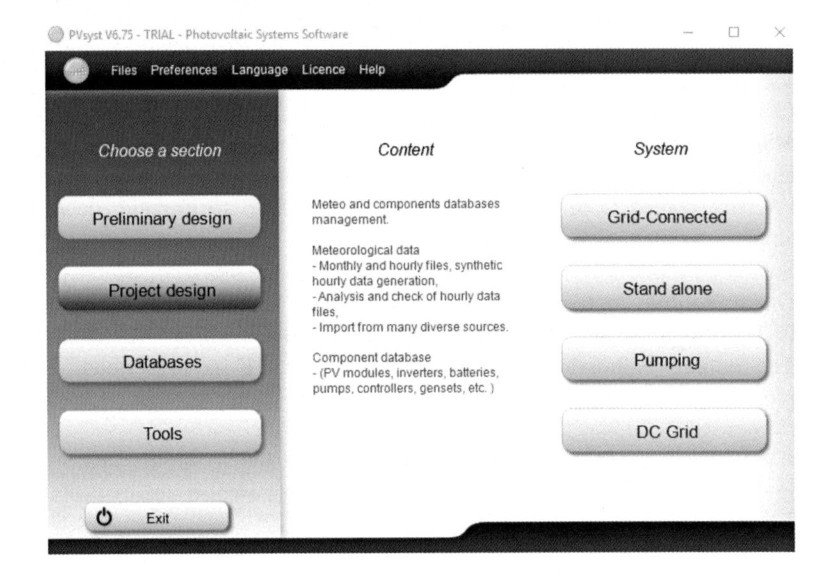

Step 2

- Provide file name
- In *Sit File* select the required place. In component choice select *New site*. In this Geographical Site Parameters page we can provide latitude/longitude value or select *Show Map* for selection of place through GOOGLE map.

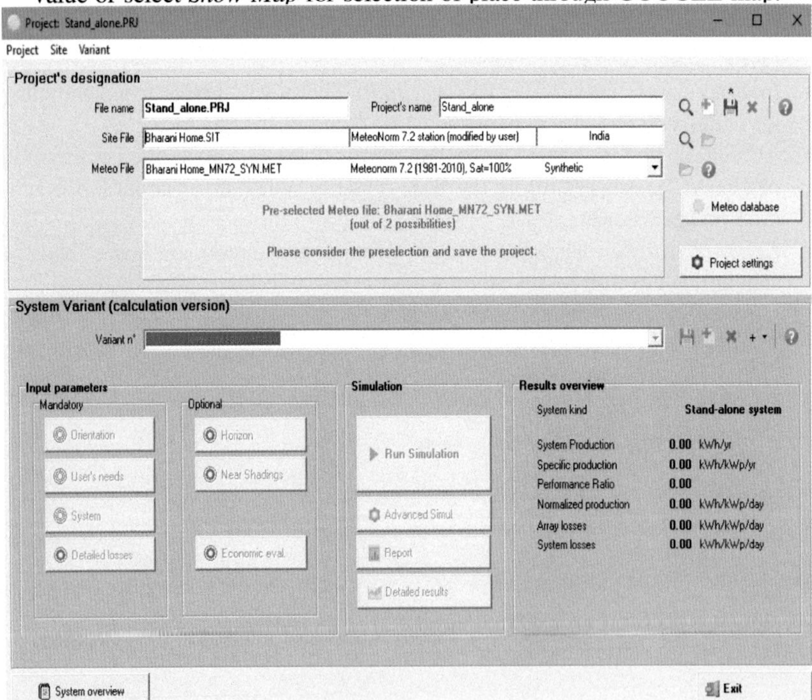

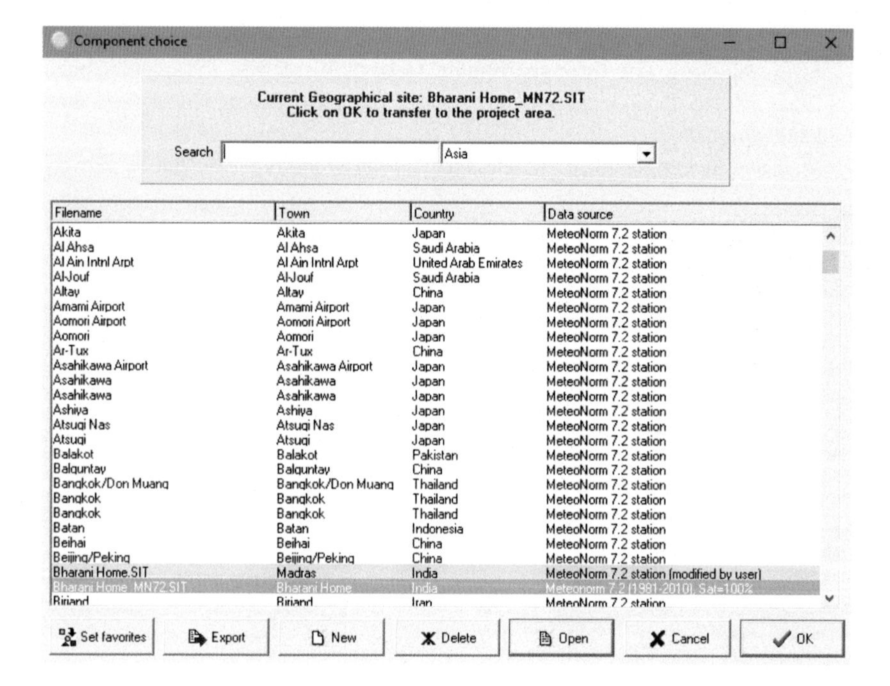

Filename	Town	Country	Data source
Akita	Akita	Japan	MeteoNorm 7.2 station
Al Ahsa	Al Ahsa	Saudi Arabia	MeteoNorm 7.2 station
Al Ain Intnl Arpt	Al Ain Intnl Arpt	United Arab Emirates	MeteoNorm 7.2 station
Al-Jouf	Al-Jouf	Saudi Arabia	MeteoNorm 7.2 station
Altay	Altay	China	MeteoNorm 7.2 station
Amami Airport	Amami Airport	Japan	MeteoNorm 7.2 station
Aomori Airport	Aomori Airport	Japan	MeteoNorm 7.2 station
Aomori	Aomori	Japan	MeteoNorm 7.2 station
Ar-Tux	Ar-Tux	China	MeteoNorm 7.2 station
Asahikawa Airport	Asahikawa Airport	Japan	MeteoNorm 7.2 station
Asahikawa	Asahikawa	Japan	MeteoNorm 7.2 station
Asahikawa	Asahikawa	Japan	MeteoNorm 7.2 station
Ashiya	Ashiya	Japan	MeteoNorm 7.2 station
Atsugi Nas	Atsugi Nas	Japan	MeteoNorm 7.2 station
Atsugi	Atsugi	Japan	MeteoNorm 7.2 station
Balakot	Balakot	Pakistan	MeteoNorm 7.2 station
Balguntay	Balguntay	China	MeteoNorm 7.2 station
Bangkok/Don Muang	Bangkok/Don Muang	Thailand	MeteoNorm 7.2 station
Bangkok	Bangkok	Thailand	MeteoNorm 7.2 station
Bangkok	Bangkok	Thailand	MeteoNorm 7.2 station
Batan	Batan	Indonesia	MeteoNorm 7.2 station
Beihai	Beihai	China	MeteoNorm 7.2 station
Beijing/Peking	Beijing/Peking	China	MeteoNorm 7.2 station
Bharani Home.SIT	Madras	India	MeteoNorm 7.2 station (modified by user)
Bharani Home_MN72.SIT	Bharani Home	India	Meteonorm 7.2 (1991-2010), Sat=100%
Birjand	Birjand	Iran	MeteoNorm 7.2 station

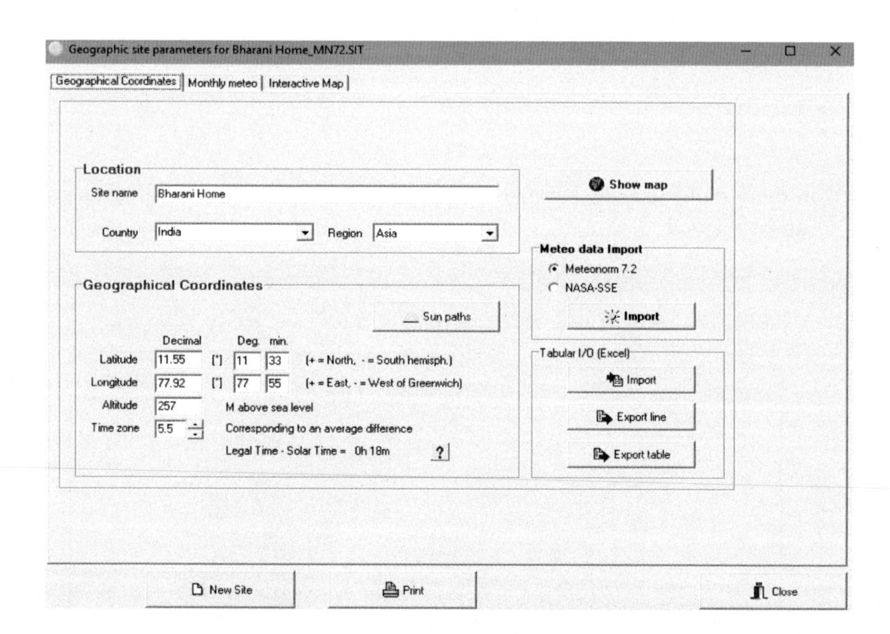

Step 3

- After selecting site through Map in Geographical Site Parameters page select Meteo Data Import either from *Meteomorn7.2* or *NASA-SSE* then select *Import*
- After that in Geographical Site Parameters page *Monthly Meteo* can be viewed

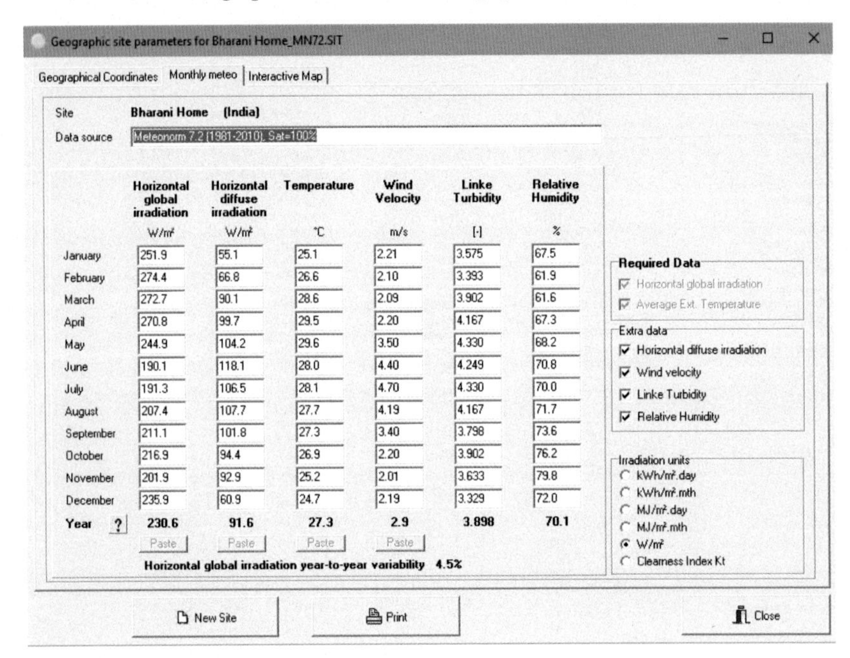

- In Geographical Site Parameters page we can view *Sun Path* diagram can be viewed

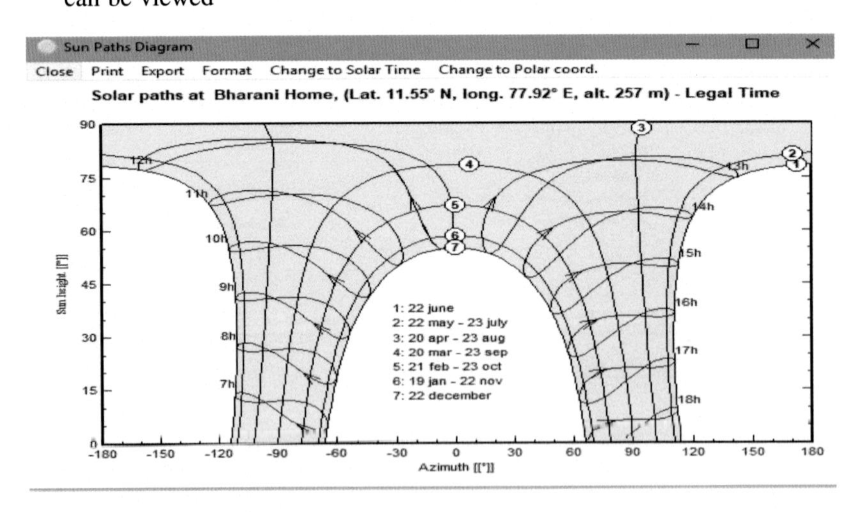

- After selecting *Site File* save the project

Step 4

- In *Input Parameters* select *Orientation* and select *Yearly Irradiation Yield*. In this we can provide *Azimuth angle* and *Plane Tilt* angle

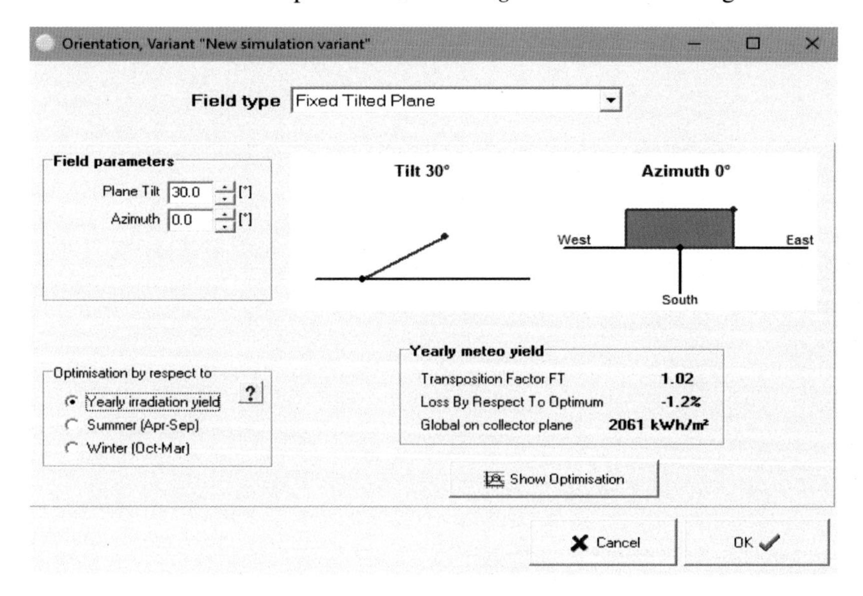

Step 5

- In *Input Parameters* select *Users Needs* we can select type *of Load, Power rating* and *Duration*. In consumption definition by we can select year, season, month for which calculation has to be done.

Daily use of Energy, Variant "New simulation variant"

Definition of Daily Household consumptions for Summer (Jun-Aug)

Consumptions | Hourly distribution

Daily consumptions

Number	Appliance	Power		Daily use		Hourly distrib	Daily energy
3	Lamps (LED or fluo)	10	W/lamp	4.0	h/day	OK	120 Wh
1	TV / PC / Mobile	75	W/app.	3.0	h/day	OK	225 Wh
1	Domestic appliances	200	W/app.	3.0	h/day	OK	600 Wh
1	Fridge / Deep-freeze	0.80	kWh/day	24.0	h/day	OK	799 Wh
0	Dish- & Cloth-washers	0.0	W aver.	0.0	h/day		0 Wh
0	Other uses	0	W/app.	0.0	h/day		0 Wh
0	Other uses	0	W/app.	0.0	h/day		0 Wh
	Stand-by consumers	6	W tot	24 h/day			144 Wh

? Appliances info

Total daily energy 1888 Wh/day
Total monthly energy 56.6 kWh/month

Consumption definition by: Year / Seasons / Months ?

Week-end or Weekly use: Use only during 7 days in a week

Display Values of: Summer / Autumn / Winter / Spring — Copy Values

Model: Load — Save

Other profile — Cancel — OK

Step 8

- Run the simulation detailed report can be obtained

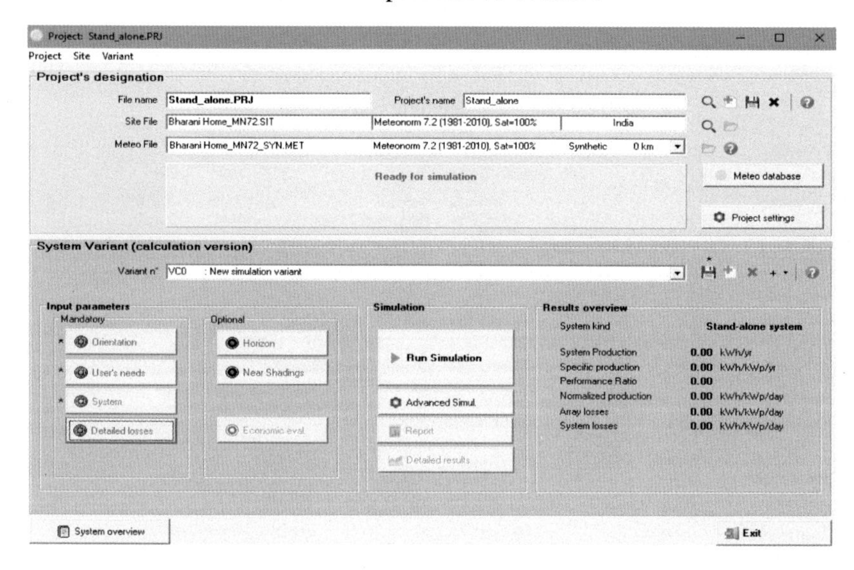

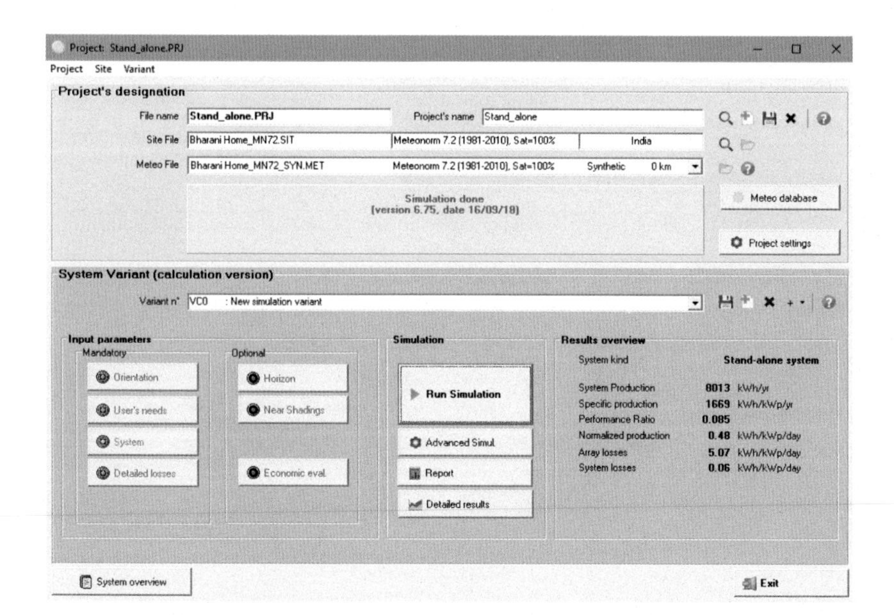

- The report is generated.

Report:

PVSYST V6.75		16/09/18	Page 1/4

Stand Alone System: Simulation parameters

Project : **Stand_alone**

Geographical Site	**Bharani Home**		Country	**India**
Situation	Latitude	11.55° N	Longitude	77.92° E
Time defined as	Legal Time	Time zone UT+5.5	Altitude	257 m
	Albedo	0.20		
Meteo data:	**Bharani Home**	Meteonorm 7.2 (1981-2010), Sat=100% - Synthetic		

Simulation variant : **New simulation variant**

Simulation date 16/09/18 11h12

Simulation parameters	System type	**Stand-alone system**		
Collector Plane Orientation	Tilt	30°	Azimuth	0°
Models used	Transposition	Perez	Diffuse	Perez, Meteonorm

PV Array Characteristics

PV module Si-poly	Model	**TP 320L BZp**		
Original PVsyst database	Manufacturer	Tata Power Solar System		
Number of PV modules	In series	3 modules	In parallel	5 strings
Total number of PV modules	Nb. modules	15	Unit Nom. Power	320 Wp
Array global power	Nominal (STC)	**4800 Wp**	At operating cond.	4309 Wp (50°C)
Array operating characteristics (50°C)	U mpp	101 V	I mpp	43 A
Total area	Module area	29.1 m²	Cell area	26.3 m²

PV Array loss factors

Thermal Loss factor	Uc (const)	20.0 W/m²K	Uv (wind)	0.0 W/m²K / m/s
Wiring Ohmic Loss	Global array res.	40 mOhm	Loss Fraction	1.5 % at STC
Serie Diode Loss	Voltage Drop	0.7 V	Loss Fraction	0.6 % at STC
Module Quality Loss			Loss Fraction	-0.4 %
Module Mismatch Losses			Loss Fraction	1.0 % at MPP
Strings Mismatch loss			Loss Fraction	0.10 %

Incidence effect (IAM): User defined IAM profile

0°	40°	50°	60°	70°	75°	80°	85°	90°
1.000	1.000	0.990	0.970	0.890	0.820	0.690	0.450	0.000

System Parameter	System type	**Stand Alone System**		
Battery	Model	**Battery module Li-Ion, 26V 180 Ah**		
	Manufacturer	Generic		
Battery Pack Characteristics	Voltage	26 V	Nominal Capacity	540 Ah
	Nb. of units	3 in parallel		
	Temperature	Fixed (20°C)		
Controller	Model	Universal controller with MPPT converter		
	Technology	MPPT converter	Temp coeff.	-5.0 mV/°C/elem.
Converter	Maxi and EURO efficiencies	97.0 / 95.0 %		
Battery Management control	Threshold commands as	SOC calculation		
	Charging	SOC = 0.90 / 0.75		
	Discharging	SOC = 0.20 / 0.45		

User's needs :	Daily household consumers	Seasonal modulation
	average	2.3 kWh/Day

PVSYST V6.75		16/09/18	Page 2/4

Stand Alone System: Detailed User's needs

Project : **Stand_alone**

Simulation variant : **New simulation variant**

Main system parameters	System type	**Stand-alone system**		
PV Field Orientation	tilt	30°	azimuth	0°
PV modules	Model	TP 320L BZp	Pnom	320 Wp
PV Array	Nb. of modules	15	Pnom total	**4800 Wp**
Battery	Battery module	Li-Ion, 26V 180 Ah	Technology	Lithium-ion, LFP
Battery Pack	Nb. of units	3	Voltage / Capacity	**26 V / 540 Ah**
User's needs	Daily household consumers	Seasonal modulation	Global	**832 kWh/year**

Daily household consumers, Seasonal modulation, average = 2.3 kWh/day

Summer (Jun-Aug)

	Number	Power	Use	Energy
Lamps (LED or fluo)	3	10 W/lamp	4 h/day	120 Wh/day
TV / PC / Mobile	1	75 W/app	3 h/day	225 Wh/day
Domestic appliances	1	200 W/app	3 h/day	600 Wh/day
Fridge / Deep-freeze	1		24 Wh/day	799 Wh/day
Stand-by consumers			24 h/day	144 Wh/day
Total daily energy				1888 Wh/day

Autumn (Sep-Nov)

	Number	Power	Use	Energy
Lamps (LED or fluo)	6	10 W/lamp	5 h/day	300 Wh/day
TV / PC / Mobile	1	75 W/app	4 h/day	300 Wh/day
Domestic appliances	1	200 W/app	4 h/day	800 Wh/day
Fridge / Deep-freeze	1		24 Wh/day	799 Wh/day
Stand-by consumers			24 h/day	144 Wh/day
Total daily energy				2343 Wh/day

Winter (Dec-Feb)

	Number	Power	Use	Energy
Lamps (LED or fluo)	6	10 W/lamp	6 h/day	360 Wh/day
TV / PC / Mobile	1	75 W/app	6 h/day	450 Wh/day
Domestic appliances	1	200 W/app	4 h/day	800 Wh/day
Fridge / Deep-freeze	1		24 Wh/day	799 Wh/day
Stand-by consumers			24 h/day	144 Wh/day
Total daily energy				2553 Wh/day

Spring (Mar-May)

	Number	Power	Use	Energy
Lamps (LED or fluo)	6	10 W/lamp	5 h/day	300 Wh/day
TV / PC / Mobile	1	75 W/app	4 h/day	300 Wh/day
Domestic appliances	1	200 W/app	4 h/day	800 Wh/day
Fridge / Deep-freeze	1		24 Wh/day	799 Wh/day
Stand-by consumers			24 h/day	144 Wh/day
Total daily energy				2343 Wh/day

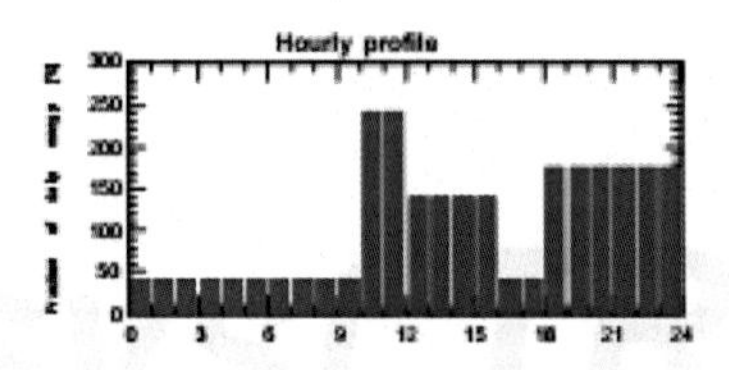

PVSYST V6.75		16/09/18	Page 3/4

Stand Alone System: Main results

Project : **Stand_alone**

Simulation variant : **New simulation variant**

Main system parameters	System type	**Stand-alone system**		
PV Field Orientation	tilt	30°	azimuth	0°
PV modules	Model	TP 320L BZp	Pnom	320 Wp
PV Array	Nb. of modules	15	Pnom total	**4800 Wp**
Battery	Battery module	Li-Ion, 26V 180 Ah	Technology	Lithium-ion, LFP
Battery Pack	Nb. of units	3	Voltage / Capacity	**26 V / 540 Ah**
User's needs	Daily household consumers	Seasonal modulation	Global	832 kWh/year

Main simulation results				
System Production	**Available Energy**	**8.01 MWh/year**	Specific prod.	1669 kWh/kWp/year
	Used Energy	0.83 MWh/year	Excess (unused)	7.13 MWh/year
	Performance Ratio PR	8.47 %	Solar Fraction SF	100.00 %
Loss of Load	Time Fraction	0.0 %	Missing Energy	0.00 MWh/year

Normalized productions (per installed kWp): Nominal power 4800 Wp

Performance Ratio PR and Solar Fraction SF

New simulation variant

Balances and main results

	GlobHor kWh/m²	GlobEff kWh/m²	E Avail MWh	EUnused MWh	E Miss MWh	E User MWh	E Load MWh	SolFrac
January	187.4	232.9	0.920	0.834	0.000	0.079	0.079	1.000
February	184.4	208.7	0.818	0.742	0.000	0.071	0.071	1.000
March	202.9	200.7	0.790	0.713	0.000	0.073	0.073	1.000
April	195.0	170.1	0.674	0.600	0.000	0.070	0.070	1.000
May	182.2	143.3	0.577	0.500	0.000	0.073	0.073	1.000
June	136.9	107.6	0.445	0.385	0.000	0.057	0.057	1.000
July	142.3	113.1	0.462	0.400	0.000	0.059	0.059	1.000
August	154.3	131.8	0.537	0.475	0.000	0.059	0.059	1.000
September	152.0	141.5	0.571	0.497	0.000	0.070	0.070	1.000
October	161.4	168.6	0.674	0.597	0.000	0.073	0.073	1.000
November	145.4	185.0	0.667	0.593	0.000	0.070	0.070	1.000
December	175.5	221.0	0.879	0.785	0.000	0.079	0.079	1.000
Year	2019.6	2003.0	8.013	7.131	0.000	0.832	0.832	1.000

Legends:
GlobHor Horizontal global irradiation
GlobEff Effective Global, corr. for IAM and shadings
E Avail Available Solar Energy
EUnused Unused energy (full battery) loss
E Miss Missing energy
E User Energy supplied to the user
E Load Energy need of the user (Load)
SolFrac Solar fraction (EUsed / ELoad)

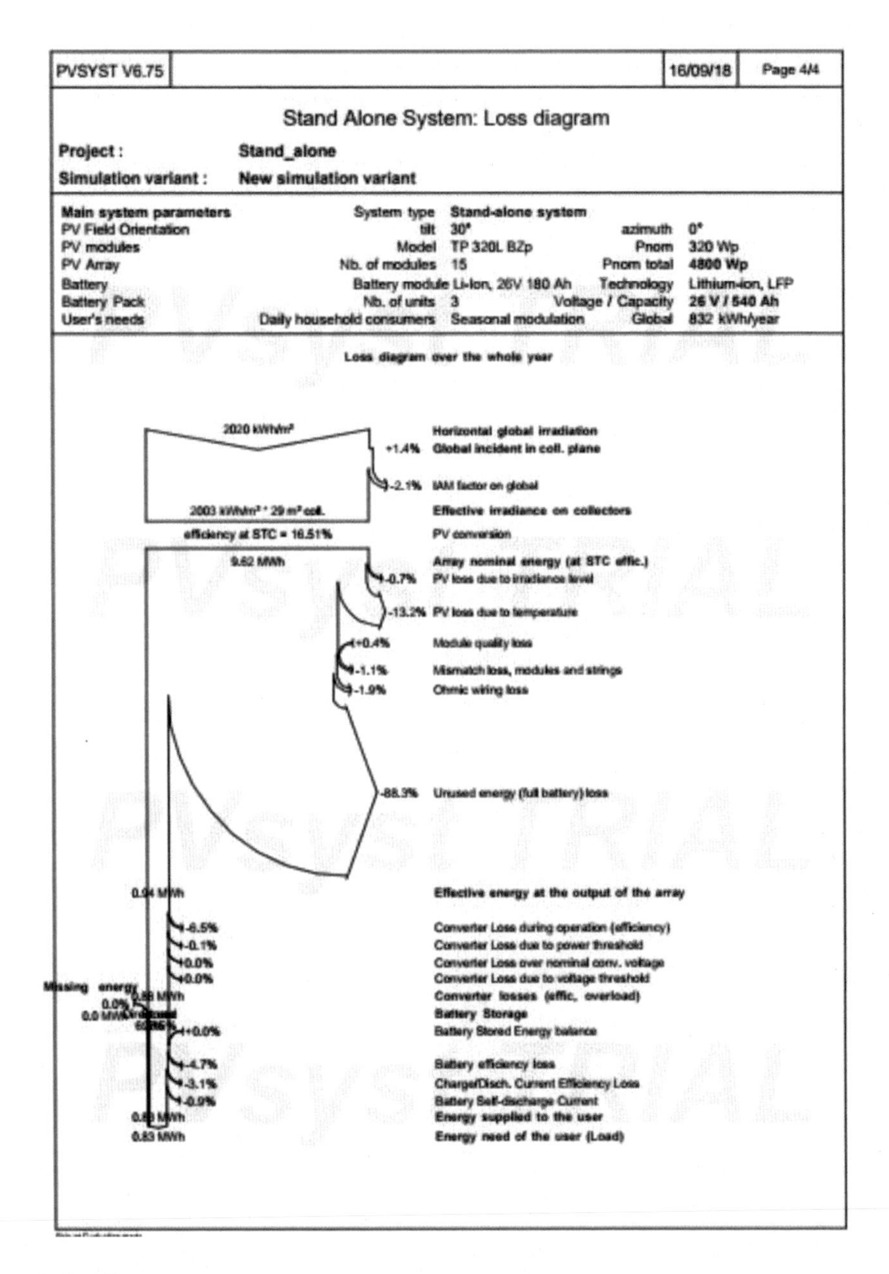
PVSYST V6.75 | 16/09/18 | Page 4/4

Stand Alone System: Loss diagram

Project : Stand_alone

Simulation variant : New simulation variant

Main system parameters	System type	Stand-alone system		
PV Field Orientation	tilt	30°	azimuth	0°
PV modules	Model	TP 320L BZp	Pnom	320 Wp
PV Array	Nb. of modules	15	Pnom total	4800 Wp
Battery	Battery module	Li-Ion, 26V 180 Ah	Technology	Lithium-ion, LFP
Battery Pack	Nb. of units	3	Voltage / Capacity	26 V / 540 Ah
User's needs	Daily household consumers	Seasonal modulation	Global	832 kWh/year

9.4 Summary

At the end of the chapter, the reader will learn to design the solar PV system using PVSYST Software. Apart from this, the learner will understand the importance of solar power plant and the design steps and components in the solar power plant.

9.5 Review questions

- How do you calculate the azimuth angle and zenith angle for the solar PV system?
- What is the importance of the combiner box?
- Brief the importance of BoS for the system integrator.
- What is the purpose of a bypass the diode and blocking diode?
- Differentiate string inverter and central inverter?
- Write the effects of shadowing on a solar PV system.
- What is meant by the dust derating factor?
- Which direction you face the solar PV panel in the Northern hemisphere and why?
- What is meant by module integrated inverter?
- What is the effect of temperature on the panel?

Index

Note: Page numbers followed by "*f*" and "*t*" refer to figures and tables, respectively.

N

O

P

R

S

Printed in the United States
By Bookmasters

Step 6

- In *Input Parameters* select *System*. In *Storage* that we can select *Battery Set*. Type of battery like Li-ion, lead acid. Capacity and voltage ratings can be provided

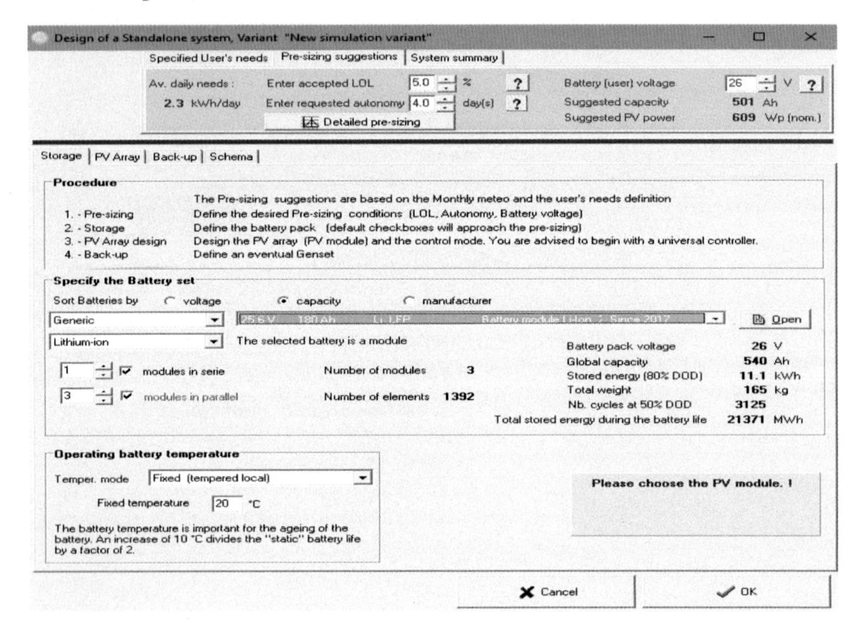

Step 7

- In *System* page *PV array* tag we can select type of *PV module* to be used. Then *Control Mode and Controllers* can be selected. Number of modules to be connected in *Series* and No. of *Strings* can be selected